EU REGULATION OF CROSS-BORDER CARBON CAPTURE AND STORAGE

EU REGULATION OF CROSS-BORDER CARBON CAPTURE AND STORAGE

Legal Issues under the Directive
on the Geological Storage of CO_2
in the light of Primary EU Law

Marijn HOLWERDA

Cambridge – Antwerp – Portland

Intersentia Publishing Ltd.
Trinity House | Cambridge Business Park | Cowley Road
Cambridge | CB4 0WZ | United Kingdom
Tel.: +44 1223 393 753 | Email: mail@intersentia.co.uk

Distribution for the UK:
NBN International
Airport Business Centre, 10 Thornbury Road
Plymouth, PL6 7 PP
United Kingdom
Tel.: +44 1752 202 301 | Fax: +44 1752 202 331
Email: orders@nbninternational.com

Distribution for the USA and Canada:
International Specialized Book Services
920 NE 58th Ave. Suite 300
Portland, OR 97213
USA
Tel.: +1 800 944 6190 (toll free)
Email: info@isbs.com

Distribution for Austria:
Neuer Wissenschaftlicher Verlag
Argentinierstraße 42/6
1040 Wien
Austria
Tel.: +43 1 535 61 03 24
Email: office@nwv.at

Distribution for other countries:
Intersentia Publishing nv
Groenstraat 31
2640 Mortsel
Belgium
Tel.: +32 3 680 15 50
Email: mail@intersentia.be

The Energy & Law Series
The Energy & Law Series is published in parallel with the Dutch series Energie & Recht.
Members of the editorial committee are:
Prof. Dr. Martha M. Roggenkamp, University of Groningen and Simmons & Simmons, Rotterdam (editor in chief)
Prof. Dr. Kurt Deketelaere, Institute of Environmental and Energy Law, University of Leuven
Prof. Dr. Leigh Hancher, Allen & Overy, Amsterdam and Tilburg University, Tilburg and Council Member, WRR
Dr. Tom Vanden Borre, Chief Counsellor, Commission for the Regulation of Electricity and Gas (CREG) and University of Leuven

EU Regulation of Cross-Border Carbon Capture and Storage. Legal issues under the Directive on the geological storage of CO_2 in the light of primary EU law
Marijn Holwerda

Cover photograph © Nikolay Sereda – Dreamstime.com

ISBN 978-1-78068-190-0
D/2014/7849/32
NUR 828

British Library Cataloguing in Publication Data. A catalogue record for this book is available from the British Library.

FOREWORD

Those who have written a PhD thesis know what a trying process it can be. Having started with my thesis in the fall of 2009, I am extremely happy to have finished in the fall of 2013. At the beginning of these four years, I sometimes got the impression that a lot of people see the writing of a PhD thesis as an unavoidable never-ending horror story or, even worse, a mission impossible. Needless to say that such an approach is not particularly encouraging for the new PhD researcher. I hope that this thesis shows young PhD researchers that it is far from a mission impossible and can perhaps even be fun.

It is only because of the help of a lot of people that I have been able to bring this PhD project to a good end. Many of these people I have already thanked or will thank in person soon. However, I would like to expressly thank a few here.

First of all, I would like to thank RWE AG and Essent B.V. and all the people from both companies who were involved in my project for their support. Without the (financial) support provided by RWE and Essent, this PhD project would not have existed in the first place. Second, I would like to thank the Dutch national research and development programme for CO_2 capture, transport and storage (CATO-2) for letting me be part of such a stimulating and successful research programme. Third, I would like to express my deep gratitude to my supervisors, Professors Martha Roggenkamp and Hans Vedder, for supervising my PhD and helping me bring this project to a good end. Fourth, I would like to thank the members of the PhD reading committee for reading my work, despite their busy schedules and summer holidays: Professors Hans Christian Bugge, Jan Jans and John Paterson. Fifth, I would like to thank my university roommate Hannah for four years of academic discussions, mind challenging and support. Finally, I would like to thank my love and life, Marieke, for her endless support and patience.

Marijn Holwerda
Groningen, September 2013

CONTENTS

INTRODUCTION

In its fight against climate change the EU foresees an important role for carbon capture and storage (CCS) technology. CCS consists of the capture of carbon dioxide (CO_2) from industrial installations, its transport to a storage site and subsequent injection into a suitable underground geological formation for the purpose of permanent storage.[1] According to preliminary estimates by the European Commission (Commission), seven million tonnes of CO_2 could be stored in the EU by 2020, and up to 160 million tonnes by 2030, assuming a 20% reduction in greenhouse gas emissions by 2020 and provided that CCS obtains private, national and EU support and proves to be an environmentally safe technology.[2] In order to speed up the development of CCS technology, the Commission set out an action plan for CCS in 2007, including the creation of a regulatory framework for CCS deployment in the EU.[3] In 2009, this regulatory framework came into being with the entry into force of Directive 2009/31/EC on the geological storage of carbon dioxide[4] (CCS Directive).

THE CCS DIRECTIVE[5]

The CCS Directive, which is part of the EU climate and energy package,[6] is one of the first dedicated legal frameworks for the deployment of CCS worldwide. Its

---

[1]   Recital 4 of the preamble to European Parliament and Council Directive 2009/31/EC of 23 April 2009 on the geological storage of carbon dioxide and amending Council Directive 85/337/EC, European Parliament and Council Directives 2000/60/EC, 2001/80/EC, 2004/35/EC, 2006/12/EC and 2008/1/EC, and Regulation (EC) No 1013/2006 [2009] OJ L140/114. For further information on the concept and technology of CCS, see Chapter I of this book.

[2]   Ibid, recital 5.

[3]   Ibid, recital 7.

[4]   See n 1.

[5]   For a more comprehensive overview of the CCS Directive, see Chapter II of this book.

[6]   The climate and energy package outlines the EU's contribution to the global fight against climate change in the coming years and in essence consists of three Directives and a Decision: European Parliament and Council Directive 2009/28/EC of 23 April 2009 on the promotion of the use of energy from renewable sources [2009] OJ L140/16, European Parliament and Council Directive 2009/29/EC of 23 April 2009 amending Directive 2003/87/EC so as to improve and extend the greenhouse gas emission allowance trading scheme of the community [2009] OJ L140/63, European Parliament and Council Directive 2009/31/EC of 23 April 2009 on the geological storage of carbon dioxide [2009] OJ L140/114 and Decision 406/2009/EC of the European Parliament and of the Council of 23 April 2009 on the effort of member states to reduce their greenhouse gas emissions to meet the Community's greenhouse

---

legal basis is ex Article 175(1) of the Treaty Establishing the European Community (TEC), now Article 192(1) of the Treaty on the Functioning of the EU (TFEU). As an alleged enabling legal framework,[7] it is intended to ensure the environmentally safe geological storage of CO_2 in the EU.[8] The CCS Directive is to make sure that those EU Member States (Member States) that seek to deploy CCS technology do so in an environmentally safe manner.[9]

As its name suggests, the Directive on the geological storage of CO_2 is primarily a Directive on the *storage* of CO_2. In drafting its proposal for the CCS Directive, the Commission took a conservative approach.[10] The default option for regulating each of the components of the CCS value chain (capture [including compression], transport and storage) was the existing legal framework, regulating activities of a similar risk.[11]

Since CO_2 capture, in its view, represented risks similar to those of the chemical/power generation sector, the Commission considered Directive 96/61/EC (Integrated Pollution Prevention and Control Directive)[12] a suitable regulatory framework for the activity.[13] Likewise, since it deemed CO_2 transport to present similar risks as natural gas transport, the Commission chose to regulate it in the same manner.[14]

---

gas emission reduction commitments up to 2020 [2009] OJ L140/136. For more information, see the Commission's website on the package <ec.europa.eu/environment/climat/climate_action.htm> accessed 8 March 2010.

[7]   See, for instance, Jos Delbeke, 'Enabling Legal Framework for Carbon Dioxide Capture and Geological Storage' <http://ec.europa.eu/clima/events/0006/jd_presentation_en.pdf> accessed 1 October 2012.

[8]   See Article 1(1) of the Directive which provides that the Directive 'establishes a legal framework for the environmentally safe geological storage of Carbon Dioxide (CO_2) to contribute to the fight against climate change'. Article 1(2) of the CCS Directive states that 'the purpose of environmentally safe geological storage of CO_2 is permanent containment in such a way as to prevent, and where this is not possible, eliminate as far as possible, negative effects and any risk to the environment and human health'. See also Commission, 'Accompanying Document to the Proposal for a Directive of the European Parliament and of the Council on the Geological Storage of Carbon Dioxide: Impact Assessment' COM (2008) 18 final 10 and 35.

[9]   As underlined by the European Council in 2008, the objective of proposing a regulatory framework for CCS was 'to ensure that this novel technology would be deployed in an environmentally safe way'. See recital 9 of the preamble to the CCS Directive.

[10]   Commission (n 8) 2.

[11]   Ibid.

[12]   Council Directive 96/61/EC of 24 September 1996 concerning integrated pollution prevention and control (Integrated Pollution Prevention and Control Directive) [1996] OJ L257. In 2011, the Integrated Pollution Prevention and Control Directive was replaced by European Parliament and Council Directive 2010/75/EU of 24 November 2010 on industrial emissions (integrated pollution prevention and control) (Recast) [2010] OJ L334/17 (Industrial Emissions Directive), with effect from January 2011.

[13]   Commission (n 8) 2.

[14]   Ibid. In this regard, see also Article 31 of the CCS Directive, which amends Annexes I and II of Council Directive 85/337/EC of 27 June 1985 on the assessment of the effects of certain

By contrast, in relation to CO_2 storage, the Commission deemed the existing legal framework (Integrated Pollution Prevention and Control Directive and EU waste legislation)[15] not to be well adapted to regulating the risks.[16] According to it, the kind of controls required differed from those under the Integrated Pollution Prevention and Control Directive, and EU waste legislation had not been designed to cover the particular risks in question.[17] Therefore, the Commission decided to develop a free-standing legal framework for CO_2 storage and to remove CCS deployment under the CCS Directive from the scope of EU waste legislation.[18]

Furthermore, the Commission chose to give itself a large role in the application of the Directive. The Commission, inter alia, reviews draft CO_2 storage permits[19] and draft decisions to transfer responsibility for the storage site to the competent authorities (post-closure),[20] and has the option to issue guidelines on various matters.[21] Given the limited experience with the large-scale deployment of CCS technology, the Commission wanted to ensure a consistent application of the CCS Directive in the various Member States, as well as to help enhance public confidence in the safety of CCS.[22]

A striking characteristic of the CCS Directive is the lack of precision and conclusiveness of some of its provisions. Article 12(1), for instance, provides that

public and private projects on the environment [1985] L175 (Environmental Impact Assessment Directive). An environmental impact assessment is a procedure that ensures that the environmental implications of decisions are taken into account before the decisions are made. See the European Commission's website on environmental impact assessment <http://ec.europa.eu/environment/eia/home.htm> accessed 2 October 2012. On this issue, see further section 3.2.1 of Chapter III of this book on CO_2 stream purity and Member States' scope to impose stricter norms.

15   The latter refers to European Parliament and Council Directive 2006/12/EC of 5 April 2006 on waste [2006] L114/9 (Waste Directive) and European Parliament and Council Regulation (EC) 1013/2006 of 14 June 2006 on shipments of waste [2006] L190 (Shipments of Waste Regulation).

16   Commission (n 8) 2.

17   Ibid 3.

18   Ibid. See Articles 35 and 36 of the CCS Directive, amending the Waste Directive and the Shipments of Waste Regulation, as a consequence of which captured CO_2 for storage can be said to no longer be considered a waste under EU waste legislation.

19   Article 10 of the CCS Directive.

20   Article 18 of the CCS Directive.

21   By virtue of Article 12(2), the Commission can, for instance, issue guidelines to help identify the conditions applicable on a case-by-case basis for respecting the CO_2 stream acceptance criteria in Article 12(1). I further address the issue of CO_2 stream purity in Chapter III on CO_2 stream purity and Member States' scope to impose stricter norms. In relation to possible Commission guidelines, see also Article 18(2) of the Directive.

22   See recital 33 of the preamble to the CCS Directive and Commission, 'Questions and Answers on the Proposal for a Directive on the Geological Storage of Carbon Dioxide', MEMO/08/36 <http://europa.eu/rapid/pressReleasesAction.do?reference=MEMO/08/36&format=HTML&aged=0&language=EN&guiLanguage=en> accessed 2 October 2012.

'a CO_2 stream shall consist *overwhelmingly* of CO_2'.[23] Likewise, Article 4(4) states that a geological formation may only be selected as a storage site if 'there is no *significant* risk of leakage, and if no *significant* environmental or health risks exist.[24] In addition, the Directive leaves a number of important choices to the Member States.[25] The latter, for instance, have to determine the (precise) regime for third-party access to CO_2 transport and storage infrastructure,[26] as well as the (precise) criteria for the financial security and mechanism required under Articles 19 and 20. What is more, Member States need to design their own liability regime for third-party damages.[27]

As is evidenced by the long list of issues to be assessed in the planned review of the CCS Directive in 2015 (Article 38),[28] the Directive has a learning-by-doing character.[29] It provides the Member States and different market players along the CCS value chain a first piece of legislation for the early deployment of CCS technologies,[30] as well as the opportunity to gain different experiences with the

---

[23]   Emphasis added. Article 12 contains the criteria related to the purity of the captured CO_2 stream for storage. It sets the boundaries for the presence in the CO_2 stream of substances other than CO_2. See further section 3.2 of Chapter III.

[24]   Emphasis added.

[25]   See also Chiara Armeni, 'An Update of the State of CCS Regulation in Europe' (2012) <www.globalccsinstitute.com/insights/authors/chiara-armeni/2012/02/13/update-state-ccs-regulation-europe> accessed 11 October 2012. In addition to the issues mentioned above, see NERA Economic Consulting, 'Developing a Regulatory Framework for CCS Transportation Infrastructure', vol 1 <www.decc.gov.uk/assets/decc/what%20we%20do/uk%20energy%20supply/energy%20mix/carbon%20capture%20and%20storage/1_20090617131338_e_@@_ccs reg1.pdf> accessed 4 January 2012. In its 2009 study analysing the regulatory options for developing CO_2 transport infrastructure in the UK, NERA Economic Consulting outlines some of the choices that Member States will have to make when setting the regulatory framework for the development and management of CO_2 transport infrastructure (e.g. the system for computing charges for network usage).

[26]   Third-party access refers to the access to infrastructure by parties who do not control the relevant infrastructure. Article 21(2) provides that third-party access 'shall be provided in a transparent and non-discriminatory manner *determined by the Member State*' (emphasis added). On third-party access to CO_2 transport and storage infrastructure, see Chapters V (Refusing access to CCS infrastructure and the general EU law principle of loyalty), VI (Refusing access to CCS infrastructure and Article 102 TFEU) and VII (The development and management of CO_2 transport infrastructure and EU antitrust law) of this thesis.

[27]   The CCS Directive only provides for rules on climate and environmental liability. On the different types of liability under the CCS Directive, see section 2.5.2.4 in Chapter II. See also recital 30 of the preamble to the CCS Directive.

[28]   Among these are, inter alia, experiences with the application of Articles 12 (CO_2 stream-purity) and 21 (third-party access to CO_2 transport networks and storage sites) of the CCS Directive.

[29]   In this regard, the CCS Directive is similar to the first EU ETS Directive, European Parliament and Council Directive 2003/87/EC of 13 October 2003 establishing a scheme for greenhouse gas emission allowance trading within the Community [2003] L275/32 (EU ETS Directive), Article 38.

[30]   By 'early deployment' I mean the deployment of CCS in the years up to around 2020.

implementation of that legal framework. As we will see in the following, adopting such a legislative approach is not without risks.

CROSS-BORDER CCS DEPLOYMENT

Considering the above-mentioned characteristics, it does not come as a surprise that the CCS Directive hardly mentions the cross-border deployment of CCS in the EU.[31] Only two provisions explicitly mention the possibility of cross-border CO_2 transport and/or storage. What is more, these provisions deal with the cross-border deployment of CCS in the EU in a very rudimentary manner. Article 22(2), among other things, provides that Member States are to consult each other in case of a cross-border dispute over access to cross-border CO_2 transport or storage infrastructure. By virtue of Article 24, Member States are to cooperate in case of cross-border CO_2 transport or cross-border storage complexes. Even though the CCS Directive acknowledges the cross-border deployment of CCS, the development of EU-wide markets for the capture, transport and storage of CO_2 is evidently not one of its declared objectives.

The limited attention given to the cross-border deployment of CCS could be explained by the Commission not having anticipated an early cross-border deployment of CCS at the time it drafted its proposal for a CCS Directive. Expectations were that CCS technologies would be widely deployed only after 2020 at the earliest.[32] As a consequence, chances of an early cross-border deployment of CCS were small.

However, recent developments, mainly related to the poor public acceptance of *onshore* geological CO_2 storage in several Member States,[33] appear to have made an early cross-border deployment of CCS in the EU (significantly) more likely than initially thought. Mikunda and others have argued that the general move towards *offshore* storage due to communication challenges with the general public in certain countries, the cost of characterising suitable offshore storage complexes, and the potential demand for CO_2 for the purposes of enhanced oil

---

[31]   For an overview of the possible factual cross-border situations, see Element Energy, 'One North Sea: A Study into North Sea Cross-border CO_2 Transport and Storage' (2010) 85 <www.regjeringen.no/upload/OED/OneNortSea_Fulldoc.pdf> accessed 11 October 2012.

[32]   See, for instance, Commission, 'Communication for the Commission to the Council and the European Parliament Sustainable Power Generation from Fossil Fuels: Aiming for Near-zero Emissions from Coal after 2020' COM (2006) 843 final 10.

[33]   On public acceptance problems in Germany and the Netherlands, see, for instance, section 3.1 of Chapter III on CO_2 stream-purity and Member States' scope to impose stricter norms. On similar developments in Denmark, see 'CO2 Storage Protests' *Jyllands-Posten* (5 August 2009) <http://jyllands-posten.dk/uknews/article4187838.ece?service=printversion> accessed 5 October 2012.

recovery[34] in the North Sea,[35] could mean that pipeline infrastructures and potentially CO_2 shipping routes will be required to cross national boundaries.[36] Likewise, Neele and others have argued that in an offshore-only scenario and in an enhanced oil recovery scenario storage abroad will already be required in 2020, since the current onshore CCS plans would need to be changed to use offshore storage sites.[37]

As we have seen, the CSS Directive was not (primarily) drafted with a view to facilitating the cross-border deployment of CCS. Quite on the contrary, several provisions in the Directive could hamper the cross-border deployment of CCS.

---

[34] Enhanced oil recovery is a form of enhanced hydrocarbon recovery (in this case oil), which refers to the recovery of hydrocarbons, in addition to those extracted, by water addition or other means. See recital 20 of the preamble to the CCS Directive.

[35] With the general move to offshore CO_2 storage in at least the North-West of the EU, the vast Norwegian offshore storage potential comes into play. In a 2009 report on European geological storage capacity, the EU GeoCapacity project estimates that out of a total of about 360,000 Mt European geological storage capacity, 200,000 Mt is located offshore Norway (about half of the total European geological storage capacity). What is more, Norway currently hosts two large-scale offshore CO_2 storage demonstration projects (Sleipner and Snøhvit), one of which is located in the North Sea (Sleipner). See Thomas Vangkilde-Pedersen and others (GeoCapacity), 'Assessing European Capacity for Geological Storage of Carbon Dioxide' (2009) 157 <www.geology.cz/geocapacity/publications/D16%20WP2%20Report%20 storage%20capacity-red.pdf> accessed 18 October 2012, and The National Mining Association, 'Carbon Capture and Storage: Status of CCS Developments' <www.nma.org/ccs/ ccsprojects.asp> accessed 18 October 2012. The GeoCapacity project was an EU co-funded (6th Framework Programme) research project assessing the European capacity for the geological storage of CO_2. For more information on the project, see its website at <www. geology.cz/geocapacity> accessed 18 October 2012.

[36] See Tom Mikunda and others (CATO2), 'Transboundary Legal Issues in CCS: Economics, Cross-border Regulation and Financial Liability of CO_2 Transport and Storage Infrastructure' (2011) 10. In this regard, see also Sonja van Renssen, 'European CCS Industry Faces Moment of Truth' *European Energy Review* (29 October 2012) <www.europeanenergyreview.eu/site/ pagina.php?id=3919> accessed 30 October 2012. CATO2 is a Dutch research and development programme which aims to facilitate and enable integrated development of CCS demonstration sites in the Netherlands. For further information, see its website at <www.co2-cato.org/> accessed 5 October 2012.

[37] Filip Neele and others (CO₂Europipe), 'Development of a Large-scale CO_2 Transport Infrastructure in Europe: Matching Captured Volumes and Storage Availability' (2010) 32–33 <www.co2europipe.eu/Publications/D2.2.1%20-%20CO2Europipe%20Report%20CCS%20 infrastructure.pdf> accessed 5 October 2012. See also Roman Mendelevitch and others, 'CO₂ Highways for Europe: Modelling a Carbon Capture, Transport and Storage Infrastructure for Europe' (2010) Centre for European Policy Studies Working Document No. 340 1 <http://aei. pitt.edu/15200/1/WD_340_CO2_Highways.pdf> accessed 27 April 2011. Mendelevitch and others argue that continued public resistance to onshore CO_2 storage can only be overcome by constructing expensive offshore storage. See also McKinsey and Company, 'Carbon Capture and Storage: Assessing the Economics' (2008) 39 <http://assets.wwf.ch/downloads/ mckinsey2008.pdf> accessed 5 October 2012. CO₂Europipe, a research project funded by the European Commission under the 7th Framework Programme, aims at paving the way towards large-scale, Europe-wide infrastructure for the transport and injection of CO_2 from zero-emission plants. For further information, see the project's website at <www.co2europipe. eu/> accessed 5 October 2012.

Article 21(2)(b), for instance, appears to give Member States the option to require the operators of CO_2 transport and storage infrastructure to refuse to grant access to the relevant infrastructure when the capacity concerned is needed to meet part of the Member State's international and EU obligations to reduce greenhouse gas emissions. As argued by Vedder, Article 21(2)(b) could effectively foreclose an entire Member State's CO_2 storage market.[38] Similarly, it is not hard to imagine the rather general wording of Article 12(1) (CO_2 stream purity) leading to the adoption of different sets of technical standards in the various Member States, in turn hindering the cross-border trade of captured CO_2 for storage.

These provisions do not, however, exist in splendid isolation. They must be placed in the broader context of primary EU law, which takes precedence over both secondary EU law, such as the CCS Directive, and national law, such as the Member States' measures implementing that Directive.[39] By setting part of the framework within which the different market players, the Member States and the Commission have to operate, primary EU law to a great extent determines whether these provisions will indeed hamper the cross-border deployment of CCS in the EU.

APPROACH

In this thesis, I outline the broader primary EU law context of a number of legal issues, related to provisions in the CCS Directive,[40] that could have far-reaching consequences for the cross-border deployment of CCS in the EU. In doing so, I will answer the question to what extent these individual issues could indeed hamper the (near-)future cross-border deployment of CCS in the EU, as well as explore possible solutions, if required. It is important to underline that not all of the legal issues explored are exclusively relevant for the *cross-border* deployment of CCS. Yet, considering that they could have extensive consequences for the latter, all merit discussion.

In addition to examining these legal issues, I intend to assess, from the point of view of the cross-border deployment of CCS, the effectiveness of the overall

---

38   Hans Vedder, 'An Assessment of Carbon Capture and Storage under EC Competition Law' (2008) 29 European Competition Law Review 586, 598.
39   In the hierarchy of EU law, primary EU law, consisting of the provisions in the Treaties (the Treaty on European Union (TEU) and the Treaty on the functioning of the European Union (TFEU)), takes precedence over secondary EU law and national law. On this issue, see further section 5.3. of Chapter V of this thesis on refusal of access to CO_2 transport and storage infrastructure and the general EU law principle of loyalty.
40   Except for Chapter IX, all chapters are related to a provision in the CCS Directive. Chapter IX deals with an issue that is of great relevance for the cross-border deployment of CCS: the development of CO_2 transport and storage infrastructure and the public funding thereof.

legislative approach chosen by the Commission. The underlying premise is that, particularly in a cross-border setting, rules are more effective (in facilitating technology deployment) if they are clear and uniform (across Member States). As indicated earlier, the CCS Directive appears to have a learning-by-doing character, setting a general framework for the early deployment of CCS in the EU.[41] This framework is to be further specified using the different experiences with the early deployment of CCS, as well as with the application of the CCS Directive. As such, it does not yet take into account the near-future cross-border deployment of CCS. In addition, the Commission's legislative approach could potentially endanger the consistency of rules in the different Member States.[42] In this work, I explore the question of to what extent such legislative approach, taking account of the primary EU law context, is generally likely to lead to problems for the cross-border deployment of new technologies such as CCS.

The legal issues that I examine in this thesis were selected by means of an analysis of the CCS Directive in light of the likely expansion of cross-border deployment of CCS in the EU. This analysis revealed several legal issues that could present barriers to the growth of cross-border CCS if not addressed. These issues directly affect the business case for the cross-border deployment of CCS in the EU by influencing transaction costs[43] and operating costs.

The selected issues are related to three overarching, recurring themes that generally are of great significance for the cross-border deployment of any technology in the EU: availability of and access to relevant (cross-border) infrastructure (Chapters V–IX), (differential) product standards (Chapter III) and taxation of products transported cross-border (Chapter IV). Although I do not claim to give an exhaustive overview of cross-border legal issues related to the CCS Directive, I believe that this work covers the most important overarching themes related to the cross-border deployment of CCS in the EU.

---

41   Interestingly, the EU seems to have proposed a somewhat similar approach for the inclusion of CCS in the Clean Development Mechanism (CDM) during the United Nations Framework Convention on Climate Change (UNFCCC) negotiations on a post-Kyoto international climate change regime. See 'Submission by Slovenia on Behalf of the European Community and its Member States: Carbon Dioxide Capture and Storage as Clean Development Mechanism Activities' (2008) <www.ccs-info.dk/euccscdm.pdf> accessed 9 October 2012.

42   In a 2009 report on EU CCS deployment, the Global CCS Institute underlined the importance of an effective and consistent transposition of the CCS Directive for the success of any cross-border CCS projects. See Global CCS Institute, 'Strategic Analysis of the Global Status of Carbon Capture and Storage – Report 3: Country Studies, the European Union' (2009) 1 <http://cdn.globalccsinstitute.com/sites/default/files/publications/8517/strategic-analysis-global-status-ccs-country-study-european-union.pdf> accessed 10 October 2012.

43   Transaction costs can be described as the costs of transferring resources between individuals. See William Jack and Tavneet Suri, 'Risk Sharing and Transaction Costs: Evidence from Kenya's Mobile Money Revolution' 1 <www.mit.edu/~tavneet/Jack_Suri.pdf>.

Finally, in this thesis, I do not make a distinction between the onshore and offshore deployment of CCS technologies in the EU. The assumption is that the conclusions of this thesis hold as long as EU law applies. In the *Salemink* case, the EU Court of Justice held that the territorial applicability of EU law extends to the continental shelves of the Member States.[44]

CONTRIBUTION

To date, legal (and policy) research on CCS in the EU has predominantly focused on issues such as the long-term liability for CO_2 storage,[45] the financial security and mechanism required under Articles 19 and 20 of the CCS Directive,[46] and, in relation to the international legal framework, the cross-border movement of CO_2 under the London Protocol.[47, 48] This focus can be explained by the interest

---

[44] Case C-347/10 *Salemink* [2012] not yet reported, paras 30–37. In the remainder of this book, the EU Courts will, for the sake of recognisability and consistency, be referred to as the European Court of Justice (ECJ) and the Court of First Instance (CFI). Together they are referred to as the Courts.

[45] The term long-term liability refers to the post-closure liability (environmental, climate and civil) for a storage site. On this issue, see e.g. Nkaepe Etteh, 'Carbon Capture and Storage: Liability Implications' (2009) University of Dundee Centre for Energy, Petroleum and Mineral Law and Policy Annual Review 2009/10 <www.dundee.ac.uk/cepmlp/gateway/index.php?news=31264> accessed 11 October 2012; Avelien Haan-Kamminga, 'Long-term Liability for Geological Storage in the European Union' (2011) University of Groningen Centre of Energy Law Working Paper <http://papers.ssrn.com/sol3/papers.cfm?abstract_id=1858631> accessed 11 October 2012; and Ian Havercroft, 'Long-term Liability of CCS Business' (presentation at the Global CCS Institute's regional members' meeting in Tokyo, 2012) <www.slideshare.net/globalccs/ian-havercroft-global-ccs-institute-longterm-liability-of-ccs-business> accessed 11 October 2012.

[46] See, for instance, Kristofer Hetland, 'Financial Security from an EU Perspective' (presentation held for the International Energy Agency (IEA) CCS Regulators Network 2010) <www.iea.org/media/workshops/2010/financialmech/Hetland.pdf> accessed 11 October 2012; David Holyoake and Carla Hill (ClientEarth), 'Final Hurdles: Financial Security Obligations under the CCS Directive' (2012) <www.clientearth.org/reports/ccsa-report.pdf> accessed 11 October 2012; European CCS Demonstration Project Network, 'Thematic Report: Regulatory Development Session May 2012' (2012) <www.ccsnetwork.eu/uploads/publications/thematic_report_-_regulatory_development_session_-_may_2012.pdf> accessed 11 October 2012; and Chiara Armeni, 'Key Legal Issues Arising from the Transposition of the EU CO_2 Storage Directive' (presentation held at the 4th IEA International CCS Regulatory Network Meeting, Paris, 2012) <www.iea.org/media/workshops/2012/ccs4thregulatory/Chiara_Armeni.pdf> accessed 11 October 2012.

[47] The London Protocol is a protocol to the 1972 Convention on the Prevention of Marine Pollution by Dumping of Wastes and Other Matter (London Convention). Under the London Protocol all dumping is prohibited except for a limited number of wastes or other matter that may be considered for dumping, listed in Annex I to the Protocol.

[48] On the latter, see e.g. IEA, 'Carbon Capture and Storage and the London Protocol – Options for Enabling Transboundary CO_2 Transfer' (2011) IEA Working Paper <www.iea.org/publications/freepublications/publication/CCS_London_Protocol.pdf> accessed 11 October 2012; Tom Mikunda and others, 'Towards a CO2 Infrastructure in North-Western Europe: Legalities, Costs and Organizational Aspects' (2011) 4 Energy Procedia, 2409, 2409–16; and

that (potential) CCS market players have shown in these issues. They generally view the long-term liability and financial security/mechanism requirements under the CCS Directive as significant obstacles to the large-scale commercial deployment of CCS in the EU. The same goes for the prohibition on the cross-border transport of captured CO_2 for storage in Article 6 of the London Protocol. In recent years, some research has been conducted on legal issues associated with the cross-border deployment of CCS in the EU.[49] Yet, these studies mainly dealt with issues similar to those addressed in previous legal studies on CCS in the EU.

In this thesis, I examine an area that, to my knowledge, has not been explored so far. In analysing the interaction between on the one hand a number of cross-border legal issues related to the CCS Directive and, on the other, their primary EU law context, I sketch part of the legal framework within which (future) market players, the Member States, and the Commission will have to operate. In doing so, I outline the relevant legal boundaries and provide concrete lessons and recommendations for all three of these parties. The thesis is structured as follows.

STRUCTURE

In Chapter I, I briefly explore the concept and the technology of CCS. This provides the reader with the technological background to the regulatory issues addressed in the following chapters by briefly dealing with the concept of CCS and the different capture, compression, transport and storage technologies.

In Chapter II, I give a brief overview of the CCS Directive with a focus on the Directive's provisions on CO_2 storage. The short overview gives the reader a coherent impression of the way in which CO_2 capture, transport and storage are primarily regulated in the EU.

The first two chapters provide a background to the chapters in the second part of this thesis, in which I analyse the selected cross-border legal issues related to the CCS Directive.

IEA and Global CCS Institute, 'Tracking Progress in Carbon Capture and Storage – International Energy Agency/Global CCS Institute Report to the third Clean Energy Ministerial' (2012) <www.iea.org/publications/freepublications/publication/IEAandGlobal CCSInstituteTrackingProgressinCarbonCaptureandStoragereporttoCEM3FINAL-1.PDF> accessed 11 October 2012.

[49] See, for instance, Element Energy (n 31); Mikunda and others, 'Transboundary Legal Issues in CCS' (n 35); and FT Blank and others (CATO2), 'Permitting Cross-border Networks in Relation to Monitoring, Verification and Accounting under EU-ETS' (2011) <www.co2-cato. org/publications/publications/permitting-cross-border-networks-in-relation-to-monitoring-verification-and-accounting-under-eu-ets> accessed 11 October 2012.

In Chapter III, I investigate the possibilities for Member States to adopt CO_2 stream-purity criteria that are more stringent than those in Article 12 of the CCS Directive. As we have seen above, Article 12(1) provides that the 'CO_2 stream shall consist overwhelmingly of carbon dioxide'. Governments in Member States that have shown an interest in accommodating early CCS demonstration projects, are increasingly under pressure to guarantee the safety of early CCS deployment. These Member States could be tempted to increase the safety standard in Article 12. Since the legal basis of the CCS Directive is ex Article 175(1) of the TEC, Member States would be allowed to adopt stricter CO_2 stream-purity criteria, subject to these standards being compatible with the Treaties.

Yet, it is not clear whether such criteria would be compatible with the free movement provisions in the TFEU. The development of different national sets of CO_2 stream-purity criteria could hinder the cross-border transport and storage of captured CO_2, since it would likely increase costs along the CCS value chain. In this chapter, I explore the question of to what extent there is scope under EU law for Member States to adopt more stringent CO_2 stream-purity criteria.

In Chapter IV, I deal with Articles 19 and 20 of the CCS Directive and the compatibility of Member States' financial security/mechanism charges with Article 110 TFEU. Under Article 19 (financial security), Member States are to require applicants for a CO_2 storage permit to prove that they have the financial means to meet all future obligations, including closure and post-closure requirements. By virtue of Article 20 (financial mechanism), Member States are to require the storage operator to make a financial contribution available to the competent authority before responsibility for the storage site is transferred to the latter.

Member States' financial security/mechanism charges under Articles 19 and 20 could form a hindrance to the free movement of captured CO_2 within the EU in that they could make it more difficult for 'foreign' CO_2 streams to be stored than for 'domestic' CO_2 streams. Article 110 TFEU prohibits Member States from imposing on imported products any internal taxation that is in excess of that imposed on similar domestic products or that protects competing domestic products. In this chapter, I answer the question of in what way Member States should design the financial security/mechanism charge for it not to breach Article 110 TFEU.

In Chapter V, I examine Article 21(2)(b) of the CCS Directive, which appears to give Member States the possibility to require the operators of CO_2 transport and storage infrastructure to refuse to grant access to the relevant infrastructure when the capacity concerned is needed to meet part of the Member State's international and EU obligations to reduce greenhouse gas emissions. As a consequence, CO_2 transport and storage capacity in some Member States could (partly) be unavailable to third parties in other Member States, possibly leading to problems for the latter Member States in trying to meet their national targets for the reduction of greenhouse gas emissions.

However, Article 21(2)(b) appears to be at odds with Article 4(3) of the Treaty on European Union (TEU) in conjunction with Article 194(1)(c) TFEU. The former requires Member States to assist each other in carrying out tasks which flow from the Treaties (principle of loyalty), while the latter calls for secondary EU climate and energy legislation to be drafted 'in a spirit of solidarity between Member States'. In Chapter V, I address the questions of to what extent Article 21(2)(b) of the CCS Directive is compatible with Article 4(3) TEU in conjunction with Article 194(1)(c) TFEU and in what way both provisions could narrow the Member States' scope for implementing Article 21(2)(b).

In Chapter VI, as in Chapter V, I analyse third-party access to CO_2 transport and storage infrastructure. Article 21(2)(c) of the CCS Directive would seem to provide the operators of CO_2 transport and storage infrastructure the possibility to refuse to grant access to the relevant infrastructure when the technical specifications of the specific CO_2 stream are incompatible with the required technical standard and the incompatibility 'cannot reasonably be overcome'. As differences in national technical standards for CCS infrastructure are likely to appear, this provision has the potential to form an obstacle to the cross-border trade of captured CO_2.

Yet, the question is whether a refusal to grant third-party access on technical grounds is compatible with Article 102 TFEU, which prohibits the refusal of access to infrastructure which is considered indispensable to enter a certain market. In Chapter VI, I examine the question of to what extent there is scope under Article 102 TFEU for CO_2 transport and storage operators to refuse to grant access to their infrastructure on technical grounds.

In Chapter VII, I explore the development and management of CO_2 transport infrastructure in relation to EU antitrust law. Article 21 of the CCS Directive is silent on the development and management (capacity allocation and congestion management)[50] of CO_2 transport infrastructure in relation to (future) cross-

---

[50] The term capacity allocation refers to the mechanisms that are used by the network operators to allocate new capacity on EU gas and electricity transport markets. For an overview of the types of capacity allocation mechanisms in EU gas and electricity transport markets, see Commission, 'Commission Staff Working Document on Capacity Allocation and Congestion Management for Access to the Natural Gas Transmission Networks Regulated under Article 5 of Regulation (EC) No 1775/2005 on Conditions for Access to the Natural Gas Transmission Networks' SEC(2007) 822. Congestion management procedures are instruments used to

border capacity requirements. This absence of regulation of the development and management of CO_2 transport infrastructure in relation to (future) cross-border capacity requirements could hinder the cross-border deployment of CCS in the EU.

Nevertheless, by virtue of a number of EU competition cases, the operators of CO_2 transport infrastructure could be under several obligations related to the cross-border capacity allocation and congestion management on such infrastructure. In these cases, the Commission addressed the development and management of gas transport infrastructure under Article 102 TFEU. In Chapter VII, I answer the question in what way the relevant cases (also known as the 'gas foreclosure cases') impose requirements to the cross-border capacity allocation and congestion management on CO_2 transport infrastructure.

In Chapter VIII, I examine Article 4(1) of the CCS Directive and the selection of CO_2 storage locations. Article 4(1) of the CCS Directive provides that Member States retain the right to determine the areas from which storage sites are selected, including the right not to allow any CO_2 storage in parts or in the whole of their territory. As a consequence of the little onshore storage capacity that has become available in the first few years of EU CCS demonstration,[51] it is conceivable that the Commission will try and obtain a more central role in the selection of CO_2 storage locations. Recent developments in EU nuclear law show that the poor availability of final disposal capacity – in the case of CCS CO_2 storage capacity – can lead to the adoption of EU legislation obliging Member States to create final disposal facilities.[52]

As part of the review of the CCS Directive in 2015, the Commission could try to amend Article 4(1) in order for the provision to allow it to force Member States to accept CO_2 storage in (parts of) their territory. In such a scenario, it would make sense for the Commission to designate a number of storage locations to be used for the storage of CO_2 captured in several Member States. This could arguably facilitate the cross-border deployment of CCS since it would create opportunities for Member States with little or some of their own storage capacity.

---

allocate and re-allocate capacity allocated by the transmission system operator to market participants in a transparent, fair and non-discriminatory manner to prevent or remedy a situation of scarcity of network capacity. See ERGEG, 'ERGEG Principles: Capacity Allocation and Congestion Management in Natural Gas Transmission Networks' (2008) 17.

51   See, for instance, the Introduction to Chapter III on CO_2 stream-purity and Member States' scope to impose stricter norms.

52   See n 5 Chapter VIII. While leaving some flexibility as to the dates a disposal facility is put into operation, Articles 1(1), 2, 4(3)(c), 11 and 12 of Directive 2011/70/Euratom of 19 July 2011 on the responsible and safe management of spent fuel and radioactive waste [2011] OJ L199/48, oblige the Member States to initialise without undue delay the process towards the planning and the realisation of disposal of spent fuel and radioactive waste.

In Chapter VIII, I sketch the EU legal framework with which an amendment of Article 4(1) for these purposes will have to comply. More specifically, I answer the questions of to what extent Articles 192 and 194 TFEU would provide proper legal bases for such an amendment of Article 4(1) and to what extent an amended Article 4(1) would be compatible with Article 345 TFEU on EU rules and Member States' system of property ownership.

In Chapter IX, I deal with the public funding of CO_2 transport and storage infrastructure in view of EU state aid law. As we have seen above, recent developments have (strongly) increased the chances of early CCS projects having to cross Member States' borders. For the cross-border deployment of CCS, it is of great importance that sufficient early CO_2 transport and storage infrastructure is developed. An obvious way to advance the development of such infrastructure is to provide public funding.[53] Yet, when Member States decide to support the development of CO_2 transport and storage infrastructure, Article 107(1) TFEU comes into play. Article 107(1) TFEU prohibits Member States from granting aid that distorts competition and affects trade between Member States.

To avoid the uncertainty and loss of time caused by the notification and standstill procedure under Article 108(3) TFEU, which could delay the development of CO_2 transport and storage infrastructure, Member States would be wise to make sure that public support for the development of CO_2 transport and storage infrastructure does not constitute state aid under Article 107(1) TFEU in the first place. One way to do so is to have the public funding meet the conditions listed in the *Altmark* case, as a consequence of which compensation for the discharge of a service of general economic interest[54] does not constitute state aid.[55] The most difficult and indeterminate of the *Altmark* criteria, arguably,

---

[53] At the EU level, preparatory steps for the construction of CCS infrastructure with public means have already been taken. Under the new entrants reserve of the EU greenhouse emission trading scheme (EU ETS), a scheme that sets aside emissions allowances, 300 million allowances are available for the co-financing of 12 commercial CCS demonstration projects as well as renewable energy demonstration projects. See Article 10(a)(8) of the EU ETS Directive.

[54] The Commission has defined services of general economic interest as 'market services which the Member States subject to specific public service obligations by virtue of a general interest criterion'. See Commission, 'Communication from the Commission – Services of General Interest in Europe' [2001] C 17/04, 23.

[55] Case C-280/00 *Altmark* [2003] ECR I-07747. In *Altmark*, the ECJ held that for the compensation for the discharge of a service of general economic interest to escape classification as state aid, (1) the recipient undertaking has to have clearly defined public service obligations, (2) the parameters for calculating the compensation must have been objectively and transparently established, (3) the compensation must not exceed all or part of the costs of discharging the public service obligations, taking into account the relevant receipts and a reasonable profit, and (4) the public service provider must be selected by means of a public procurement procedure or, alternatively, the level of compensation for the discharge of the service of general economic interest/public service obligation has to be based

is the fourth criterion, also known as the 'efficiency criterion'. In Chapter IX, I explore the question of in what way Member State authorities should design public funding of CO_2 transport and storage infrastructure for it to be compatible with the *Altmark* efficiency criterion and thus to escape classification as state aid, provided that the other three *Altmark* criteria are met.

Finally, in the conclusions, I answer the questions of to what extent the relevant legal issues could indeed hamper the (near-)future cross-border deployment of CCS in the EU (and explore possible solutions if required) and in what way the legislative approach chosen by the Commission could generally lead to problems from the point of view of the cross-border application of new technologies such as CCS. Also, I make a number of concrete recommendations, both in relation to individual issues explored in the different chapters, as well as on the Commission's regulatory approach towards the cross-border deployment of CCS.

The research for this book is based on a study of the relevant case law of the EU Courts, the Commission's decision practice, legal literature and literature on CCS technology, economics and policy, as well as documents from the EU institutions (predominantly from the Commission). The relevant case law and decision practice of the EU Courts and the Commission was selected by identifying the applicable provision(s) from the TFEU and TEU and subsequently searching for secondary and primary sources on these provisions (the latter by means of search terms in EUR-Lex and the Commission's competition cases search engine). The law is stated as at 1 December 2012. Wherever possible, subsequent developments have been taken into account. All referencing is in accordance with the fourth edition of the Oxford University Standard for the Citation of Legal Authorities (OSCOLA).

---

on the costs of a typical, well-run undertaking adequately provided with the means to meet the necessary public service requirements (efficiency criterion).

CHAPTER I

CARBON CAPTURE AND STORAGE: CONCEPT AND TECHNOLOGY

Even though the technology has been applied for several decades in the (US) oil and gas industry, CCS has only recently come to the fore as an instrument to reduce greenhouse gas emissions, both in the EU and the rest of the world. Technological developments are proceeding rapidly, in turn increasing the chances of mature EU markets for the capture, transport and storage of CO_2 developing and of CCS becoming a realistic option for reducing greenhouse gas emissions. In the following sections, I will give a brief overview of the concept and technology of CCS.[1] The intention is to provide the reader with the technological background to the regulatory issues that are addressed in the following chapters.

Before discussing the different capture, compression, transport and storage technologies/concepts, it is important to stress that CCS is not a single technology, but a combination of processes that involves technologies which successively capture, compress, transport, inject (into suitable storage reservoirs), and monitor CO_2.[2] While all the component technologies exist today, and a complete CCS system can be assembled from them, the state of development of the component technologies is at different points on the maturity scale.[3] Moreover, the state of development of a fully-integrated CCS system is less than some of its separate components.[4] The industry so far has very limited experience with projects that integrate capture, transport, and long-term storage of CO_2 at a scale to produce commercial base load power.[5] At the time of

---

[1]   This chapter draws heavily on the United Nations Intergovernmental Panel on Climate Change (IPCC) 2005 Special report on CCS. Even though the report dates from 2005, it still provides one of the most complete and authoritative assessments of the different CCS technologies. More recent reports by other institutions, such as the International Energy Agency (IEA) and the Carbon Sequestration Leadership Forum (CSLF) have been used to add to the information provided by the IPCC report.

[2]   Hiranya Fernando and others, 'Capturing King Coal: Deploying Carbon Capture and Storage Systems in the U.S. at Scale' (World Resources Institute 2008) 10 <http://pdf.wri.org/capturing_king_coal.pdf> accessed 18 March 2010.

[3]   Ibid.

[4]   Ibid.

[5]   Ibid.

writing, there are only five fully-integrated, commercial-scale CCS projects in operation, in Norway, Algeria and North America.[6]

1.1 CAPTURE TECHNOLOGIES[7]

CO_2 capture requires separating CO_2 from industrial or energy-related emissions into relatively pure streams, which can then be pressurised for transport. The main application of CCS is likely to be at large point sources[8] of CO_2 emissions such as fossil fuel power plants, fuel processing plants and other industrial plants, particularly for the manufacture of iron, steel, cement and bulk chemicals.[9] There are four basic methods for capturing CO_2 from use of fossil fuels:

1) capture from industrial process streams;
2) post-combustion capture;
3) oxy-fuel combustion capture;
4) pre-combustion capture.

1.1.1 CAPTURE FROM INDUSTRIAL PROCESS STREAMS

CO_2 has been captured from industrial process streams for 80 years.[10] Current examples of CO_2 capture from process streams are purification of natural gas and production of hydrogen-containing synthesis gas for the manufacture of ammonia, alcohols and synthetic liquid fuels.[11] CO_2 capture from industrial processes is a mature technology, but is not expected to contribute to significant abatement of emissions, since high concentration industrial sources represent only a limited share of the sector's total emissions (3% to 4%).[12] Moreover, single production units tend to be smaller point sources than power plants, which

---

[6]   This was the state of play in March 2010. See International Energy Agency, 'Technology Roadmap: Carbon Capture and Storage' (2009) 10.

[7]   For a clear illustration of the different capture technologies see the CCS-dedicated website of energy utility Vattenfall at <www.vattenfall.com/en/ccs/capture-of-co2.htm> accessed 2 May 2010.

[8]   A point source of pollution is a single identifiable localised source of air, water, thermal, noise or light pollution. In this case, it is a source of CO_2 emissions.

[9]   IPCC, 'IPCC Special Report on Carbon Dioxide Capture and Storage. Prepared by Working Group III of the Intergovernmental Panel on Climate Change' (Bert Metz and others (eds) Cambridge University Press 2005) 108.

[10] Ibid.

[11] Ibid.

[12] Fernando and others (n 2) and IEA, 'CO_2 Capture and Storage: A Key Carbon Abatement Option' (2008) 66.

increases the capital cost of CO_2 capture per unit of output.[13] Nevertheless, CO_2 capture from industrial processes may still be significant, since it could provide early examples of solutions that can be applied on larger scale elsewhere.[14] In the iron and steel industry for example, the largest energy-consuming manufacturing sector in the world,[15] a number of early opportunities exist for the capture of CO_2 emissions.[16] According to recent IEA estimates, CCS used together with oxygen injection could result in a reduction of 85% to 95% of the CO_2 emissions attributable to the core production process (up to 75% of total emissions).[17]

1.1.2 POST-COMBUSTION CAPTURE

Capture of CO_2 from flue gases produced by combustion of fossil fuels and biomass in air is referred to as post-combustion capture.[18] Post-combustion capture is a commercially available[19] mature technology used at many locations around the world in a wide range of industrial manufacturing processes, refining and gas processing.[20] The same capture technologies can also be applied to power plants, but have yet to be fully demonstrated at commercial-scale[21] power plants.[22] Increasing the technology to this size is generally not considered to be a major problem.[23]

In post-combustion capture technology, flue gas, instead of being discharged directly to the atmosphere, is passed through equipment which separates most of

---

[13]   IEA, 'Key Carbon Abatement' (n 12).
[14]   IPCC (n 9) 113.
[15]   10–15% of total industrial consumption, i.e. 10% of worldwide CO_2 emissions. See IEA, 'Key Carbon Abatement' (n 12) 67.
[16]   IPCC (n 9) 112–13.
[17]   IEA, 'Key Carbon Abatement' (n 12) 68. This concurs with IPCC estimates. See IPCC (n 9) 113.
[18]   IPCC (n 9) 109.
[19]   The term 'commercially available' is frequently used in various IEA (and other) reports on CCS, but rarely defined. It does, however, seem to refer to the final phase of the 'technology development cycle', in which the technology operates in a mature market and is distributed on a large scale. In its 2005 special report on CCS, the IPCC states that 'mature market' means that the technology is in operation with multiple replications of the commercial-scale technology worldwide. Before a technology becomes market mature, it is usually confined to niche markets and used only in selected commercial applications. See IPCC (n 9) 5. For more information on the technology development cycle see the IPCC website <www.ipcc.ch/publications_and_data/ar4/wg3/en/tssts-ts-2-6-technology-research.html> accessed 9 June 2010.
[20]   IEA, 'Key Carbon Abatement' (n 12) 48 and IEA, 'Technology Roadmap' (n 6) 25.
[21]   The IEA does not indicate what it considers to be a commercial-scale power plant, i.e. up to what capacity (MW) post-combustion capture technologies have been fully demonstrated at power plants.
[22]   IEA, 'Technology Roadmap' (n 6) 25.
[23]   IEA Greenhouse Gas R&D Programme, 'Capturing CO_2' (2007) 2.

the CO_2.[24] The CO_2 is fed to a storage reservoir and the remaining flue gas is discharged to the atmosphere.[25] There are several commercially available process technologies which can in principle be used for CO_2 capture from flue gases.[26] However, absorption processes based on chemical solvents[27] are currently the preferred option for post-combustion CO_2 capture.[28] These processes are proven and commercially available at present.[29]

Considering that man-made CO_2 emissions from stationary sources come mostly from combustion systems such as power plants, furnaces in industries and iron and steel production plants, the strategic importance of post-combustion capture systems becomes evident.[30] Any attempt to mitigate CO_2 emissions from stationary sources on a relevant scale using CCS will have to address CO_2 capture from combustion systems.[31] Various alternative capturing processes (e.g., the use of novel solvents) are currently being researched.[32] Before post-combustion technologies can be deployed in large-scale power plants (around 500 MW), several challenges remain such as scaling up proven technologies to 20–50 times[33] that of current unit capacities and reducing the gross energy penalty.[34, 35]

1.1.3 OXY-FUEL COMBUSTION CAPTURE

The oxy-fuel combustion process eliminates N_2 (nitrogen) from the flue gas by combusting a hydrocarbon or carbonaceous fuel in either pure oxygen or a mixture of pure oxygen and a CO_2-rich recycled flue gas (instead of air).[36] This results in a flue gas that is mainly CO_2 and H_2O.[37] After cooling to condense water vapour, the net flue gas contains about 80–98% CO_2, depending on the fuel

---

[24] IPCC (n 9) 109.
[25] Ibid.
[26] Ibid 114.
[27] In absorption processes, CO_2 is separated from other flue gases by using a chemical solvent that reacts with CO_2 in an absorption process.
[28] IPCC (n 9) 114.
[29] Ibid 121.
[30] Ibid 113–14.
[31] Ibid 114.
[32] Ibid 118–20.
[33] Ibid 121.
[34] Post-combustion capture processes based on absorption have a high energy requirement, reducing the efficiency of power cycles. This is primarily due to the required heating of the solvent (to break chemical bonds between the solvent and CO_2) and to a lesser extent to processes such as the compression of the CO_2 product. See IPCC (n 9) 117.
[35] IEA, 'Technology Roadmap' (n 6) 25–26.
[36] IPCC (n 9) 122.
[37] Ibid 109.

type used and the particular oxy-fuel combustion process.[38] The high concentration of CO_2 makes it easy to further concentrate the flue gas to an almost pure CO_2 stream, ultimately enhancing capture efficiency.

At present, a number of different oxy-fuel combustion technologies exist, with different systems to supply the heat for combustion. The CO_2 capture efficiency in oxy-fuel combustion capture systems is very close to 100%.[39] Although elements of the technology are in use in the aluminium, iron, steel and glass melting industries today, they have yet to be deployed on a commercial scale.[40] However, the key separation step in most oxy-fuel capture systems (O_2 from air) is an existing technology.[41] The energy penalty for producing oxygen is by far the most important cause for reduced efficiency in an oxy-fuel cycle compared to a conventional power plant[42] and one of the several challenges that have to be met before the technique can be deployed on a large scale.[43]

1.1.4 PRE-COMBUSTION CAPTURE

Pre-combustion capture involves the removal of CO_2 from the primary fuel before combustion takes place. The fuel is reacted with air or oxygen to produce a fuel that contains CO (carbon monoxide) and H_2 (hydrogen).[44] This synthesis gas is then reacted with steam to produce a mixture of CO_2 and H_2, after which the CO_2 is separated through an absorption process and transported for storage.[45] The hydrogen is combusted in a gas turbine to generate power, resulting in a flue gas consisting only of water vapour.[46]

It is possible to envisage two applications of pre-combustion capture.[47] The first is in producing a fuel (hydrogen) that is essentially carbon-free. Secondly, pre-combustion capture can be used to reduce the carbon content of fuels, with the excess carbon being made available for storage. Pre-combustion technologies are commercially used in various industrial applications and all the components of the process have been tested at pilot plant scale.[48] For large scale application,

---

[38]   IPCC (n 9) 122.
[39]   Ibid.
[40]   Ibid.
[41]   Ibid.
[42]   Ibid 130.
[43]   IEA, 'Technology Roadmap' (n 6) 26.
[44]   IEA GHG R&D Programme (n 23) 3.
[45]   The Parliament of the Commonwealth of Australia, 'Between a Rock and a Hard Place: The Science of Geosequestration' (2007) 29.
[46]   Ibid.
[47]   IPCC (n 9) 130.
[48]   IEA, 'Key Carbon Abatement Option' (n 12) 51.

pre-combustion capture is still in the developmental stages, but there are a number of promising technologies.[49]

Because of their potential to deliver on a large scale and at high thermal efficiencies, a suitable mix of electricity, hydrogen and lower carbon-containing fuels or chemical feedstocks in an increasingly carbon-constrained world, pre-combustion technologies are considered to be of high strategic importance.[50] Critical elements that need further development are for instance the hydrogen turbines, where the hydrogen is used for electricity generation.[51]

1.1.5 CARBON CAPTURING AND POWER PLANT EFFICIENCY

Capturing and compressing CO_2 requires energy and thus reduces the thermal efficiency of power plants and increases the amount of fossil fuel used to achieve a given power generation output.[52] In 2007, the IEA examined the thermal efficiency of a number of power plants, both with and without CO_2 capture.[53] Several factors contribute to the efficiency reductions for CO_2 capture. These factors vary depending on the fuel and technology used for combustion. Examples are power and steam required for CO_2 capture, CO_2 compression/ purification and the production of oxygen.

1.2 COMPRESSION TECHNOLOGIES

Before CO_2 is transported, it is compressed into a supercritical fluid[54] to make transport more efficient and thus cheaper. CO_2 compression uses the same equipment as natural gas compression, a mature, well-developed technology in the natural gas industry, with some modifications to suit the properties of CO_2.[55] There is substantial operating experience in compressing and handling

---

49 Parliament of Australia (n 45) 29. For an overview of these promising technologies see IPCC (n 9) 130–38.
50 IPCC (n 9) 141.
51 IEA, 'Key Carbon Abatement Option' (n 12) 51.
52 IEA GHG R&D Programme (n 23) 5–7 and China-UK Near Zero Emissions COAL (NZEC) Initiative, 'Summary Report' (2009) 8 <www.nzec.info/en/assets/Reports/China-UK-NZEC-English-031109.pdf> accessed 22 March 2010.
53 IEA GHG R&D Programme (n 23) 5–7.
54 A supercritical fluid is at a pressure and temperature above its critical point, i.e. the phase boundary between liquid and gas terminates and the properties of the gas and liquid phases approach one another, resulting in only one phase at the critical point. The fluid in question will then have the properties of both a gas and a liquid.
55 The Delphi Group and Alberta Research Council, 'Building Capacity for CO_2 Capture and Storage in the APEC Region: A Training Model for Policy Makers and Practitioners' (2005)

CO_2 in large-scale applications (e.g., for enhanced oil recovery operations [EOR][56]).[57] The amount of energy required for pressurisation depends on the transportation distance and the pressure of the underground reservoir (which depends on its depth).[58] According to the IEA, typical pressurisation of CO_2 reduces plant efficiency by 4–5%.[59]

Past experience has shown that the development of novel compression technologies takes a very long time.[60] Avoiding corrosion and hydrate formation[61] are the main operational issues when dealing with CO_2.[62] Since CO_2 dissolves in water and forms carbonic acid ($CO_2 + H_2O \rightarrow H_2CO_3$), which is corrosive, minimising the water content in the CO_2 stream ('dry CO_2') is essential for safe operation of the compressor.[63] Because of the high energy requirements needed for compression, CO_2 compression is still considered to be a major obstacle in addressing capture needs.[64]

1.3   TRANSPORT TECHNOLOGIES

CO_2 is transported in three states: gas, liquid and solid.[65] Transporting CO_2 as a solid is currently not cost-effective or feasible from an energy usage point of view.[66] There are many ways to transport CO_2, such as by pipeline, marine tanker, train, truck or compressed gas cylinder. However, at the moment only transport by pipeline and tanker are considered commercially reasonable

---

4–2   <http://canmetenergy-canmetenergie.nrcan-rncan.gc.ca/fichier/78892/apec_training_2005.pdf> accessed 22 March 2010.

56   The term enhanced oil recovery refers to a number of technologies that can be used to increase the amount of oil that can be extracted from fields by pumping e.g. water or CO_2 into the reservoir. Alongside enhanced oil recovery, there are other enhanced hydrocarbon recovery technologies such as enhanced gas recovery (EGR) and enhanced coal bed methane (a gas extracted from coal) recovery (ECBM).

57   Fernando and others (n 2) 12.

58   IEA, 'Key Carbon Abatement Option' (n 12) 65.

59   Ibid.

60   Rajat Suri, 'CO_2 Compression for Capture-Enabled Power Systems' (MSc thesis, Massachusetts Institute of Technology 2009) 152.

61   A hydrate is a compound which is formed when water or its elements are attached to another molecule.

62   Delphi Group and Alberta Research Council (n 55) 4–2.

63   Fernando and others (n 2) 12.

64   Suri (n 60) 50. In a recent stakeholder survey of the UK-China Near Zero Emissions Coal Initiative (NZEC), a joint UK-China Initiative to demonstrate CCS in China and the EU by 2020, the high energy penalty and corresponding costs of compression technologies were seen as an important barrier to CCS. See NZEC (n 52) 41.

65   IPCC (n 9) 181.

66   IEA, 'Key Carbon Abatement Option' (n 12) 81.

options for the large quantities of CO_2 associated with centralised collection hubs or point source emitters such as large power plants.[67]

Pipelines have been used for several decades to transport CO_2 obtained from natural underground or other sources to oil fields for enhanced oil recovery purposes.[68] To avoid corrosion and hydrate formation, CO_2 should be dried before it is transported by pipeline. If the CO_2 cannot be dried, it may be necessary to build the pipeline of a corrosion-resistant alloy ('stainless steel').[69] Field experience indicates very few problems with transportation of high-pressure dry CO_2 in carbon steel pipelines.[70]

Depending on the source of the flue gas, the CO_2 stream recovered may contain concentrations of substances such as H_2S (hydrogen sulphide), SO_x (sulphur oxide), NO_x (nitrogen oxide), O_2, N_2 (nitrogen gas) and Ar (argon).[71] These impurities might have an impact on the physical state of the CO_2 stream and hence the operation of the compressors, pipelines and storage tanks.[72] This can be solved by purification of the CO_2 stream. Since CO_2 is transported in a supercritical state (ten times denser than methane), and since the assumed average distance between booster stations would be 200 km (compared to between 120 km and 160 km for natural gas), transporting CO_2 will require less energy than transporting natural gas over the same distance.[73]

On the basis of its BLUE Map scenario (50% reduction of greenhouse gases by 2050 – 1/5 of which to be achieved by CCS),[74] the IEA estimates that between 70,000km to 120,000km and 200,000km to 360,000km of CO_2 pipeline will be needed globally in 2030 and 2050, respectively.[75] The US, China and OECD regions will need about half of this network.[76] By 2050, according to the IEA scenario, the pipeline network in these regions will have to be of a size that exceeds the total expected length of China's natural gas pipelines in 2015

---

[67]   Carbon Sequestration Leadership Forum (CSLF), 'Technology Roadmap' (2009) 8.
[68]   Ibid.
[69]   IPCC (n 9) 182.
[70]   Ibid.
[71]   Delphi Group and Alberta Research Council (n 55) 4–7.
[72]   Ibid.
[73]   IEA, 'Key Carbon Abatement Option' (n 12) 82–83.
[74]   A 50% reduction of greenhouse gas emissions by 2050 (compared to 1990 levels) is in line with the EU's aim to keep global warming below 2°C above the pre-industrial temperature, equivalent to below 1.2°C above today's level. See Commission, 'Building a post-2012 global climate regime: the EU's contribution' (2010) <http://ec.europa.eu/environment/climat/future_action.htm> accessed 23 March 2010.
[75]   IEA, 'Technology Roadmap' (n 6) 30.
[76]   Ibid.

(100,000km).[77] By 2050, the CO_2 network in the US would need to transport a mass equivalent to three times the total amount of gas transported in natural gas pipelines.[78] The considerable size of the required network is necessitated by the fact that storage sites often are located some distance away from the CO_2 source.[79]

The above does not, however, mean that all required pipeline capacity needs to be constructed anew. In a 2008 report commissioned by the Netherlands Ministry of Economic Affairs and the Netherlands Oil and Gas Exploration and Production Association (NOGEPA), the potential for reuse of existing gas transport, injection and storage infrastructure was researched in relation to CO_2 storage in depleted gas fields on the Dutch continental shelf.[80] As rebuilding a completely new infrastructure for CO_2 storage will be expensive and technically challenging, reuse of existing gas infrastructure is an attractive option.[81] The report draws two relevant conclusions.[82] First, price developments in the oil and gas markets make the 'out of operation' or abandonment year, and thereby the availability, of existing gas pipelines very uncertain. Second, as long as the properties of the transported CO_2 stream are such that corrosion is avoided, there will probably not be any technical impediments to the use of existing gas pipelines. Nevertheless, to preserve the existing gas infrastructure for CO_2 transport and storage, maintenance will be required.

While CO_2 transport by *ship* is an established technology on a smaller scale, the quantities of CO_2 involved in CCS are enormous and present an altogether different challenge in terms of shipping capacity.[83] Moreover, CO_2 is continuously captured at the plant on land, but the cycle of ship transport is discrete, and so a marine transportation system requires temporary storage on

---

[77]   Oilandgaspress.com, 'China's gas pipeline length to reach 100,000 km in 2015' (22 January 2010) <www.oilandgaspress.com/wp/2010/01/22/chinas-gas-pipeline-length-to-reach-100000-km-in-2015/> accessed 23 March 2010.

[78]   IEA, 'Key Carbon Abatement Option' (n 12) 83.

[79]   In the Netherlands, for example, only a limited number of industrial installations and power plants are responsible for the majority of CO_2 emissions (resp. 12 for 73% and 20 for 81% in 2004). However, the Netherlands is surrounded by several areas with a large number of big point sources (UK, Ruhr area), which together are responsible for about half of the yearly European CO_2 emissions (1500 Mt). This, in combination with the availability of substantial storage capacity, has been thought to make transport of CO_2 to the Netherlands an attractive and potentially economically viable option. See Platform Nieuw Gas, Werkgroep CO_2-opslag/ Schoon fossiel, 'Beleidsrapport Schoon Fossiel' 19.

[80]   NOGEPA, 'Potential for CO_2 storage in depleted gas fields on the Dutch continental shelf' (2008) <www.nogepa.nl/LinkClick.aspx?fileticket=G11SzzTQuFc%3d&tabid=546&language=en-GB> accessed 23 June 2010.

[81]   Ibid 7 and 15.

[82]   Ibid 30.

[83]   Fernando and others (n 2) 13.

land and a loading facility.[84] The advantages of transport by ship are that it offers increased flexibility in routes and that it may be cheaper than pipelines (reduction in infrastructure capital costs),[85] particularly for longer distance transportation.[86] As a short-term measure, ship and train transport present viable options, particularly for regions that have low prospective CO_2 storage capacities.[87]

1.4 STORAGE TECHNOLOGIES

Captured CO_2 can be stored in certain types of geological formations, by injecting it into the ocean[88] and in a number of other ways, such as through *mineral carbonation*,[89] *biofixation*,[90] and various *industrial uses*.[91] At the time of writing, none of the storage options, except geological storage of CO_2, seem to be particularly promising in terms of storage capacity and duration, and public acceptance. In the following, I will therefore address the geological storage of CO_2 only.

1.4.1 GEOLOGICAL STORAGE OF CO_2

In geological storage, CO_2 is injected under high pressure into deep, stable rocks in which there are countless tiny pores that trap natural fluids.[92] The first engineered injection of CO_2 into subsurface geological formations took place in Texas (US) in the early 1970s, as part of an enhanced oil recovery project. Ever since, geological storage of CO_2 has taken place at several locations. The world's

---

[84]   IPCC (n 9) 186.
[85]   IEA, 'Key Carbon Abatement Option' (n 12) 84.
[86]   CSLF, 'Technology Roadmap' (n 67) 9.
[87]   IEA, 'Technology Roadmap' (n 6) 30.
[88] CO_2 ocean storage is not further discussed here as Article 2(4) of the CCS Directive explicitly prohibits the storage of CO_2 in the water column. Article 3(2) of the Directive defines the water column as 'the vertically continuous mass of water from the surface to the bottom sediments of a water body'.
[89]   Mineral carbonation is based on the reaction of CO_2 with metal oxide bearing materials such as calcium and magnesium to form mineral carbonates. See IPCC (n 9) 322.
[90]   Biofixation or algal bio-sequestration is a technique for the production of biomass using CO_2 and solar energy, typically employing microalgae or cyano-bacteria. See CSLF, 'Technology Roadmap' (n 67) 15.
[91]   *Industrial uses* of CO_2 are numerous. CO_2 is for example used in food and beverages, horticulture, welding and safety devices and the fertiliser industry.
[92]   IEA Greenhouse Gas R&D Programme, 'Geological Storage of Carbon Dioxide: Staying Safely Underground' (2008) <www.co2crc.com.au/dls/external/geostoragesafe-IEA.pdf> accessed 24 March 2010.

first large-scale CO_2 storage project was initiated by Statoil and its partners at the Sleipner Gas Field in the North Sea in 1996.

To geologically store CO_2, it must first be compressed, usually to a supercritical fluid state. Depending on the rate that temperature increases with depth, the density of CO_2 will increase (and volume will decrease) with depth, until at about 800m or deeper, the injected CO_2 will be in a dense supercritical state.[93] Geological storage of CO_2 can be undertaken in a variety of geological settings in sedimentary basins.[94] Oil fields, depleted gas fields, deep coal seams and saline formations[95] are all possible storage formations.[96] These formations are generally considered to have the greatest potential capacity for CO_2 storage.[97]

In addition to storage in sedimentary formations, storage in natural and man-made caverns, basalt and organic-rich shales has been considered.[98] However, storage in caverns and mines cannot, realistically, be expected to make a significant contribution to tackling climate change.[99] The majority of mines are not leak proof, especially at pressures much greater than atmospheric.[100] Moreover, those mines that are leak proof have alternative uses.[101] Subsurface geological storage is possible both onshore and offshore, with offshore sites accessed through pipelines from the shore or from offshore platforms.[102] The continental shelf and some adjacent deep-marine sedimentary basins are potential offshore storage sites, but the majority of sediments of the abyssal deep ocean floor are too thin and impermeable to be suitable for geological storage.[103]

1.4.1.1 Underground CO_2 behaviour

Injection of CO_2 into deep geological formations is achieved by pumping the CO_2 down into a well. The equipment and practices for injection are widely used in the oil and gas industry.[104] Injection equipment is fully commercial, although

---

93   IPCC (n 9) 200.
94   Ibid.
95   A saline formation is a sediment or rock body containing brackish water or brine. See CO_2 Capture Project, 'Glossary' <www.co2captureproject.com/glossary.html> accessed 24 March 2010.
96   IPCC (n 9) 200.
97   CSLF, 'Technology Roadmap' (n 67) 10.
98   IPCC (n 9) 200.
99   British Geological Survey (BGS), 'Sequestration: The Underground Storage of Carbon Dioxide' in EJ Moniz (ed), *Climate change and energy pathways for the Mediterranean: Workshop proceedings, Cyprus* (Springer 2007) 65.
100   Ibid.
101   Ibid.
102   IPCC (n 9) 200.
103   Ibid.
104   IEA GHG R&D Programme (n 92) 14.

more advanced technologies for CO_2 injection are also being developed.[105] Once the CO_2 is injected into the reservoir, the processes of migration and trapping begin. Various flow and transport mechanisms control the spread of CO_2.

The rate of fluid flow depends on the number and properties of the fluid phases present in the formation.[106] Supercritical CO_2 is much less viscous than water and oil, creating buoyancy forces that drive CO_2 upwards. In natural gas reservoirs, CO_2 is denser than natural gas, with CO_2 migrating downwards. In saline formations and oil reservoirs, the buoyant plume of injected CO_2 migrates upwards, but not evenly. This is because permeability barriers within the reservoir cause the CO_2 to migrate laterally.[107] As the density of CO_2 increases and its volume decreases when it is injected at depths between approximately 500m and 1,000m, CO_2 occupies much less space in the subsurface than at the surface.[108] This makes storage of large masses of CO_2 in shallow reservoir rocks impractical, since the physical conditions at shallow depths underground mean that relatively small masses of CO_2 would occupy relatively large volumes of pore space.[109]

1.4.1.2 CO₂ storage mechanisms

1.4.1.2   CO$_2$ storage mechanisms

A combination of physical and geochemical trapping mechanisms determines the effectiveness of geological CO_2 storage. The most effective storage sites are those where CO_2 is immobile because it is trapped permanently under a thick, low-permeability seal, converted to solid minerals, or adsorbed on the surfaces of coal micro pores or through a combination of physical and chemical trapping mechanisms.[110]

The physical trapping of CO_2 below low-permeability seals ('cap rocks'), such as very low-permeability shale or salt beds, is the principal means of storing CO_2 in geological formations.[111] Physical trapping mechanisms range from mechanical trapping under impermeable rocks (*stratigraphic* and *structural* trapping) to processes whereby CO_2 dissolves in the formation water and then migrates with the groundwater (*hydrodynamic* trapping) or is trapped in the tiny pores of rocks[112] by the capillary pressure of water (*residual* trapping). Eventually, the

[105]   IEA GHG R&D Programme (n 92) 14.
[106]   IPCC (n 9) 205.
[107]   Ibid 206.
[108]   BGS (n 99) 66.
[109]   Ibid.
[110]   IPCC (n 9) 208.
[111]   Ibid.
[112]   After the injection of CO_2 stops, water from the surrounding rocks will begin to move back into the pore spaces containing the CO_2. This pressure from the water ('capillary pressure') may immobilise significant amounts of CO_2 (5–30%). See IPCC (n 9) 206.

injected CO_2 will be confined by geochemical trapping mechanisms, i.e. a sequence of geochemical interactions with the rock and the formation water. As a first step, much of the injected CO_2 will eventually dissolve in the saline water or in the oil that remains in the rock (*solubility* trapping).[113]

Next, depending on the rock formation, the dissolved CO_2 may react chemically with the surrounding rocks to form stable minerals (*mineral* trapping).[114] This is a slow process which takes place over thousands of years, but it provides the most secure form of storage.[115]

1.4.1.3 Storage formations

For a geological formation to be suitable for CO_2 storage, it needs to have (1) adequate capacity and injectivity,[116] (2) a satisfactory sealing cap rock or confining unit, and (3) a sufficiently stable geological environment to avoid compromising the integrity of the storage site.[117] The latter criterion makes basins formed in mid-continent locations or near the edge of stable continental plates excellent targets for long-term CO_2 storage.[118] Two important attributes of possible storage or seal rock are porosity[119] and permeability[120] of the rock.[121] Rocks suitable for storage typically have high porosity to provide space for the CO_2 and high permeability for the CO_2 to move into that space.[122] By contrast, cap rocks typically have low porosity and permeability to trap the fluids stored below.[123]

a) Oil and gas fields

In depleted oil and gas fields, CO_2 fills the pores in the rocks that were once filled with oil or natural gas.[124] For a number of reasons, depleted oil and gas reservoirs

---

113 IEA GHG R&D Programme (n 92) 8.
114 Ibid.
115 Ibid.
116 Injectivity refers to the rate at which the CO_2 can be injected into a storage reservoir formation. Typically, the CO_2 must be injected at much the same rate as it is captured from the sources. See IEA GHG R&D Programme (n 92) 8.
117 IPCC (n 9) 213.
118 Ibid.
119 Porosity is a measure of the space in the rock for storing fluids. It usually reduces with depth, which reduces storage capacity and efficiency. See IEA GHG R&D Programme (n 92) 8 and IPCC (n 9) 214.
120 Permeability is a measure of the ability of the rock to allow fluid flow, which is strongly affected by the shape, size and connectedness of the spaces in the rock. See IEA GHG R&D Programme (n 92) 8.
121 IEA GHG R&D Programme (n 92) 8.
122 Ibid.
123 Ibid.
124 Ibid 9.

are considered to be prime candidates for CO_2 storage.[125] First, the oil and gas that originally accumulated in traps (structural and stratigraphic) did not escape (in some cases for many millions of years), which is said to demonstrate their integrity and safety. Second, the geological structure and physical properties of most oil and gas fields have been extensively studied and characterised. Third, computer models have been developed in the oil and gas industry to predict the movement, displacement behaviour and trapping of hydrocarbons. Finally, some of the infrastructure and wells already in place may be used for handling CO_2 storage operations. However, those same wells of often variable quality and integrity also pose a risk, since they could constitute leakage pathways for the stored CO_2.[126] Care must be taken to ensure that exploration and production operations have not damaged the reservoir or seal (especially in the vicinity of the wells), and that the seals of shut-in wells[127] remain intact.[128]

b)   Saline formations

Saline formations[129] are widespread, including in areas with no appreciable oil and gas production, and contain enormous quantities of water which are unsuitable for agriculture or human consumption.[130] Saline formations meet all the necessary criteria to provide long-term storage, although their quality and capacity to store CO_2 varies depending on their geological characteristics.[131] Deep saline formations provide by far the largest potential volumes for geological storage of CO_2.[132] However, exact estimates are for a number of reasons very difficult to make.[133] Apart from being abundant, saline formations also have the advantage that they do not have the problem of abandoned oil and gas production wells. The Sleipner project[134] in the Norwegian sector of the North Sea was the first demonstration of CO_2 storage in a deep saline formation designed specifically in response to climate change mitigation.[135] Following Sleipner, several other comparable large-scale storage projects in deep saline

---

[125]   IPCC (n 9) 215.
[126]   CSLF, 'Technology Roadmap' (n 67) 11.
[127]   A shut-in well is a well that is closed off.
[128]   CSLF, 'Technology Roadmap' (n 67) 11.
[129]   See n 95. This includes 'deep saline aquifers', i.e. deep underground rock formations composed of permeable materials and containing highly saline fluids. See IPCC (n 9) 404.
[130]   IPCC (n 9) 217.
[131]   IEA GHG R&D Programme (n 92) 10 and CSLF (n 67) 11.
[132]   CSLF, 'Technology Roadmap' (n 67) 10.
[133]   IPCC (n 9) 222. The storage mechanisms in saline formations, for example, operate both simultaneously and on different time scales, making capacity estimates highly time-bound.
[134]   In 1996, Norwegian Statoil began with the injection of approximately 1 Mt CO_2 per year into the Utsira formation at a depth of about 1,000m below the sea floor. Over the lifetime of the project a total of 20 Mt CO_2, equalling about 10% of the total yearly Netherlands' greenhouse gas emissions is expected to be stored.
[135]   CSLF, 'Technology Roadmap' (n 67) 11.

formations have taken off, including the In Salah gas project in Algeria and the Snøhvit LNG project (Barents Sea).

c)   Unmineable coal beds and other storage options

Unmineable coal beds[136] could also be used to store CO_2.[137] CO_2 pumped into a coal seam is not only stored by becoming adsorbed onto the coal, it may also displace any CH_4 (methane) at the adsorption sites (CO_2 has a greater affinity to be adsorbed onto coal than methane).[138] The adsorbed CO_2 is likely to be stored for geological time[139] and remain in place even without cap rocks, as long as the coal is not mined or depressurised.[140] CO_2 storage in coal is, however, limited to a relatively narrow depth range.[141] The likely future fate of a coal seam is a key determinant of its suitability for storage.[142] Conflicts between mining and CO_2 storage are possible, particularly for shallow coals. The methane in coal represents only a small proportion of the energy value of the coal, and the remaining energy would be sterilised if the coal was used as a CO_2 storage reservoir, i.e. the coal could not be mined or gasified without releasing the CO_2 into the atmosphere.[143]

In the US, Burlington Resources have injected over 100 $KtCO_2$ into coal since 1996.[144] Recent results have shown that some US low-rank coals could store 5 to 10 times as much CO_2 as the methane they originally contained.[145] *Other geological storage options* such as salt caverns, oil shales, mines and basalts are in the early stages of development, and appear to have limited capacity except possibly as local niche options for geological storage of CO_2.[146]

1.4.1.4   Global geological storage capacity estimates

Due to the lack of data, the fact that frequently more than one trapping mechanism is active, and several other factors, it is quite difficult to estimate global capacity for geological storage of CO_2.[147] Global capacity estimates have

---

[136]   These are coal beds below a depth that can be economically mined.
[137]   CSLF, 'Technology Roadmap' (n 67) 12.
[138]   BGS (n 99) 75.
[139]   The IPCC special report defines geological time as 'the time over which geological processes have taken place'. See IPCC (n 9) 406.
[140]   IPCC (n 9) 219 and CSLF, 'Technology Roadmap' (n 67) 12.
[141]   CSLF, 'Technology Roadmap' (n 67) 12.
[142]   IPCC (n 9) 219.
[143]   BGS (n 99) 75.
[144]   Ibid.
[145]   IEA, 'Key Carbon Abatement Option' (n 12) 99.
[146]   CSLF, 'Technology Roadmap' (n 67) 12.
[147]   IPCC (n 9) 220.

generally been calculated by simplifying assumptions and using very simplistic methods, which makes them unreliable.[148] The uncertainties related to global storage capacity estimates are clearly outlined in some of the studies conducted so far.[149] There are large differences between what have been called 'theoretical', 'realistic' (technically viable), and 'viable' (economically viable) capacity estimates, and different projects have in the past used different reduction factors to relate these estimates.[150] In an overview of regional and worldwide estimates used by the IEA in a 2008 report on CCS e.g., global storage capacity estimates range from 100–200,000 $GtCO_2$.[151]

Nevertheless, according to the British Geological Survey, global underground storage capacity is likely to be large.[152] Given that oil and gas fields occupy only a very small part of the saline water-bearing reservoir rocks in the world's sedimentary basins, it would be highly unlikely that the storage capacity of the latter would be less than the former.[153] Thus, the British Geological Survey estimates total global storage capacity to be sufficient for at least 80 years of CO_2 storage and probably more.[154] In its 2009 Technology Roadmap on CCS the IEA indicates that in theory, global storage capacity is more than sufficient to store over 1.2 $GtCO_2$ in 2020 and 145 $GtCO_2$ in 2050.[155]

---

[148]   IPCC (n 9) 220.
[149]   See, for instance, CSLF, 'Phase II Final Review from the Taskforce for Review and Identification of Standards for CO_2 Storage Capacity Estimation' (2007); BGS (n 99); IEA, 'Key Carbon Abatement Option' (n 12) and IEA, 'Technology Roadmap' (n 6).
[150]   CSLF, 'Phase II' (n 149).
[151]   IEA, 'Key Carbon Abatement Option' (n 12) 88.
[152]   BGS (n 99) 74.
[153]   Ibid.
[154]   Ibid.
[155]   IEA, 'Technology Roadmap' (n 6) 32. By comparison: in 2007, global CO_2 emissions from fuel combustion were close to 30 $GtCO_2$. See IEA, 'CO_2 emissions from fuel combustion – Highlights' (2009) 44.

CHAPTER II

THE CCS DIRECTIVE:
A BRIEF OVERVIEW

In the following, I briefly explore the way in which the CCS Directive regulates the three parts of the CCS chain: the capture, transport and storage of CO_2. For two reasons, most attention is paid to the Directive's provisions on CO_2 storage. First, these provisions make up the bulk of the regulation in the CCS Directive. Second, a number of provisions dealing with CO_2 capture and transport are explored in more depth in the following chapters. I briefly mention these provisions below, but refer to the following chapters for a more detailed examination. Before examining the Directive's rules covering the different parts of the CCS chain, I briefly discuss the scope of the CCS Directive as well as the way in which the EU has chosen to (financially) incentivise CCS deployment. By providing a short overview of the CCS Directive's regulation of the three parts of the CCS chain, I intend to give the reader a coherent impression of the way in which CO_2 capture, transport and storage are primarily regulated in the EU.

2.1   SCOPE OF THE CCS DIRECTIVE

Article 2 provides that the CCS Directive applies to the geological storage of CO_2 in the territory of the Member States, their exclusive economic zones and their continental shelves within the meaning of the United Nations Convention on the Law of the Seas (UNCLOS). Storage beyond these areas is not allowed. This means that the CCS Directive does not allow the Member States to store their captured CO_2 in the subsoil of the high seas. Following the OSPAR Convention,[1] the CCS Directive further expressly prohibits CO_2 storage in the water column. By virtue of the definition of the term 'water column' in Article 3(2), this prohibition includes CO_2 storage in the deep water or on the

---

[1]   The OSPAR Convention is a regional treaty on the protection of the marine environment in the North-East Atlantic. By virtue of Decision 2007/1 of the OSPAR Commission, the placement of CO_2 streams in the water column or on the seabed is, with a few exceptions, prohibited. The EU, as well as 13 EU Member States, are contracting parties to the OSPAR Convention.

bottom of lakes. In addition, small-scale R&D CO_2 storage projects (< 100 kilotonnes) are excluded from the scope of the Directive.

Interestingly, recital 20 of the preamble to the CCS Directive states that the injection of captured CO_2 to recover hydrocarbons such as oil and gas (enhanced hydrocarbon recovery) is not in itself included in the Directive. It provides that where enhanced hydrocarbon recovery is combined with geological storage of CO_2, the provisions of the CCS Directive should apply. Article 3(1) of the Directive defines 'geological storage of CO_2' as 'injection accompanied by storage of CO_2 streams in underground geological formations'. Following the definition in Article 3(1), the question is whether the distinction between enhanced hydrocarbon recovery and geological CO_2 storage made in recital 20 is tenable.

In practice, both CO_2 storage projects and enhanced hydrocarbon recovery projects will likely lead to a significant amount of the injected CO_2 being geologically stored. According to the International Energy Agency (IEA), CO_2 storage projects, joint CO_2 storage/enhanced hydrocarbon recovery projects and enhanced hydrocarbon recovery projects are all likely to result in significant quantities of CO_2 being geologically stored.[2] On the basis of the definition in Article 3(1), I would therefore argue that enhanced hydrocarbon recovery activities, contrary to that alleged in recital 20,[3] would fall within the scope of the CCS Directive.

2.2 INCENTIVISING CCS DEPLOYMENT

The main incentive for CO_2 emitters to deploy CCS technologies is provided by the EU greenhouse gas emissions trading scheme (EU ETS). The idea is that every tonne of geologically stored CO_2 will count as not having been emitted under the EU ETS. Accordingly, it does not have to be covered by an emissions allowance. By storing their CO_2 emissions, EU ETS emissions permit holders, such as power plants and refineries, can either sell the emissions allowances they keep or not buy any allowances in the first place.

If the price of an emissions allowance under the EU ETS is higher than the costs of capturing, transporting and geologically storing a tonne of CO_2, EU ETS emissions permit holders will have a financial incentive to deploy CCS

---

[2] IEA, 'Carbon Capture and Storage – Model Regulatory Framework' (2010) 114 <www.iea.org/publications/freepublications/publication/model_framework.pdf> accessed 11 February 2013.
[3] In this regard, it needs to be stressed that the preamble to an EU Directive, contrary to its provisions, has no binding legal force. See, for instance, Case C-162/97 *Nilsson* [1998] ECR I-07477.

technologies. Once CCS becomes a cost-effective option to reduce CO_2 emissions, the participants in the EU ETS are expected to automatically start deploying these technologies.

In phase II of the EU ETS (2008–2012) individual CCS projects could already be recognised under the emissions trading scheme through prior approval by the Commission.[4] To provide a long-term incentive for the further development of the different technologies, CCS is fully recognised as an instrument to reduce CO_2 emissions in the revised (post-2012) EU ETS.[5] Article 12(3)(a) of the EU ETS Directive provides that no emissions allowances have to be surrendered for CO_2 emissions that are permanently stored in accordance with the CCS Directive.

In addition to the above, the EU has arranged for various forms of subsidies for CCS deployment. First, Article 10(3) of the EU ETS Directive states that Member States should use at least 50% of the revenues from the auctioning of emissions allowances for a number of climate policy instruments, including CCS.[6] Second, by virtue of Article 10(a)(8) of the EU ETS Directive up to 300 million allowances in the new entrants' reserve are available for the construction and operation of the up to 12 planned EU CCS demonstration projects as well as innovative renewable energy demonstration projects. The sale of the first 200 million allowances from the new entrants' reserve brought in €1.5 billion, €1.2 billion of which has been granted to renewable energy projects. For various reasons, no CCS projects have been awarded funding, but this is expected to be different in the second funding round. Finally, under the European Economic Recovery Plan, €1 billion has been granted to six EU CCS demonstration projects.

These subsidies are meant to bridge the gap between the costs of CCS deployment (per tonne of CO_2 emissions avoided) and the prevailing low emissions allowance prices under the EU ETS. While the different subsidies contribute to making CCS a financially more attractive instrument for reducing CO_2 emissions, they distort the EU emissions trading market. The effect of subsidising specific climate technologies under the EU ETS is a greater deployment of relatively expensive technologies, eventually leading to lower emissions allowance prices

---

[4]   See Article 24 of European Parliament and Council Directive 2003/87/EC of 13 October 2003 establishing a scheme for greenhouse gas emission allowance trading within the Community and amending Council Directive 96/61/EC [2003] OJ L275 (EU ETS Directive).

[5]   To this end, the capture, transport and geological storage of greenhouse gases (including CO_2) have been added to Annex I to the EU ETS Directive, which contains the list of activities for which an EU ETS emissions permit is required.

[6]   On the basis of the wording of Article 10(3), I would argue that the Member States, however, are not *obliged* to spend any auctioning revenues on the development of CCS technologies.

and rising overall CO_2 emissions reduction costs.[7] The EU subsidisation of CCS technologies, in other words, comes at the expense of the cost-effectiveness of emissions reductions realised under the EU ETS.

2.3 CO_2 CAPTURE

2.3.1 CO_2 STREAM PURITY

One of the most important provisions in the CCS Directive regulating CO_2 capture is Article 12. Article 12 contains the requirements regarding the purity of the CO_2 stream for storage. This refers to the substances other than CO_2 (impurities) that are allowed to be in the captured CO_2 stream. The composition of the captured CO_2 stream can vary from project to project as a consequence of differences in, for instance, fuel type, capture process and post-capture processing. Article 12 does not prescribe any concrete purity standards. It provides that a CO_2 stream for storage has to consist *overwhelmingly* of CO_2 and may not be used for the disposal of waste. In Chapter III, I further examine the requirements in Article 12 in relation to Member States' scope to impose stricter norms.

2.3.2   ENVIRONMENTAL IMPACT ASSESSMENT

Article 31 of the CCS Directive brings capture installations under the scope of Directive 85/337/EC (Environmental Impact Assessment Directive).[8] As a consequence, an environmental impact assessment has to be carried out in the capture permit process. Not all capture installations will, however, automatically be made subject to an environmental impact assessment. I further explain this procedure in section 3.2.1 of Chapter III.

---

[7]   See, for instance, Heleen Groenenberg and Heleen de Coninck, 'Effective EU and Member State Policies for Stimulating CCS' (2008) 2 International Journal of Greenhouse Gas Control, 653, 658.

[8]   Council Directive 85/337/EEC of 27 June 1985 on the assessment of the effects of certain public and private projects on the environment [1985] OJ L175 (Environmental Impact Assessment Directive). The Environmental Impact Assessment Directive provides that development consent for public and private projects which are likely to have significant effects on the environment should be granted only after prior assessment of the likely significant environmental effects of these projects has been carried out. See the preamble and Article 2 of the Environmental Impact Assessment Directive.

2.3.3 OBLIGING CO_2 CAPTURE

By virtue of Article 37 of the CCS Directive, the capture of CO_2 is brought under the scope of Directive 2008/1/EC (Integrated Pollution Prevention and Control Directive).[9] This ensures that best available techniques (BATs)[10] to improve the composition of the CO_2 stream have to be established and applied.[11] Also, combustion plants with a rated thermal input of at least 50 MW are required to meet the emission limits set by Directive 2001/80/EC (Large Combustion Plants Directive).[12] The effect of the emissions limits in the environmental permits issued under the Integrated Pollution Prevention and Control and Large Combustion Plants Directives is that CO_2 capturers will generally already have to reduce flue gas concentrations of substances such as sulphur dioxide and nitrogen oxides.

During the negotiations on the CCS Directive, one of the most debated issues was the possible introduction of an emissions performance standard[13] de facto obliging new large coal fired power plants to capture their CO_2 emissions. The European Commission (Commission) had originally proposed to include in the Large Combustion Plants Directive a provision that obliged each new large combustion plant to have suitable space on its site for capture and compression equipment and required the availability of suitable transport facilities, storage sites and the technical feasibility of retrofitting the installation with capture equipment

---

9   European Parliament and Council Directive 2008/1 of 15 January 2008 concerning integrated pollution prevention and control [2008] L24. The Integrated Pollution Prevention and Control Directive lays down measures to minimise the release of polluting emissions into air, water or soil in order to achieve a high level of protection for the environment as a whole. In 2011, the Directive was replaced by European Parliament and Council Directive 2010/75/EU of 24 November 2010 on industrial emissions (integrated pollution prevention and control) (Recast) [2010] OJ L334/17 (Industrial Emissions Directive), with effect from January 2011.

10   Article 2(10) of the Industrial Emissions Directive defines the term 'best available technique'. In essence, it boils down to the requirement that the installations concerned use the, from an environmental protection point of view, best technology for reducing emissions, as long as it is economically and technically viable to do so.

11   Recital 27 of the CCS Directive.

12   European Parliament and Council Directive 2001/80 of 23 October 2001 on the limitation of emissions of certain pollutants into the air from large combustion plants [2001] L309. This Directive sets limits on sulphur dioxide, nitrogen oxides and dust emissions. Like the Integrated Pollution Prevention and Control Directive, the Large Combustion Plants Directive was replaced by the Industrial Emissions Directive in 2011.

13   In a proposed amendment to the Commission's original proposal for the CCS Directive (discussed below), the European Parliament defined an emissions performance standard as 'the maximum permissible quantity of CO_2 that may be emitted to air per unit of electrical output, calculated as grams per kilowatt hour (g CO_2/kwh) on an annual average basis'. See European Parliament, 'Report on the proposal for a directive of the European Parliament and of the Council on the geological storage of carbon dioxide and amending Council Directives 85/337/EEC, 96/61/EC, Directives 2000/60/EC, 2001/80/EC, 2004/35/EC, 2006/12/EC and Regulation (EC) No 1013/2006 (COM(2008)0018 – C6-0040/2008 – 2008/0015(COD))' A6-0414/2008 amendment 126 <www.europarl.europa.eu/sides/getDoc.do?pubRef=-//EP//NONSGML+REPORT+A6-2008-0414+0+DOC+PDF+V0//EN> accessed 14 February 2013.

to have been assessed (capture readiness).[14] The intention was to make sure that large industrial installations initially constructed without CO_2 capture equipment could still decide to start capturing their CO_2 emissions at a later moment.

The European Parliament (Parliament) had sought to amend the proposed Article 32 of the CCS Directive by replacing the capture readiness requirement with an emissions performance standard obliging all new large power plants not to emit more than 500 g CO_2/kWh, as from January 2015.[15] Parliament's amendment, however, was not adopted in the final text of the CCS Directive. Instead, the capture readiness requirement proposed by the Commission was weakened.

The resulting Article 33 requires the operators of new large combustion plants to reserve space on the installations site for capture and compression equipment under three conditions only: a suitable storage site is available and it is technically and economically feasible to both retrofit the installation with capture equipment and develop a CO_2 transport facility. Considering the high costs associated with CO_2 capture, I would argue that the requirement of economic feasibility seriously weakens the capture readiness obligation in Article 33.

Even though Parliament and the Council of the EU (Council) could not agree on an emissions performance standard for new large power plants, the idea has not entirely been abandoned. Article 38(3) of the CCS Directive requires the Commission to examine the necessity of a later introduction of an emissions performance standard for new large power plants, when reviewing the CCS Directive in 2015.[16] Nevertheless, this assessment shall take place only if CCS has proven to be a safe and economically feasible instrument for reducing CO_2 emissions.

Besides proposing an EU emissions performance standard for new large power plants, Parliament also suggested to delete Article 9(3)(iii) of the Integrated Pollution Prevention and Control Directive.[17] This provision required permits issued under the Integrated Pollution Prevention and Control Directive not to contain an emission limit value[18] for emissions of greenhouse gases covered

---

[14] See Article 32 of the Commission's original proposal for the CCS Directive: Commission, 'Proposal for a Directive of the European Parliament and of the Council on the geological storage of carbon dioxide and amending Council Directives 85/337/EEC, 96/61/EC, Directives 2000/60/EC, 2001/80/EC, 2004/35/EC, 2006/12/EC and Regulation (EC) No 1013/2006' COM (2008) 18 final.

[15] European Parliament n 13.

[16] See Article 38 of the CCS Directive.

[17] European Parliament (n 13) amendment 124. Article 9(3)(iii) has been replaced by Article 9(1) of the Industrial Emissions Directive.

[18] Article 2(8) of the Industrial Emissions Directive defines an emission limit value as 'the mass, (...) concentration and/or level of an emission, which may not be exceeded during one or more periods of time'.

under the EU ETS, unless necessary to prevent significant local pollution. By deleting Article 9(3)(iii), Parliament sought to clear the way for Member States to introduce CO_2 emissions performance standards through the permits granted under the Integrated Pollution Prevention and Control Directive.

It could be argued that Parliament's amendment was unnecessary as Article 9(3(iii) only (explicitly) prohibited the introduction of an emission limit value and did not refer to emissions performance standards. Nevertheless, emissions performance standards can be seen as a subcategory of the prohibited emission limit values. In that light, Parliament's wish to delete the prohibition in Article 9(3)(iii) made sense. Yet, like the proposal for an EU emissions performance standard, this amendment did not make it into the final text of the CCS Directive. Article 9(1) of the Industrial Emissions Directive still prohibits the inclusion of an emission limit value for EU ETS greenhouse gases[19] in the permit granted under the Directive.[20]

2.4 CO_2 TRANSPORT

EU regulation of CO_2 transport for storage is essentially based on the idea that captured CO_2 is transported by pipeline and not, for instance, by ship. Several provisions in the CCS Directive demonstrate this point. First, Article 3(22) defines the transport network as 'the network of pipelines, including associated booster stations, for the transport of CO_2 to the storage site'. Even though shipping routes, including associated shipping terminals or CO_2 hubs used to load captured CO_2 onto tankers, could just as well be part of a CO_2 transport network, there is no mentioning of them in Article 3(22).

Second, by virtue of Directive 2009/29/EC (revising EU ETS Directive),[21] only the transport of greenhouse gases (for storage) *by pipelines* has been added to the list of activities in Annex I of the EU ETS Directive. Accordingly, no EU ETS permit is required for the transport of greenhouse gases (for storage) by ship. A

---

19   From 1 January 2013, a number of new activities responsible for emissions of nitrous oxide and perfluorocarbons have been included in the EU ETS. The permit issued under the Industrial Emissions Directive may also not contain an emission limit value for these greenhouse gases.

20   Nonetheless, a lively (practical) debate has developed on the legality of national CO_2 emissions performance standards under EU law. On this discussion, see, for instance, Lorenzo Squintani, Marijn Holwerda and Kars de Graaf, 'Regulating Greenhouse Gas Emissions from EU ETS installations: What Room is Left for the Member States?' in Marjan Peeters, Mark Stallworthy and Javier de Cendra de Larragán, *Climate Law in EU Member States: Towards National Legislation for Climate Protection* (Edward Elgar 2012) 67.

21   European Parliament and Council Directive 2009/29/EC of 23 April 2009 so as to improve and extend the greenhouse gas allowance trading scheme of the Community [2009] L140/63.

reason for this could be the quite extensive EU and international maritime safety regulation CO_2 shipping is subjected to.[22]

Third, nowhere in the CCS Directive is the option of CO_2 transport per ship mentioned. This is remarkable, considering the explicit reference to shipping as one of the two main kinds of modes of EU CO_2 transport in the Commission's impact assessment of the CCS Directive.[23]

2.4.1 THIRD-PARTY ACCESS

The most important rules on CO_2 transport for storage in the CCS Directive, arguably, are the requirements regarding third-party access to CO_2 transport infrastructure. Third-party access refers to access to the relevant infrastructure by parties who do not own or control that infrastructure. Article 21 of the CCS Directive addresses third-party access to CO_2 transport networks and storage sites. By virtue of this provision, it basically is up to each Member State to determine the manner in which to design their regime for third-party access to CO_2 transport and storage infrastructure, as long as such access is provided in a fair, open and non-discriminatory manner. The CCS Directive does not prescribe a particular third-party access regime.[24] I further examine the third-party access requirements in Article 21 in Chapters V, VI and VII.

2.5 CO_2 STORAGE

Point of departure for the geological storage of CO_2 in the EU is the full discretion of Member States to select CO_2 storage sites, including the right not to allow for any storage in their territory.[25] The provisions in the CCS Directive cover the three basic phases of any CO_2 storage project: pre-storage, operation,

---

[22] Commission, 'Accompanying Document to the Proposal for a Directive of the European Parliament and of the Council on the geological storage of carbon dioxide: Impact Assessment' COM (2008) 18 final paras 86–89.

[23] Ibid para 72.

[24] However, Brockett and Roggenkamp have both argued that the CCS Directive's provisions on third-party access to transport networks and storage sites intend to apply the principles of negotiated rather than regulated access. See Scott Brockett, 'The EU Enabling Legal Framework for Carbon Capture and Geological Storage' (2009) 1 Energy Procedia 4433, 4439; and Martha M Roggenkamp, 'The Concept of Third Party Access Applied to CCS' in Martha M. Roggenkamp and Edwin Woerdman (eds), *Legal Design of Carbon Capture and Storage* (Intersentia 2009) 273, 293.

[25] Article 4(1) of the CCS Directive. As recital 19 of the preamble to the Directive notes, Member States might want to put potential CO_2 storage reservoirs to other uses, such as natural gas storage or the production of geothermal energy. These considerations were not included in the text of recital 19 as originally proposed by the Commission.

and closure and post-closure. In the following, I briefly explore the legal requirements in all three phases of CO_2 storage.

2.5.1 PRE-STORAGE

Before any CCS project can start injecting CO_2 underground, two steps have to be taken. First, a proper storage site has to be selected. Second, all permitting procedures need to have been completed.

2.5.1.1 Storage site selection

From a safety point of view, the selection of a storage site with the appropriate geological characteristics is crucial.[26] Article 4(4) of the CCS Directive determines that a storage site may only be selected if there are no significant risks of leakage and no significant environmental or health risks. Whether no such risks exist is to be determined through a geological characterisation of the site by means of three-dimensional computer modelling, as described in Annex 1 to the Directive. This modelling is used to predict the underground behaviour of injected CO_2. Site characterisation, arguably, is the most time-consuming and costly part of the CO_2 storage site selection process.[27]

2.5.1.2 Permitting procedures

If it is necessary to obtain more information about the potential storage site by, for instance, performing drilling activities or injection tests, such activities may only be undertaken on the basis of an exploration permit.[28] Likewise, no CO_2 storage site may be operated without a storage permit, the Directive's central instrument for guaranteeing the environmentally safe geological storage of CO_2 in the EU.[29] Both permitting procedures appear to be based on the regime applied in Directive 94/22/EC (Hydrocarbon Licensing Directive).[30] Permits are exclusive, granting the holder the sole right to perform the exploration or storage activities, and are awarded through competitive procedures based on objective,

---

26 See recital 19 of the preamble to the CCS Directive.

27 CO2CRC, 'Storage Capacity Estimation, Site Selection and Characterisation for CO_2 Storage Projects' (2008) 2 <www.co2crc.com.au/dls/pubs/08–1001_final.pdf> accessed 21 February 2013.

28 Article 5(1) of the CCS Directive.

29 Article 6(1) of the CCS Directive and recital 24 of the preamble to the CCS Directive.

30 Martha M. Roggenkamp, Kars de Graaf and Marijn Holwerda, 'Afvang, Transport en Opslag van CO_2: De Implementatie van Richtlijn 2009/31/EG in Nederland' (2010) 37 Milieu en Recht 548, 553. This is European Parliament and Council Directive 94/22/EC of 30 May 1994 on the conditions for granting and using authorisations for the prospection, exploration and production of hydrocarbons [1994] OJ L164.

published and transparent criteria. Applicants have to show that they are financially sound and technically competent as well as reliable to operate and control the site. To encourage investments in exploration, exploration permit holders have priority in the storage permitting procedure.

As part of the storage permit procedure, applicants have to submit, among other things, the results of the compulsory environmental impact assessment of the potential storage site. Importantly, Article 19 requires potential storage operators to provide financial security. This financial security is to ensure that the operator can meet all obligations arising from the permit during operation, closure and post-closure, until responsibility for the storage site has been transferred to the competent authority. In case a storage operator defaults on his obligations under the Directive, the competent authority can draw on the financial security. The financial security requirements in Article 19 are controversial and have been heavily debated by industry.

To guarantee consistency between the permitting procedures in the different Member States the Commission reviews all draft storage permits.[31] It issues a non-binding opinion, from which the Member States may not depart without stating their reasons. The Commission's opinion is based on a review by a scientific panel, which assesses whether the information on the basis of which the permit decision is to be taken is comprehensive and reliable.[32] In this way, the Commission hopes to exchange information and best practices and to enhance public confidence in early CCS deployment.

2.5.2 OPERATION

2.5.2.1 Monitoring

As soon as a storage permit storage has been awarded, CO_2 injection into the storage reservoir may start. In the operating phase, all efforts are directed at ensuring that the injected CO_2 remains safely in the underground. Crucial in this regard are the monitoring requirements in Article 13 of the CCS Directive. This provision requires storage operators to monitor the injection facilities, the storage complex[33] (including the CO_2 plume) and the surrounding environment. This is to verify whether the underground behaviour of the injected CO_2 matches the modelling done in the pre-storage phase and to keep a close eye on possible

---

[31] Article 10 of the CCS Directive.
[32] Commission, 'Accompanying Document' (n 23) para 138.
[33] Article 3(6) of the CCS Directive defines the storage complex as 'the storage site and surrounding geological domain which can have an effect on overall storage integrity and security'.

adverse effects on the environment. The storage operators need to report the results of monitoring to the competent authority at least once a year.[34]

2.5.2.2 Inspections

In addition, Member States have to organise a system of routine and non-routine inspections of all storage complexes covered by the CCS Directive.[35] As part of these inspections, surface installations such as the injections facilities are visited and all relevant records kept by the storage operator are checked. The routine inspections are conducted at least once a year until three years after closure of the site or five years after the transfer of responsibility for the storage site to the competent authority. Non-routine inspections are carried out in case of problems such as leakage and serious complaints related to the environment or human health. After each inspection, the competent authority has to prepare a report on the results of the inspection, evaluating whether the operator (still) complies with the requirements of the storage permit. Under Article 11(3), it has the power to withdraw the permit in the event of the storage operator not meeting his obligations.

2.5.2.3 When something goes wrong

Article 16 of the CCS Directive deals with the measures that are to be taken in case of leakage[36] or other 'significant irregularities'.[37] In the event of leakage or a risk of leakage or damage to the environment or human health, the operator has to immediately notify the competent authority and take corrective measures. The latter basically are measures that prevent or stop injected CO_2 from leaking from the reservoir. If the competent authority decides to take any corrective measures, it will draw on the financial security required under Article 19 in order to recover any costs made.

2.5.2.4 Liabilities

Besides the obligation to take corrective measures, storage operators bear three types of liability in relation to the geological storage of CO_2 pursuant to the CCS Directive: environmental, climate and civil liability.

---

34 Article 14(1) of the CCS Directive.
35 Article 15 of the CCS Directive. This, for instance, excludes the small-scale R&D storage projects mentioned in section 2.1.
36 By virtue of Article 3(5) of the CCS Directive, leakage is 'any release of CO_2 from the storage complex'.
37 Article 3(17) of the CCS Directive defines a significant irregularity as 'any irregularity in the injection or storage operations or in the condition of the storage complex itself, which implies the risk of a leakage or risk to the environment or human health'.

The first refers to liability for damage to protected species and natural habitats, water and land. By virtue of Article 34 of the CCS Directive, liability for environmental damage is regulated by Directive 2004/35/EC (Environmental Liabilities Directive).[38] As a consequence, the Environmental Liabilities Directive's strict liability regime applies to CO_2 storage activities. This means that the operator of a storage site may be held financially liable for environmental damage caused by CO_2 storage activities, even if he is not at fault. In case of (an imminent threat of) environmental damage occurring, he is required to take (preventive and) remedial measures pursuant to Articles 5(1) and 6(1) of the Environmental Liabilities Directive.

The second type of liability arises when stored CO_2 leaks into the atmosphere. The revising EU ETS Directive adds the geological storage (as well as the capture and transport) of greenhouse gases to Annex I to the EU ETS Directive. As a consequence, storage operators need to surrender emissions trading allowances for any leaked greenhouse gas emissions. Industry has generally perceived climate liability as a great financial obstacle to deploying CCS technologies. This is mainly due to the uncertainties surrounding the size of the required compensation, which are predominantly caused by the unpredictability of EU ETS allowance prices.

Finally, civil liability arises when CO_2 storage causes damage to individuals or property. Underground storage of CO_2 could, for instance, cause minor earthquakes and ground movement, which could damage property. Recital 34 explicitly states that liabilities other than those covered by the CCS Directive, EU ETS Directive and Environmental Liabilities Directive should be dealt with at national level. Civil liability is therefore solely regulated under national law.

2.5.3 CLOSURE AND POST-CLOSURE

2.5.3.1 Closure

Under Article 17 of the CCS Directive, a storage site may be closed when:

a) the conditions in the storage permit have been met;
b) an operator's request for closure is authorised by the competent authority; or
c) the competent authority decides so after withdrawal of a storage permit

---

[38] This is European Parliament and Council Directive 2004/35/EC of 21 April 2004 on environmental liability with regard to the prevention and remedying of environmental damage [2004] L143.

2.5.3.2 Transfer of responsibility

Importantly, the operator of a storage site remains responsible for that site until responsibility is transferred to the competent authority. After closure of a site, the storage operator would normally not want to keep responsibility for the site indefinitely, since this would be too large a business risk.[39] Therefore, responsibility for the storage is to be transferred to the competent authority when:[40]

a) the stored CO_2 will likely be completely and permanently contained;
b) a minimum period of at least 20 years, or less if the competent authority so decides, has elapsed;
c) the operator has made available a financial contribution covering the costs of the competent authority after transfer of responsibility;[41] and
d) the site has been sealed and injection facilities have been removed

The transfer of responsibility to the competent authority includes environmental and climate liabilities, but excludes civil liability. The latter is, as indicated above, not regulated by the CCS Directive. Depending on national legislation, a storage operator could, therefore, transfer the environmental and climate liability for the storage site only a few years after closure, whereas he could bear civil liability for the same site for a much longer period.

It is important to underline that the transfer of responsibility, including environmental and climate liabilities, does not mean that the storage operator will not, in any event, face post-closure costs. Where there has been fault on the part of the operator, including cases of deficient data, concealment of relevant information, negligence, wilful deceit or a failure to exercise due diligence, the competent authority has to recover incurred costs from the former storage operator.[42]

In order to demonstrate the likelihood of stored CO_2 staying completely and permanently contained (first criterion in Article 18(1)), the storage operator has to demonstrate that:

a) the injected CO_2 behaves as modelled;
b) there are no detectable leakages; and
c) the store site is evolving towards a situation of long-term stability.[43]

---

[39]   What is more, the lifespan of companies is probably shorter than the (post-closure) period during which a storage site needs to be monitored and watched.
[40]   Article 18(1) of the CCS Directive.
[41]   See Article 20 of the CCS Directive. I further discuss this 'financial mechanism' below.
[42]   Article 18(7) of the CCS Directive. See also recital 35 of the preamble to the Directive.
[43]   Article 18(1) of the CCS Directive.

A consistent application of these criteria is guaranteed in two ways. First, Article 18(2) provides the Commission with the opportunity to issue guidelines on the assessment of these matters, which it did in 2011.[44] As shown by the guidelines, permanent containment of stored CO_2 is demonstrated through modelling, based on the modelling done in the pre-storage phase. Second, by virtue of Article 18(4), the Commission reviews all drafts decisions to transfer responsibility in a way similar to its review of draft storage permits.

2.5.3.3   *Financial mechanism*

Finally, just like there is a provision ensuring that the storage operator can meet all financial obligations during operation of the storage site, there is a provision that guarantees that the competent authority is able to do the same after it has become responsible for the site. Under Article 20, the storage operator has to make a financial contribution (financial mechanism) to the competent authority before responsibility for the storage site is transferred. The contribution may be used to cover the costs incurred by the competent authority in ensuring that the CO_2 is completely and permanently contained.[45] This financial mechanism should cover at least the anticipated cost of monitoring for a period of 30 years.

2.6   IMPLEMENTING THE CCS DIRECTIVE

Being a Directive, the CCS Directive needs to be implemented into national law. By virtue of Article 288 TFEU, Directives are binding as to the results to be achieved, but allow the Member States to choose the form and methods of implementation. Article 39 of the CCS Directive requires the Member States to have had transposed the Directive into national law by 25 June 2011. Spain is the only Member State to have met this transposition deadline.[46] As a consequence,

---

[44]   Commission, 'Implementation of Directive 2009/31/EC on the Geological Storage of Carbon Dioxide – Guidance Document 3: Criteria for Transfer of Responsibility to the Competent Authority' (2011) <http://ec.europa.eu/clima/policies/lowcarbon/ccs/implementation/docs/gd3_en.pdf> accessed 26 February 2013.

[45]   To help the Member States in implementing Articles 19 and 20, the Commission also adopted guidelines for these provisions in 2011. See Commission, 'Implementation of Directive 2009/31/EC on the Geological Storage of Carbon Dioxide – Guidance Document 4: Article 19 Financial Security and Article 20 Financial Mechanism' (2011) <http://ec.europa.eu/clima/policies/lowcarbon/ccs/implementation/docs/gd4_en.pdf> accessed 26 February 2013.

[46]   See Chiara Armeni, 'An Update on the State of CCS Regulation from Europe' (2012) <www.globalccsinstitute.com/insights/authors/chiara-armeni/2012/02/13/update-state-ccs-regulation-europe> (accessed 18 March 2013). On the implementation process see also the website of the Global CCS Institute on the status of the transposition of the CCS Directive in the EU Member States <www.globalccsinstitute.com/networks/cclp/legal-resources/dedicated-ccs-legislation/europe/transposition-status> accessed 19 March 2013.

in 2011, the Commission started a large number of infringement procedures under Article 258 TFEU.[47]

The extent to and speed at which Member States have complied with their obligations to transpose the CCS Directive have varied quite considerably as a result of a series of factors, including public opposition to the technology (the Netherlands and Germany), complex division of powers between the regions and central government (the UK and Germany) and complicated internal legal procedures (Poland).[48] In similar vein, the way in which the Member States have implemented the CCS Directive into national law has varied considerably.

The Netherlands, for instance, have predominantly implemented the Directive by amendment of national mining legislation, the Dutch Mining Act,[49] whereas Germany has chosen to create a separate CCS act, the Act on the Demonstration and Use of the Technology for the Capture, Transport and Permanent Storage of CO_2 (KSpG).[50] Contrary to the Dutch implementing legislation, the German law in essence is a CCS demonstration law, allowing for an annual storage of no more than 1.3 Mtonnes CO_2 and a maximum storage capacity of 4 Mtonnes CO_2/year in the whole of Germany.[51]

As I have argued in the Introduction to this thesis, a number of provisions in the CCS Directive leave a large margin of discretion to the Member States and contain ambiguities.[52] This, arguably, enhances chances of the implementation of these provisions differing in the various Member States. In the following chapter, we will explore a typical example of such a provision, Article 12 on CO_2 stream purity.

---

[47] Article 258 TFEU provides that if the Commission considers that a Member State has failed to fulfil an obligation under the Treaties, it shall deliver a reasoned opinion on the matter after giving the State concerned the opportunity to submit its observations. If the Member State does not comply with the opinion within the period laid down by the Commission, the latter may bring the matter before the Court.

[48] Armeni (n 46).

[49] Wet van 31 oktober 2002, houdende regels met betrekking tot het onderzoek naar en het winnen van delfstoffen en met betrekking tot met de mijnbouw verwante activiteiten (Mijnbouwwet).

[50] Matthias Lang and Ulrich Mutschler, 'Overview German Carbon Capture and Storage (CCS) Law' <www.germanenergyblog.de/?page_id=3061> accessed 19 March 2013.

[51] Ibid.

[52] See also Armeni (n 46).

CHAPTER III

CO$_2$ STREAM PURITY AND MEMBER STATES' SCOPE TO IMPOSE STRICTER NORMS[1]

- Member States seeking to adopt CO$_2$ stream-purity criteria that are stricter than those in Article 12 CCS Directive are advised to design these criteria in such a way that impurity levels can be determined on a case-by-case basis, in relation to the characteristics of each individual project

- Member States would likely be allowed to, for instance, adopt a decree imposing ranges of impurity levels for various categories of capture technology

3.1 INTRODUCTION

Public knowledge of CCS is still very limited. A number of early EU CCS demonstration initiatives have met with strong local opposition as people are worried over health and environmental risks. In Germany, the population of the states of Brandenburg and Schleswig-Holstein fiercely opposed plans for demonstration projects by Swedish and German utilities Vattenfall and RWE, respectively.[2] Uproar over several early demonstration projects eventually led to the German bill for the implementation of the CCS Directive to allow for only

---

1   This chapter is based on a 2011 article in Climate Law. See Marijn Holwerda, 'Deploying Carbon Capture and Storage Safely: The Scope for Member States of the EU to Adopt More Stringent CO$_2$ Stream-Purity Criteria under EU Law' (2011) 2 Climate Law 37.

2   Malte Kohls, 'Developing CCS in Germany' (presentation held at the Energy Delta Convention, Groningen, 2009) <www.rug.nl/energyconvention/EDC/Archive/EDC2009/overviewspeakers2009> accessed 10 December 2010; Nicholas Comfort, 'RWE May Build Dutch Carbon Capture Plant Amid German Opposition' *Bloomberg* (22 January 2010) <www.bloomberg.com/apps/news?pid=20601100&sid=aOZGhPaq04MY> accessed 15 April 2010; and Paul Voosen, 'Frightened, Furious Neighbours Undermine German CO2-Trapping Power Project' *New York Times* (7 April 2010) <www.nytimes.com/gwire/2010/04/07/07/greenwire-frightened-furiousneighbors-undermine-german-35436.html?scp=6&sq=paul%20voosen&st=cse> accessed 11 May 2010.

limited yearly amounts of CO_2 (1.3 Mtonne) to be stored.[3] In the Netherlands, the government decided to stop spatial planning procedures for the planned CCS demonstration project in the town of Barendrecht and to no longer strive for a large-scale onshore CCS demonstration project in the north of the Netherlands.[4] In both cases, the lack of local support for the planned project(s) played a large role in the government's decision to no longer work towards onshore CO_2 storage in the relevant areas.[5]

Governments in Member States that have shown an interest in accommodating early CCS demonstration projects, such as Germany and the Netherlands, are increasingly under pressure to guarantee the safety of early CCS deployment. Even though it has been argued that Member States only rarely adopt stricter environmental standards,[6] Member States like Germany and the Netherlands could be tempted to strengthen the safety standards in the CCS Directive. The negotiations on the CCS Directive indicate that Article 12 of the directive would in that case be a likely target. During these negotiations, the European Parliament (Parliament) long insisted on stricter CO_2 stream-purity criteria.[7] Parliament initially pushed for a minimum level of ninety per cent CO_2 in the captured CO_2 stream and later even insisted on a ninety-five per cent CO_2 level as well as on a prohibition on H_2S (hydrogen sulphide) and SO_2 (sulphur dioxide).

---

[3]   Berliner Informationsdienst, 'Einigung im Vermittlungsausschuss zur Solarförderung und CCS' (2012) <www.polisphere.eu/bid/einigung-im-vermittlungsausschuss-zur-solarforderung-und-ccs/> accessed 20 August 2012.

[4]   Dutch Ministry of Economic Affairs, Agriculture and Innovation, 'Uitwerking van de afspraken voor de individuele CO2-opslagprojecten die momenteel in voorbereiding zijn' (2010) 2 <www.rijksoverheid.nl/ministeries/eleni/documenten-en-publicaties/kamerstukken/2010/11/04/uitwerking-van-de-afspraken-voor-de-individuele-co2-opslagprojecten-die-momenteel-in-voorbereiding-zijn.html> accessed 12 December 2010; and Dutch Ministry of Economic Affairs, Agriculture and Innovation, 'CCS-projecten in Nederland' (2011) <www.rijksoverheid.nl/documenten-en-publicaties/kamerstukken/2011/02/14/ccs-projecten-in-nederland.html> accessed 26 May 2011.

[5]   Ibid.

[6]   Jan Jans and others, '"Gold plating" of European Environmental Measures?' (2009) 6 Journal for European Environmental and Planning Law 417, 434.

[7]   See European Parliament, 'Draft report on the proposal for a directive of the European Parliament and of the Council on the geological storage of carbon dioxide and amending Council Directives 85/337/EEC, 96/61/EC, Directives 2000/60/EC, 2001/80/EC, 2004/35/EC, 2006/12/EC and Regulation (EC) No 1013/2006 (COM(2008)0018 – C6–0040/2008 – 2008/0015(COD)) A6–0414/2008 amendments 18 and 44 <www.europarl.europa.eu/oeil/FindByProcnum.do?lang=en&procnum=COD/2008/0015> accessed 26 May 2011; and European Parliament, 'Report on the proposal for a directive of the European Parliament and of the Council on the geological storage of carbon dioxide and amending Council Directives 85/337/EEC, 96/61/EC, Directives 2000/60/EC, 2001/80/EC, 2004/35/EC, 2006/12/EC and Regulation (EC) No 1013/2006 (COM(2008)0018 – C6–0040/2008 – 2008/0015(COD))' A6–0414/2008 amendment 84 <www.europarl.europa.eu/sides/getDoc.do?language=EN&reference=A6–0414/2008> accessed 26 May 2011.

Since the legal basis of the CCS Directive is ex Article 175(1) of the Treaty Establishing the European Community (TEC) (now Article 192(1) of the Treaty on the Functioning of the EU (TFEU)), Member States would, in principle, be allowed to adopt stricter purity criteria than those laid down in Article 12. According to ex Article 176 TEC (now Article 193 TFEU), 'the protective measures adopted pursuant to Article 175 shall not prevent any Member State from maintaining or introducing more stringent protective measures'. More stringent protective measures, however, must be notified to the European Commission (Commission) and be compatible with the TEC,[8] including the provisions on the free movement of goods, persons, services and capital.[9] Yet, it is not clear whether stricter CO_2 stream-purity criteria would be compatible with the TFEU free movement provisions. The development of different national sets of CO_2 stream-purity criteria could hinder the cross-border transport and storage of captured CO_2, since it would likely increase costs along the CCS value chain.

In the following sections, I will try and answer the question to what extent there is scope under EU law for Member States to adopt stricter CO_2 stream-purity criteria. By doing so, I intend to outline part of the EU legal framework for the implementation and application of Article 12 by the Member States. Section 3.2 briefly examines Article 12 of the CCS Directive, as well as the European Commission's (Commission) guidelines on CO_2 stream composition. Section 3.3 addresses the classification of captured CO_2 for storage that is transported across Member States' borders, under EU free movement provisions.

Section 3.4 briefly explores two concepts of EU law which are of great importance for determining the room for Member States to adopt stricter environmental measures, the concepts of exhaustion and (minimum) harmonisation.[10] Section 3.5 addresses the relationship between Article 193 TFEU and EU secondary environmental law, which has in recent years led to a great deal of legal debate and figured in several cases before the Courts. Section 3.6 analyses the concepts of exhaustion and harmonisation in relation to the CO_2 stream-purity criteria in the CCS Directive. Finally, section 3.7 draws a number of lessons for Member States seeking to impose stricter CO_2 stream-purity criteria.

---

[8]   Unlike Article 176 TEC, Article 193 TFEU refers to the 'Treaties'. These are the Treaty on European Union (TEU) and the Treaty on the Functioning of the European Union (TFEU). See Article 1(2) TFEU.

[9]   Ex Article 176 TEC. The provisions on the free movement of goods, persons, services and capital can be found in Titles II and IV of the TFEU.

[10]   These two concepts deal with respectively the scope and the degree of harmonisation of the relevant EU legislation.

3.2 ARTICLE 12 OF THE CCS DIRECTIVE AND THE COMMISSION'S GUIDELINES

3.2.1 ARTICLE 12 OF THE CCS DIRECTIVE

Article 12 on CO_2 stream acceptance criteria and procedure provides that 'a CO_2 stream shall consist overwhelmingly of CO_2'. The term 'overwhelmingly' was first used in the London Protocol, the first international marine environment protection instrument[11] to allow for the offshore geological storage of CO_2.[12] It was deliberately chosen by the London Protocol's Scientific Group, which drafted the amendment to the protocol's Annex I, to prevent setting arbitrary levels for CO_2 stream components.[13]

Setting an arbitrary level for CO_2 stream components would not necessarily improve the security of storage and could have perverse effects by increasing the energy penalty for CO_2 capture[14] and emissions from the capture process.[15] Some storage reservoirs may, for instance, be rich in H_2S at levels far greater than present in the CO_2 stream.[16] Purity criteria that would require the amount of H_2S in the CO_2 stream to stay below these levels would increase the energy penalty for CO_2 capture as well as emissions from the capture process (both leading to higher capturing costs), while not necessarily increasing the safety of CCS deployment.[17]

The formula that CO_2 streams 'consist overwhelmingly of carbon dioxide'[18] allows for a case-by-case assessment of levels of impurity, recognising the natural variation in storage site characteristics and different transport constructions.[19] The London Protocol provides that the CO_2 stream may contain incidental

---

[11]   Shortly after the London Protocol, the OSPAR Convention was amended to allow for the offshore geological storage of CO_2. The OSPAR Convention is a regional treaty on the protection of the marine environment in the North-East Atlantic. The term 'overwhelmingly' is also used in the OSPAR Convention. See Articles 3(2)(f)(ii) and 3(3)(b) of respectively Annexes II and III to the OSPAR Convention.

[12]   The London Protocol is a protocol to the 1972 Convention on the Prevention of Marine Pollution by Dumping of Wastes and Other Matter (London Convention). Under the London Protocol all dumping is prohibited except for a limited number of wastes or other matter that may be considered for dumping, listed in Annex I to the Protocol.

[13]   Tim Dixon and others, 'International Marine Regulation of CO_2 Geological Storage. Developments and Implications of London and OSPAR' (2009) 1 Energy Procedia 4503, 4506.

[14]   See n 34 in Chapter I.

[15]   Dixon and others (n 13) 4506.

[16]   Ibid.

[17]   Likewise, setting the required CO_2 content too high could also exclude certain CCS technologies such as oxy-fuel combustion capture (see section 1.1.3 of Chapter I). See Dixon and others (n 13) 4506.

[18]   Article 4(2) of Annex I to the amended London Protocol.

[19]   Dixon and others (n 13) 4506.

associated substances derived from the source material and the capture and sequestration processes used and that no wastes or other matter may be added for the purpose of disposing of those wastes or other matter.[20]

The CCS Directive goes a bit further than the London Protocol by providing that incidental substances and substances added for the monitoring and verification of CO_2 migration ('tracers') shall be below levels that either (a) adversely affect the integrity of the storage site or the relevant transport infrastructure, (b) pose a significant risk to the environment or human health, or (c) breach the requirements of applicable Community legislation.[21]

The criteria in (a) and (b) are not very clear. The Directive does not specify when incidental substances or tracers adversely affect the integrity of transport and storage infrastructure (criterion (a)). Furthermore, the Directive's definition of the crucial term in criterion (b), i.e. 'significant risk', is vague: 'a combination of a probability of occurrence of damage and a magnitude of damage that cannot be disregarded without calling into question the purpose of this Directive for the storage site concerned'.[22] The first part of this definition refers to a formula for risk that is commonly used in risk management, whereas the second part is to define the meaning of the adjective 'significant'.

Yet, when does disregarding a foreseeable risk run contrary to the purpose of the Directive for the storage site concerned? Article 1 of the CCS Directive provides that the Directive establishes a legal framework for the environmentally safe geological storage of CO_2, the purpose of which is the permanent containment of CO_2. This would seem to suggest that incidental substances and tracers in the CO_2 stream should stay below levels that raise doubts as to whether the CO_2 to be injected will be permanently contained in the storage site concerned after injection.

Finally, the criterion in (c) refers to the requirements imposed on the operators of capture installations by several EU environmental Directives. First of all, Article 31 of the CCS Directive amends Directive 85/337 (Environmental Impact

---

[20] See Article 4(2) and (3) of Annex I to the amended London Protocol. Similar provisions can be found in Article 3(2)(f)(ii) and (iii) of Annex II to the amended OSPAR Convention.
[21] Article 12(1) CCS Directive.
[22] Article 3(13) CCS Directive. The criterion in (b) resembles the fourth criterion for allowing offshore CO_2 storage under the OSPAR Convention. Under the amended Annexes I and III to the OSPAR Convention, the fourth criterion only allows for offshore storage of captured CO_2 if CO_2 streams 'are intended to be retained in these formations permanently and will not lead to significant adverse consequences for the marine environment, human health and other legitimate uses of the maritime area'. See Articles 3(2)(f)(iv) and 3(3)(d) of respectively Annexes II and III to the OSPAR Convention.

Assessment Directive)[23] so as to include capture installations in the latter Directive. This means that an environmental impact assessment has to be carried out in the capture permit process.[24]

Not all capture installations will automatically be made subject to an environmental impact assessment. The Environmental Impact Assessment Directive differentiates between projects which are automatically to be subjected to an environmental impact assessment (listed in Annex I to the Directive) and projects for which Member States may either through (a) a case-by-case examination or (b) self-set thresholds or criteria or a combination of the both determine whether the project shall be made subject to an environmental impact assessment (listed in Annex II to the Directive). Article 31 of the CCS Directive amends the Environmental Impact Assessment Directive and adds to the list of projects under Annex I (automatically subjected to environmental impact assessment):

> [I]nstallations for the capture of CO_2 streams for the purposes of geological storage pursuant to Directive 2009/31/EC from installations covered by this Annex, or where the total yearly capture of CO_2 is 1,5 megatonnes or more.

Annex I lists installations such as crude-oil refineries, power stations and combustion installations with a heat output of at least 300 MW, and integrated chemical installations.

Capture installations from installations which are not on the Annex I list, but which capture at least 1,5 Mt CO_2 a year are also automatically subject to an environmental impact assessment. On the basis of this criterion alone,[25] fourteen out of a total of twenty-five EU large-scale integrated CCS projects (LSIPs)[26] mentioned in a 2010 global status update by the Global CCS Institute would automatically be subjected to an environmental impact assessment.[27] Article 31(2)(a) of the CCS Directive amends Annex II to the Environmental

---

[23] Council Directive 85/337/EEC of 27 June 1985 on the assessment of the effects of certain public and private projects on the environment [1985] OJ L175 (Environmental Impact Assessment Directive). In 2011, Directive 85/337/EEC was replaced by European Parliament and Council Directive 2011/92/EC of 13 December 2011 on the assessment of the effects of certain public and private projects on the environment [2012] OJ 26/1.

[24] Recital 27 CCS Directive. Article 31 of the CCS Directive amends the Environmental Impact Assessment Directive and brings CO_2 pipelines, including associated booster stations, CO_2 storage sites and CO_2 capture installations under the Environmental Impact Assessment Directive.

[25] Notwithstanding the possibility that capture installations fall under one of the categories of installations mentioned in Annex I to the Environmental Impact Assessment Directive.

[26] Including a number of projects in non-EU Member State Norway.

[27] These 14 projects are either capturing or plan to capture $\geq$ 1,5 Mt CO_2 a year. See Global CCS Institute, 'The Status of CCS Projects – Interim Report 2010' (2010) 18–25 <www.

Impact Assessment Directive so as to have all capture installations from installations not covered by Annex I to the latter Directive fall under the Directive's Annex II.

The competent authority in each Member State thus has a certain margin of discretion as to whether these installations should be subjected to a (usually costly environmental impact assessment. Considering the 1,5 Mt CO_2/year threshold for automatically being subject to an environmental impact assessment, it can be expected that not too many capture installations will fall under the Directive's Annex II, at least not once the CCS demonstration phase has ended (probably after 2020).

Alongside the above, the CCS Directive also amends Directive 2008/1 (Integrated Pollution Prevention and Control Directive).[28] Article 37 adds the capture of CO_2 by installations covered by the Integrated Pollution Prevention and Control Directive to the list of activities listed in Annex I to the latter Directive. This helps ensure that best available techniques (BATs)[29] to improve the composition of the CO_2 stream are be established and applied.[30] In addition, combustion plants with a rated thermal input of at least 50 MW are required to meet the emission limit values set by Directive 2001/80 (Large Combustion Plants Directive).[31, 32]

The amounts and proportions of various components extracted from a raw flue gas stream before CO_2 capture will affect the relative concentrations of components remaining in the gas stream.[33] Removal of air pollutants from a raw flue gas will generally already be required in order to comply with the emission limit values set in the environmental permits issued under both

---

globalccsinstitute.com/downloads/general/2010/The-Status-of-CCS-Projects-Interim-Report-2010.pdf> accessed 14 September 2010.

28   European Parliament and Council Directive 2008/1 of 15 January 2008 concerning integrated pollution prevention and control [2008] L24. The Integrated Pollution Prevention and Control Directive lays down measures to minimise the release of polluting emissions into air, water or soil in order to achieve a high level of protection for the environment as a whole. See recital 9 and Article 1 of the Integrated Pollution Prevention and Control Directive.

29   Article 2(12) of the Integrated Pollution Prevention and Control Directive defines the term 'best available technique'. In essence it boils down to the requirement that the installations concerned use the, from an environmental protection point of view, best technology, as long as it is economically and technically viable to do so.

30   Recital 27 CCS Directive.

31   European Parliament and Council Directive 2001/80 of 23 October 2001 on the limitation of emissions of certain pollutants into the air from large combustion plants [2001] L309.

32   The Large Combustion Plants Directive sets limits on SO_2, NO_x and dust emissions.

33   Commission, 'Implementation of Directive 2009/31/EC on the Geological Storage of Carbon Dioxide –Guidance Document 2: Characterisation of the Storage Complex, CO_2 Stream Composition, Monitoring and Corrective Measures' (2011) 60 <http://ec.europa.eu/clima/policies/lowcarbon/docs/gd2_en.pdf> accessed 26 May 2011.

Directives.[34] The Integrated Pollution Prevention and Control and Large Combustion Plants Directives have been recast into the Industrial Emissions Directive, which entered into force in January 2011.[35]

Article 12(3) of the CCS Directive provides for the commitment to observe the criteria set out in Article 12(1). Accordingly, the storage operator is to accept and inject a CO_2 stream only if an analysis of its composition and a risk assessment have been carried out, and only if the latter has shown that containment levels are in line with the criteria of Article 12(1). The storage operator is required to keep a register of the quantities and properties of the CO_2 streams delivered and injected, including their composition.

3.2.2 THE COMMISSION'S GUIDELINES ON CO_2 STREAM COMPOSITION

Article 12(2) of the CCS Directive allows the Commission to adopt guidelines to help identify the conditions applicable, on a case by case basis, for respecting the criteria in Article 12(1). At the beginning of 2011, the Commission published a set of guidance documents, one of which partly deals with CO_2 stream composition.[36] The purpose of the legally non-binding documents[37] is to assist stakeholders in the implementation of the CCS Directive in order to promote a coherent implementation of the directive throughout the EU.[38]

The part of the guidance document which deals with CO_2 stream composition does not cover hazards associated with CO_2 itself.[39] The guidelines on CO_2 stream composition start with the requirements imposed by the CCS Directive, as outlined in the section above. Thereafter, the two central terms in Article 12(1)

---

[34]   Commission, 'Guidance Document 2' (n 33) 61.

[35]   European Parliament and Council Directive 2010/75/EU of 24 November 2010 on industrial emissions (integrated pollution prevention and control) [2010] OJ L334 (Industrial Emissions Directive).

[36]   Commission, ' Guidance Document 2' (n 33) 58.

[37]   Article 288 TFEU lists the legal acts of the Union. Accordingly, Regulations, Directives and Decisions are binding, while recommendations and opinions are not. Guidelines are, however, administratively binding on the Commission itself. See to that extent e.g. Case C-351/98, *Spain v. Commission* [2002] ECR I-08031, para 53. They form rules of practice from which the administration may not depart in an individual case without giving reasons that are compatible with the principle of equal treatment. See Joined Cases C-189/02 P, C-202/02 P, C/205–02 P to C/208–02 P and C/213–02 P *Dansk Rørindustri* [2005] ECR I-05425, para 209.

[38]   Commission, 'Guidance Document 2' (n 33) 2.

[39]   Ibid 59.

are defined: 'incidental substances' and 'trace substances'.[40] According to the Commission, the first are

> [S]ubstances that are present in the CO$_2$ stream as a result of being (a) naturally in the feedstock (i.e. coal, gas, oil, biomass, coal-biomass mixtures etc.), (b) picked up in the capture process; or (c) incidentally entrained or intentionally added to prevent hazards during the transportation and injection processes.

The guidelines characterize trace substances as those substances 'added to assist in monitoring and verification of CO$_2$ migration in the storage complex'.

The composition of the CO$_2$ stream can vary due to the specific components in the feedstock, the type of process that is used to convert the feedstock into usable energy, the capture process and any post-capture processing.[41] The guidelines describe the influence of the different kinds of CO$_2$ capture processes[42] on the composition of the CO$_2$ stream and list indicative compositions of CO$_2$ streams from the main capture technologies.[43] In the remainder of the guidelines on CO$_2$ stream composition, the three criteria mentioned in Article 12(1) of the CCS Directive are discussed by respectively addressing pipeline impacts, storage integrity and health and environmental hazards.[44]

On the basis of a number of reports by scientists, government bodies and consultancies, the guidelines identify several issues related to CO$_2$ stream composition for the above-mentioned three areas. In each area, suggested limits for the various substances are given, as well as possible actions by the competent authority. In the case of the integrity of transport infrastructure, for instance, limitations on H$_2$O and O$_2$ content are mentioned as means to reduce pipeline corrosion.[45] Furthermore, storage site integrity requires both the competent authority and the storage operator to pay particular attention to acid interaction[46] with the geological formation.[47]

Finally, the guidelines propose a certain approach for the competent authority to follow in determining the concentration limits of incidental substances, starting

---

40   Commission, 'Guidance Document 2' (n 33) 59.
41   Ibid 60.
42   These are post-combustion capture, pre-combustion capture, oxy-fuel combustion and CO$_2$ streams from industrial processes. For more information on these capture technologies, see section 1.1 of Chapter I.
43   Commission, 'Guidance Document 2' (n 33) 61–68.
44   Ibid 68–82.
45   Ibid 82. On the issue of corrosion, see also section 1.2 of Chapter I.
46   Acid gases present in the CO$_2$ stream could interact with formation water in the storage site, resulting in reduced storage integrity. See Commission, 'Guidance Document 2' (n 33) 74.
47   Ibid 82.

with a proposal for CO_2 stream composition by the operator of the site in question and followed by a review of certain key parameters for respectively pipeline safety, human/environmental safety, storage integrity, and well integrity.[48]

The limited scientific experience with CCS and the need for operating flexibility are clearly reflected in the guidelines for CO_2 stream composition. There still is debate over the effects of different substance concentrations and the proposed indicative limits show a range of possibilities. In the coming years, the first large-scale EU CCS demonstration projects will probably play an important role in the further development of these suggested requirements for CO_2 stream composition.

3.3 QUALIFYING CAPTURED CO_2 FOR STORAGE UNDER EU INTERNAL MARKET PROVISIONS[49]

When captured CO_2 for storage is transported across Member States' borders, the TFEU free movement provisions may apply, even when the CO_2 was captured in a non-EU country. The question is whether in that case the provisions on the free movement of services or those on the free movement of goods apply. In other words: do the transporters and storage operators handling captured CO_2 for storage provide a *service* to the capturer or are the three parties along the chain instead engaged in the trading of a *good*?

At first sight, it seems logical to assume that the TFEU provisions on the free movement of services apply. In the case of the geological storage of CO_2, both the transporter and the storage operator will not acquire the captured CO_2 for their own use. Rather, they will help the capturer dispose of a substance that represents a cost to him. CO_2 captured for storage represents a cost to the capturer as the company will generally participate in the EU's greenhouse gas emissions trading scheme (EU ETS), which requires it to surrender an emissions allowance for each tonne of CO_2 emitted into the atmosphere. Captured CO_2 can even represent a cost to the transport or storage operator as both the transport by pipelines (with a view to storage) and the storage of greenhouse gases have been added to the activities covered by Directive 2003/87/EC (EU ETS Directive).[50] Any leakage of CO_2 during transport or

---

48   Commission, 'Guidance Document 2' (n 33) 60.
49   For a more detailed examination of the status of captured CO_2 for storage under EU internal market provisions, see section 4.3 of the next chapter.
50   See Annex I of the revised EU ETS Directive. The transport of CO_2 by ship has not been added to the Annex I of the EU ETS Directive. Considering the obvious (jurisdictional) question

storage will have to be compensated through the surrender of the equivalent in emission allowances.

However, case law seems to suggest that the provisions on the free movement of services are unlikely to apply. In *Walloon Waste*, the European Court of Justice (ECJ) indicated that waste constitutes a good within the meaning of the TFEU provisions on the free movement of goods.[51] It stated that 'objects' that are transported cross-border 'in order to give rise to a commercial transaction' are covered by the TFEU provisions on the free movement of goods 'irrespective of the nature of those transactions'.[52]

Even though captured and transported CO_2 for storage is, as a consequence of Articles 35 and 36 of the CCS Directive, excluded from the scope of the Waste Directive (Directive 2006/12/EC)[53] and thus the Shipments of Waste Regulation (Regulation 1013/2006),[54] the ECJ's ruling in *Walloon Waste* is relevant for the classification of captured CO_2 for storage under the TFEU free movement provisions; the point in common is that both non-recyclable waste and captured CO_2 for storage are disposed of, even though the latter is formally no longer classified as waste under EU environmental legislation.

The EU provisions on the free movement of goods are contained in Articles 34 to 36 TFEU. Articles 34 and 35 TFEU prohibit restrictions on imports and exports between Member States. Article 36 TFEU contains a limitative list of grounds justifying restrictions on imports and exports, among which is the protection of the health and life of humans, animals, an plants. In the case of *Cassis de Dijon*, the ECJ created a second, open category of derogations, namely the 'mandatory requirements'.[55] Since the *Danish Bottles* case, this category includes the protection of the environment.[56]

---

which competent authority should in the case of cross-border CO_2 transport by ship issue the greenhouse gas emissions permit, this is perhaps not surprising. Yet it is somewhat remarkable that CO_2 transport by ship is not covered by the EU ETS Directive and the jurisdictional issue as such does not seem insurmountable.

51   Case C-2/90 *Commission v. Belgium* [1992] ECR I-04431.

52   Ibid para 26.

53   European Parliament and Council Directive 2006/12/EC of 5 April 2006 on waste [2006] OJ L114 (Waste Directive). This Directive has now been replaced by European Parliament and Council Directive 2008/98/EC of 19 November 2008 on waste [2008] OJ L312/3.

54   European Parliament and Council Regulation (EC) 1013/2006 of 14 June 2006 on shipments of waste [2006] L190 (Shipments of Waste Regulation).

55   Case 120/78 *Cassis de Dijon* [1979] ECR 00649, para 8.

56   Case 302/86 *Danish Bottles* [1988] ECR 04607, para 9.

3.4 ASSESSING THE SCOPE FOR STRICTER ENVIRONMENTAL PROTECTION MEASURES

When assessing the scope for Member States to take more stringent environmental protection measures, two questions are of relevance. The first is whether the relevant subject matter has been exhaustively regulated by EU legislation.[57] The second is to what extent the applicable EU legislation harmonises national legislation. In the following, the EU law concepts of, respectively, exhaustion and (minimum) harmonisation are briefly addressed.

3.4.1 EXHAUSTION

The concept of exhaustion is generally ill-defined and often confused with total harmonisation, a form of harmonisation which is discussed in section 3.4.2, below.[58] Basically, the question is whether the relevant area is 'occupied by the EU'. The drafting of legislation demands that various interests, often private as against public interests, are weighed and balanced. When assessing whether a specific area has been exhaustively regulated by EU legislation, the question is whether the specific weighing and balancing of interests that a Member State intends to do has already been done by the EU legislature (usually Parliament and Council). If this is the case, the subject matter has been exhaustively regulated and a Member State can no longer take unilateral legislative action (unless the EU legislation allows it to do so). When a specific matter has not been exhaustively regulated by EU legislation, legislative action (unilateral) by a Member State is not pre-empted.

In a long line of case law, the ECJ has held that a Member State cannot rely on (possible) Treaty derogations (Article 36 TFEU)[59] to justify national legislation on a matter which has been exhaustively regulated by EU legislation.[60]

---

[57] See also S Weatherill, 'Pre-emption, Harmonisation and the Distribution of Competence to Regulate the Internal Market' in Catherine Barnard and Soanne Scott (eds), *The Law of the Single European Market, Unpacking the Premises* (Hart Publishing 2002) 41, 51–52. Weatherill identifies four questions in relation to particular (harmonising) instruments of secondary legislation like Directives and Regulations, of which the question of the scope of coverage of the secondary legislation is the first.

[58] Piet Jan Slot, 'Harmonisation' (1996) 21 European Law Review 378, 389. The ECJ, for instance, often refers to 'exhaustive harmonisation'. See e.g. Case C-374/05 *Gintec* [2007] ECR I-09517, para 34; Case C-205/07 *Gysbrechts* [2008] ECR I-09947, para 33; Case C-165/08 *Commission v. Poland* [2009] ECR I-06843, para 34; and Case C-428/08 *Monsanto* [2010] ECR I-06765, paras 60 and 90.

[59] These cases all relate to Article 36 TFEU.

[60] See Case 148/78 *Ratti* [1979] ECR 01629, Case C-29/87 *Dansk Denkavit* [1988] ECR 02965, Case C-169/89 *Gourmetterie* [1990] ECR I-02143, para 8, Case C-5/94 *Hedley Lomas* [1996]

Moreover, in a number of cases the ECJ has indicated that a national measure on a matter which has been exhaustively regulated must be assessed in the light of the provisions of the relevant harmonising measure and not those of the Treaty (including Article 34 TFEU).[61] The case of *DaimlerChrysler* makes clear that when a national measure on an exhaustively regulated matter meets the requirements of the applicable EU legislation, a review of its compatibility with the Treaty provisions on the free movement of goods is not necessary.[62] Any such measure must, however, still comply with primary EU law, including the provisions on the free movement of goods.[63]

3.4.2 (MINIMUM) HARMONISATION

Once it has been determined that a specific matter has been exhaustively regulated by EU legislation, the next question is to what extent the applicable EU legislation harmonises national legislation. In other words: what harmonisation method has been chosen for the applicable Directive or Regulation? Several forms of harmonisation exist in EU law, ranging from minimum harmonisation to total harmonisation.

In case of minimum harmonisation, minimum EU standards are set and Member States are in principle free to adopt stricter standards. The applicable EU legislation sets a floor, the Treaty itself sets a ceiling and the Member States are free to pursue an independent domestic policy between the two.[64] In the case of total harmonisation, national legislation is (fully) replaced by EU legislation and Member States may not adopt stricter measures, unless provided for in the relevant EU legislation.[65] In other words, exhaustive

---

ECR I-2553, para 18, Case C-1/96 *Compassion in World Farming* [1998] I-01251, paras 47 and 64, and Case C-102/96 *Commission v. Germany* [1998] ECR I-06871, paras 21–22.

[61] Case C-150/88 *Parfümerie-Fabrik* [1989] ECR 3891, para 28, Case C-37/92 *Vanacker and Lesage* [1993] ECR I-4947, para 9, Case C-324/99 *DaimlerChrysler* [2001] ECR I-9897, para 32, Case C-99/01 *Linhart and Biffl* [2002] ECR I-9375, para 18, Case C-322/01 *Deutscher Apothekerverband* [2003] ECR I-14887, para 64, and Case C-205/07 *Gysbrechts* (n 58) para 33.

[62] Case C-324/99 *DaimlerChrysler* (n 61) paras 44–46.

[63] The latter provided that the measure applies to imported or exported goods that originate or are in free circulation in the EU. See Joanne Scott, 'Flexibility in the Implementation of EC Environmental Law' in Han Somsen (ed), *Yearbook of European Environmental Law* (Oxford University Press 2000) 37, 41.

[64] Michael Dougan, 'Minimum Harmonization and the Internal Market' (2000) 37 Common Market Law Review 853, 880 and Scott (n 63) 855.

[65] Safeguard clauses can permit Member States under certain circumstances to prohibit the marketing of products conforming to the applicable environmental Directive. Such national measures can most likely be characterised as more stringent protective measures, but they do allow a Member State to deviate from the common standard in the Directive. See Nicolas de Sadeleer, 'The Impact of the Registration, Evaluation and Authorization of Chemicals (REACH) Regulation on the Regulatory Powers of the Nordic Countries' in Nicolas de

legislation in principle allows Member States to adopt stricter standards in case of minimum harmonisation, but prevents them from doing so in case of total harmonisation.[66]

In the field of EU environmental policy, minimum harmonisation measures can be enacted under either Article 193 TFEU (ex Article 176 TEC) or Article 114 TFEU (ex Article 95 TEC). Like Article 193, Article 114 requires Member States to notify stricter protective measures to the Commission.[67] However, whereas the obligation to notify under Article 193 TFEU does not, within the national legal order, affect the validity of the rules not notified and their enforcement as against private individuals,[68] national provisions notified under Article 114(4) or (5) have to be approved by the Commission.[69] Member States are allowed to apply such measures only once they have been approved by the Commission.[70]

3.5   ARTICLE 193 TFEU AND SECONDARY EU ENVIRONMENTAL LAW

As indicated above, exhaustive legislation in principle allows Member States to adopt stricter standards where there is minimum harmonisation, but prevents them from doing so for total harmonisation. In recent years, an intense debate has developed among EU environmental lawyers as to whether the content of secondary EU legislation as such can prevent Member States from invoking Article 193 TFEU.[71] Put differently: can Article 193 TFEU prevent the EU legislature from adopting environmental legislation that leaves Member States no scope for adopting more stringent national measures?[72] This is a particularly pertinent question, since Article 193 TFEU provides Member States with a Treaty base to adopt stricter national environmental standards.

Sadeleer (ed), *Implementing the Precautionary Principle: Approaches from the Nordic Countries, EU and USA* (Earthscan Ltd 2006), 344.

[66] Harrie Temmink, 'From Danish Bottles to Danish Bees: The Dynamics of Free Movement of Goods and Environmental Protection – A Case Law Analysis' in H Somsen (ed), *Yearbook of European Environmental Law* (Oxford University Press 2000), 68.

[67] Article 114(4) and (5) TFEU.

[68] Dougan (n 64) 880 and Scott (n 63) 39.

[69] Article 114(6) TFEU.

[70] See Case C-41/93 *French Republic v. Commission* [1994] ECR I-1829, para 30 and Case T-69/08 *Poland v. European Commission* [2010] ECR II-05629, para 57.

[71] Jan Jans and Hans Vedder, *European Environmental Law* (3rd edn, Europa Law Publishing 2008) 107. The same question can, and has been asked in relation to Article 114 TFEU. See e.g. Temmink (n 66) 69 and Damian Chalmers, Gareth Davies and Giorgio Monti, *European Union Law* (2nd edn, Cambridge University Press 2010) 701.

[72] See e.g. Robert Schutze, 'Cooperative Federalism Constitutionalised: The Emergence of Complementary Competences in the EC Legal Order' (2006) 31 European Law Review 167, 174.

There are basically two views. One is that the content of secondary EU environmental legislation can indeed prevent Member States from invoking Article 193 TFEU to adopt more stringent national measures. The main argument is that Article 193 TFEU must be considered to be a codification of the predominant, but not exclusive, EU environmental legislative practice before the Single European Act.[73] The introduction of the predecessor of Article 193 TFEU (Article 130t of the Treaty Establishing the European Economic Community – TEEC), allegedly, was solely designed to give the practice of using minimum harmonisation clauses a Treaty base and 'not really intended to have legal consequences'.[74] Accordingly, Article 193 TFEU is said to express the principle that, in general, decision making under Article 192 TFEU takes the shape of minimum harmonisation, without limiting the competence of the Council to decide to what extent Member States are allowed to adopt more stringent national protective measures.[75]

According to the second view, Member States can always adopt more stringent environmental measures following harmonisation under Article 192 TFEU. This view is supported by a number of arguments. First, there is the very wording of Article 193 TFEU itself.[76] Second, from a teleological point of view the aim of achieving a high level of environmental protection within the EU[77] would be better served by allowing Member States to go beyond the political compromise of most EU environmental legislation.[78] Third, it is argued that the decreased use of minimum harmonisation clauses in recent environmental Directives illustrates the role of Article 193 TFEU as a direct constitutional limit for every EU environmental measure.[79] Finally, there is the general principle that primary EU law at all times takes precedence over secondary EU legislation.[80]

The relationship between Article 193 TFEU and secondary EU environmental legislation has figured in several cases before the ECJ and the Court of First Instance (CFI).[81] Even though both Courts did not explicitly answer the

---

[73]   Jans and Vedder (n 71)108.
[74]   Ibid.
[75]   Ibid.
[76]   See e.g. Schutze (n 72) 175, and Pål Wennerås, 'Towards an Ever Greener Union? Competence in the Field of the Environment and Beyond' (2008) 45 Common Market Law Review 1645, 1665.
[77]   See Article 191(2) TFEU.
[78]   Schutze (n 72) 175. According to Krämer, this is what the system of shared competence for environmental protection aims at. See L Krämer, 'Environmental Protection and Article 30 EEC Treaty' (1993) 30 Common Market Law Review 111, 114–15.
[79]   Schutze (n 72) 175.
[80]   Jans and Vedder (n 71) 107 and Temmink (n 66) 70.
[81]   Among these are Case C-203/96 *Dusseldorp* [1998] ECR I-04075, Case C-318/98 *Fornasar* [2000] ECR I-04785, Case C-510/99 *Tridon* [2001] ECR I-07777, Case C-6/03 *Deponiezweckverband*, Case T-387/04 *EnBW* [2007] ECR II-01195, Case C-219/07 *Nationale*

question of whether the content of secondary EU environmental legislation as such can prevent Member States from invoking Article 193 TFEU, they did implicitly rule on the matter in these cases. In the following, I chronologically examine the relevant cases.

3.5.1   CASE C-318/98 *FORNASAR*

In *Fornasar*,[82] the ECJ was asked whether Directive 91/689[83] prevented Member States from classifying as hazardous waste that did not feature on the list of hazardous waste laid down by Decision 94/904.[84] The question was whether the relevant secondary legislation prevented Member States from adopting more stringent protective measures.[85] In assessing the degree of harmonisation of the relevant list of hazardous waste, the ECJ crucially noted that:

> [I]n that connection, it must be observed that the Community rules do not seek to effect complete harmonisation in the area of the environment. Even though Article 130r of the Treaty (now Article 191 TFEU) refers to certain Community objectives to be attained, both Article 130t of the EC Treaty (now Article 193 TFEU) and Directive 91/689 allow the Member States to introduce more stringent protective measures.[86]

According to Jans and Vedder, the ECJ's ruling seemed to have settled the 'doctrinal dispute' over the relationship between Article 193 TFEU and secondary EU environmental legislation: secondary EU environmental legislation cannot prevent Member States from invoking Article 193 TFEU, since EU environmental measures 'do not seek to effect complete harmonisation'.[87]

Schutze has argued that the ECJ's ruling in *Fornasar* is clearly ambivalent.[88] The first part of the ruling can accordingly be characterised as signalling the ECJ's support for the above-mentioned second view: Article 193 TFEU allows Member

---

    Raad van Dierenkwekers [2008] ECR I-04475, Case C-100/08 *Commission v. Belgium* [2009] ECR I-00140, Case T-16/04 *Arcelor* [2010] II-00211, Case C-378/08 *ERG-I* [2010] ECR I-01919, and Case C-82/09 *Kritis* [2010] ECR I-03649.

82   Case C-318/98 *Fornasar* (n 81).

83   Council Directive 91/689/EEC of 12 December 1991 on hazardous waste [1991] OJ L377 (Hazardous Waste Directive).

84   Council Decision 94/904/EC of 22 December 1994 establishing a list of hazardous waste pursuant to Article 1(4) of Council Directive 91/689/EEC on hazardous waste [1994] OJ L356.

85   Case C-318/98 *Fornasar* (n 81) para 34.

86   Ibid para 46.

87   Jans and Vedder (n 71) 109.

88   Schutze (n 72) 178.

States to always adopt more stringent protective measures.[89] Yet Schutze contends that this is contradicted by the second part of the ruling which contains a specific analysis of the applicable secondary EU legislation and an explicit reference to the minimum harmonisation clause in Directive 91/689.[90] The latter expressly allowed Member States to supplement the Community list of hazardous waste.[91]

It is argued here that perhaps neither of the two is a correct interpretation of the ECJ's ruling in *Fornasar*. True, the ECJ's remark that EU environmental measures do not seek to effect complete harmonisation, as well as its reference to Article 130t TEC allowing Member States to introduce more stringent measures, could be read as supporting the view that EU environmental legislation cannot prevent Member States from adopting more stringent protective measures under Article 193 TFEU. Nevertheless, were the ECJ truly convinced of the latter, it could have stopped just there.

Instead, as we have seen, it went on to refer to the relevant secondary legislation by stating that 'both Article 130t of the EC Treaty *and Directive 91/689* [emphasis added] allow the Member States to introduce more stringent protective measures'. Moreover, by referring to the minimum harmonisation clause in Directive 91/689 (Article 1(4)(ii)) in its assessment of the scope for Member States to adopt more stringent protective measures, the ECJ signified the importance it attaches to the wording of the relevant secondary legislation. It can thereby be said that the ECJ has not ruled out the possibility of secondary EU environmental legislation preventing Member States from invoking Article 193 TFEU. After all, were the ECJ convinced that this would be impossible, the precise wording of the applicable secondary legislation (including possible minimum harmonisation clauses) would have been irrelevant. In that case, Member States would always, on the basis of Article 193 TFEU, be allowed to adopt stricter environmental measures, regardless of the content of the relevant harmonising legislation.

Nevertheless, it can also be argued that the ECJ, by referring to both Article 193 TFEU and the relevant harmonising legislation, merely sought to firmly argue its point. Accordingly, the reference to Article 193 TFEU and the relevant harmonising legislation does not indicate that the ECJ considered it necessary for both to allow for more stringent protective measures.

---

[89]   Schutze argues that this is illustrated by the following phrase in para 46: 'community rules do not seek to effect complete harmonisation in the area of the environment'.

[90]   That is, Article 1(4), second indent.

[91]   According to Article 1(4)(ii) of Directive 91/689 for the purposes of that Directive 'hazardous waste' means 'any other waste which is considered by a Member State to display any of the properties displayed in Annex III'.

Yet the clear difference between the ECJ's ruling and AG Cosmas' Opinion in the case suggests that the ECJ did not refer to both Article 193 TFEU and the relevant harmonising legislation just to be on the safe side.[92] Contrary to the ECJ, Cosmas explicitly and forcefully argued that EU environmental legislation could not prevent Member States from adopting stricter protective measures, stating that the relevant system of waste classification could:

> in no way exclude the competence of Member States, based directly on Article 130t, first to maintain or introduce their own autonomous definition of hazardous waste and, secondly, to take the necessary measures for the protection of the environment from hazardous waste. It suffices that (...) the measures concerned achieve a more stringent protection of the environment that complies with the other Treaty provisions.[93]

The ECJ clearly did not follow AG Cosmas' line of reasoning. Unlike the AG, it took the content of the relevant harmonising legislation into account in assessing the scope for Member States to adopt stricter protective measures. Considering the outspoken line taken by AG Cosmas, the ECJ must have been aware of the message it would convey by departing from the AG's Opinion. In this light, it seems difficult to maintain that the ECJ referred to both Article 193 TFEU and the relevant harmonising legislation merely for reasons of emphasis.

3.5.2   CASE C-510/99 *TRIDON*

In the case of *Tridon*,[94] the ECJ seemed to depart from its ruling in *Fornasar*. In this case, the Regional Court of Grenoble (France) had referred to the ECJ two questions on the interpretation of Articles 30 and 36 TEC (Articles 34 and 36 TFEU), Regulations 3626/82 and 338/97 (both on trade in endangered species)[95] and the Convention on international trade in endangered species of wild fauna and flora (CITES). The regional Court had asked the ECJ whether CITES as well as Regulations 3626/82 and 338/97 and Articles 30 and 36 TEC had to be interpreted as allowing a Member State to take or maintain domestic measures prohibiting the commercial use of captive born and bred specimens of wild species occurring in its territory.

---

[92]   Case C-318/98 *Fornasar* [2000] ECR I-04785, Opinion of AG Cosmas.
[93]   Para 40.
[94]   Case C-510/99 *Tridon* (n 81).
[95]   Council Regulation (EEC) 3626/82 of 3 December 1982 on the implementation in the Community of the Convention on international trade in endangered species of wild fauna and flora [1982] OJ L384 (CITES Regulation) and Council Regulation (EC) 338/97 of 9 December 1996 on the protection of species of wild fauna and flora by regulating trade therein [1997] OJ L61.

With regard to Regulations 3626/82 and 338/97, the ECJ contended that neither relevant Regulation precluded Member States from taking stricter measures in compliance with the provisions of the Treaty. It stated that

> [T]he introduction or maintenance of such measures is provided for, as regards Regulation No 3626/82, in Article 15 thereof, and, as regards Regulation No 338/97, which was adopted on the basis of Article 130s(1) of the EC Treaty (Article 192(1) TFEU), in Article 130t of the EC Treaty (Article 193 TFEU), which provides that the protective measures adopted pursuant to Article 130s are not to prevent any Member State from maintaining or introducing more stringent protective measures which must be compatible with the Treaty.[96]

According to the ECJ, the domestic measures protecting wild species were 'adopted in a field in which secondary Community law does not preclude a Member State from taking measures stricter than those provided for by that law'.[97] The ECJ thereby clearly indicated that EU environmental legislation could not, in its view, prevent a Member State from taking stricter protective measures under Article 193 TFEU.

3.5.3 CASE C-6/03 *DEPONIEZWECKVERBAND*

In a second preliminary reference ruling on waste disposal,[98] the ECJ seemed to distance itself from the line taken in *Tridon* and to return to the approach it chose in *Fornasar*. In *Deponiezweckverband*,[99] the ECJ was asked whether Directive 1999/31 on the landfill of waste precluded more stringent national waste legislation.[100] The relevant German legislation contained, inter alia, stricter demands on the content of waste going to landfills.

In a preliminary remark, the ECJ noted, as it did in *Fornasar*, that 'the Community rules do not seek to effect complete harmonisation in the area of the environment'.[101] In reference to its ruling in *Fornasar*, the ECJ went on to state that 'Article 176 TEC allows the Member States to introduce more stringent protective measures' and that the latter article 'makes such measures subject only to the conditions that they should be compatible with the Treaty and that

---

[96]   Case C-510/99 *Tridon* (n 81) para 45.

[97]   Ibid para 53.

[98]   Article 267 TFEU contains the preliminary reference procedure by which the ECJ can give guidance to national courts on the interpretation of the Treaties and the validity and interpretation of acts.

[99]   Case C-6/03 *Deponiezweckverband* (n 81).

[100]   Council Directive 1999/31/EC of 26 April 1999 on the landfill of waste [1999] OJ L182 (Landfill Waste Directive).

[101]   Case C-6/03 *Deponiezweckverband* (n 81) para 27.

they should be notified to the Commission'.[102] The ECJ then briefly analysed the content of the relevant provisions of Directive 1999/31, after which – and again referring to its ruling in *Fornasar* – it concluded that

> [T]he wording and the broad logic of those provisions make it clearly apparent that they set a minimum reduction to be achieved by the Member States and they do not preclude the adopting by the latter of more stringent measures (...) It follows that Article 176 EC *and the Directive* allow the Member States to introduce more stringent protection measures that go beyond the minimum requirements fixed by the Directive.[103]

In assessing the scope for Member States to take more stringent protective measures, the ECJ, like it did in *Fornasar*, referred to both to Article 193 TFEU and the relevant secondary legislation, signalling the importance it attached to the wording and the 'broad logic' of the latter. Likewise, the ECJ's ruling in *Deponiezweckverband* seemed to indicate that it did not rule out the possibility of EU environmental legislation preventing Member States from adopting stricter protective measures under Article 193 TFEU.

3.5.4 OTHER CASES (2007–2010)

Several more recent cases confirm the above picture of the ECJ being ambivalent about the relationship between Article 193 TFEU and secondary EU environmental legislation.[104] In the majority of these cases, the ECJ and the CFI seem to have abandoned the line chosen in *Fornasar* and *Deponiezweckverband* and to have returned to the kind of reasoning displayed in *Tridon*.[105] These cases suggest that the ECJ and CFI consider Article 193 TFEU to provide sufficient basis for Member States to take more stringent protective measures, implying that the content of secondary EU environmental legislation cannot prevent Member States from invoking Article 193.

---

[102] Case C-6/03 *Deponiezweckverband* (n 81) para 27. Some have read the phrase that Article 176 TEC 'makes such measures subject only to the conditions that they should be compatible with the Treaty and that they should be notified to the Commission' as illustrating the ECJ's support for the view that EU environmental legislation cannot prevent Member States from taking more stringent protective measures under Article 193 TFEU. See Wennerås (n 78) 1665. These words could, however, also be read differently, namely that the ECJ merely states that Article 193 TFEU sets no further conditions than those mentioned in the article itself.

[103] Case C-6/03 *Deponiezweckverband* (n 81) paras 31 and 32, emphasis added.

[104] Case T-387/04 *EnBW* (n 81), Case C-219/07 *Nationale Raad voor Dierenkwekers* (n 81), Case C-100/08 *Commission v. Belgium* (n 81), Case T-16/04 *Arcelor* (n 81), Case C-378/08 *ERG-I* (n 81), Case C-82/09 *Kritis* (n 81), and Case C-2/10 *Franchini Sarl* [2011] not yet reported.

[105] See Case T-387/04 *EnBW* (n 81) para 112, Case C-100/08 *Commission v. Belgium* (n 81) paras 62 and 63, Case T-16/04 *Arcelor* (n 81) para 179, Case C-82/09 *Kritis* (n 81) para 24, and Case C-2/10 *Franchini Sarl* (n 104) paras 48–50.

However, in cases like *Nationale Raad van Dierenkwekers*[106] and *ERG-I*[107] the ECJ appears to have stuck to the approach it chose in *Fornasar* and *Deponiezweckverband*. In these cases, it again referred to the relevant harmonising legislation when assessing the scope for Member States to take stricter protective measures. In contrast to the '*Tridon*-like cases', *Nationale Raad van Dierenkwekers* and *ERG-I* seem to imply that the ECJ does not rule out the possibility of secondary EU environmental legislation preventing Member States from invoking Article 193 TFEU.

The above shows that the jurisprudence of the ECJ and CFI on this matter is inconclusive and unclear. To date, both Courts have not provided clarity as to whether they consider secondary EU environmental legislation capable of preventing Member States from adopting stricter protective measures under Article 193 TFEU. The possibility that secondary EU environmental legislation can prevent Member States from invoking Article 193 TFEU to adopt more stringent protective measures has neither been ruled in nor ruled out by the Courts.

3.6 MEMBER STATES' SCOPE TO ADOPT STRICTER CO$_2$ STREAM-PURITY CRITERIA

3.6.1 EXHAUSTION AND CO$_2$ STREAM PURITY UNDER EU LAW

As we have seen, the first question when assessing the scope for Member States to take more stringent environmental protection measures is whether the specific weighing and balancing of interests that a Member State intends to do, has already been done by the EU legislature. Member States are likely to adopt more stringent CO$_2$ stream-purity criteria to try to defuse public concerns over the safety and security of CCS. This would involve the weighing and balancing of several (competing) interests. One the one hand, there is the 'public' interest of trying to appease public concerns over CCS in order to realise CCS demonstration.[108] On the other hand, there is the private interest in having a sufficient degree of operating flexibility when deploying CCS.

In order to answer the question of whether the purity of the CO$_2$ stream has been exhaustively regulated under EU law, it is necessary to examine whether the

---

[106]   Case C-219/07 *Nationale Raad voor Dierenkwekers* (n 81) para 14.

[107]   Case C-378/08 *ERG-I* (n 81) para 68.

[108]   The ultimate public interest, of course, being the mitigation of climate change through a reduction of greenhouse gas emissions.

above interests have been weighed in relation to the EU CO_2 stream-purity criteria. This requires a closer look at the objectives of respectively the CCS Directive and Article 12 on CO_2 stream purity.

Article 1(1) of the CCS Directive provides that the Directive 'establishes a legal framework for the environmentally safe geological storage of Carbon Dioxide (CO_2) to contribute to the fight against climate change'. Article 1(2) states that 'the purpose of environmentally safe geological storage of CO_2 is permanent containment of CO_2 in such a way as to prevent and, where this is not possible, eliminate as far as possible negative effects and any risk to the environment and human health'. From the wording of Article 1 of the CCS Directive, which covers the subject matter and the purpose of the Directive, it can be inferred that the objective of the Directive is to minimise the risks to the environment and human health from CCS deployment. The Directive provides a legal framework not just for the geological storage of CO_2, but for the *environmentally safe* geological storage of CO_2.[109] As underlined by the European Council in 2008, the objective of proposing a regulatory framework for CCS was 'to ensure that this novel technology would be deployed in an environmentally safe way'.[110]

As for the objective of Article 12, recital 27 of the CCS Directive's preamble provides that

> [I]t is necessary to impose on the composition of the CO_2 stream constraints that are consistent with the primary purpose of geological storage, which is to isolate CO_2 emissions from the atmosphere, and that are based on the risks that contamination may pose to the safety and security of the transport and storage network and to the environment and human health.[111]

Recital 27 indicates that the purity criteria in Article 12, taking due account of the risks from contamination, are meant to ensure that captured CO_2 remains isolated from the atmosphere. The CO_2 stream-purity criteria are to prevent the captured CO_2 (and other substances in the stream) from leaking into the atmosphere and harming the environment and human health, by guaranteeing the safety and security of the transport and storage network.

However, while seeking to minimise the risks from contamination of the CO_2 stream, Article 12 provides industry with a significant degree of operating

---

[109] See also recital 49 of the preamble to the CCS Directive, which mentions 'the establishment of a legal framework for the *environmentally safe* storage of CO_2 (emphasis added)' as the objective of the CCS Directive.

[110] Recital 9 of the preamble to the CCS Directive.

[111] Even though the preamble to a Directive is not legally binding, it often reveals quite clearly the legislature's intentions behind the relevant legislation.

flexibility. The purity criteria in Article 12 are flexible and do not contain absolute limit-values for substance impurity.[112] As argued above, the formula that 'a CO$_2$ stream shall consist *overwhelmingly* of CO$_2$',[113] allows for a case-by-case assessment of levels of impurity, recognising the natural variation in storage site characteristics and different transport constructions.[114]

At first sight, a Member State adopting stricter purity criteria to appease public concerns over the safety of CCS seems to weigh and balance interests which are different from those weighed and balanced by the EU legislature in the drafting of Article 12 of the CCS Directive. For in drafting Article 12, the EU legislature balanced the private interest of sufficient operating flexibility with the public interest of minimising the risks from contamination of the CO$_2$ stream.

A Member State adopting stricter purity criteria to defuse public concerns over the safety of CCS would balance the private interest of sufficient operating flexibility with the public interest of appeasing public concerns over a CCS demonstration project. The primary objective of stricter purity criteria would in that case not be to minimise risks from contamination of the CO$_2$ stream, that is, to protect the environment and human health, but rather to maintain public order. The conclusion could therefore be that Member States would be free to unilaterally adopt stricter purity criteria to appease public concerns over the safety of CCS, as this particular public interest has not been specifically taken into account by the EU legislature when drafting Article 12 of the CCS Directive.

Nevertheless, in the case of *Compassion in World Farming*, the ECJ ruled that when interests not covered by the relevant harmonising legislation, such as public policy or public morality, are secondary to and derive from an interest which *is* covered by that legislation, these apparently 'non-harmonised' interests must be considered to have been taken into consideration by the EU legislature when drafting the relevant harmonising legislation.[115] In other words, the subject matter has in that case been exhaustively regulated, even though the specific balancing that a Member State intends to do has only *indirectly* been done by the EU legislature.

Following the ECJ's ruling in *Compassion in World Farming*, it seems likely that the purity of the CO$_2$ stream must be considered to have been exhaustively regulated by the CCS Directive, even in relation to the public policy interest of

---

112   The guidelines on CO$_2$ stream composition do contain absolute limit values, but these are indicative and not legally binding. See (n 33).
113   Emphasis added.
114   Dixon and others (n 13) 4506.
115   Case C-1/96 *Compassion in World Farming* (n 60) paras 65–66.

appeasing public concerns over the safety of CCS. After all, public concerns over the safety of CCS to a large extent consist of worries over the consequences of CCS deployment for human health and the environment.[116] This includes worries over the consequences of 'impure' substances in the CO_2 stream, such as H_2S. People are afraid of captured CO_2 (and other substances in the stream) leaking from pipelines or reservoirs and subsequently causing illness to or even the deaths of humans, animals and plants. The public policy interest of appeasing public concerns over the safety of CCS deployment derives from the public interest of minimising risks from CO_2 stream impurities to the environment and human health.

As a consequence, the purity of the CO_2 stream for storage seems to have been exhaustively regulated under Article 12 of the CCS Directive, even in relation to the public interest of appeasing societal concerns over CCS. This means that Member States seeking to adopt stricter CO_2 stream-purity criteria in order to appease public concerns over the safety of CCS can no longer rely on the exceptions in Article 36 TFEU or on one of the mandatory requirement to justify such (trade restricting) measures. The sole framework for assessing whether Member States are allowed to adopt stricter CO_2 stream-purity criteria is the relevant harmonising legislation, that is, the CCS Directive.

In the following section, the second variable, i.e. the degree of harmonisation of the EU criteria on CO_2 stream purity, is analysed.

3.6.2 THE DEGREE OF HARMONISATION OF ARTICLE 12 OF THE CCS DIRECTIVE

In several of the cases mentioned in section 3.5, the ECJ and CFI paid close attention to the wording and the broad logic of the provisions in question as well as to the structure of the relevant Directive. Below, I will first examine the wording of Article 12 of the CCS Directive, after which I will turn to the provision's broad logic as well as to the structure/spirit of the CCS Directive.

The wording of Article 12 reveals, first, that the article does not contain a minimum harmonisation clause, that is, there is no wording explicitly allowing Member States to adopt stricter CO_2 stream-purity criteria. The fact that Article 12 does not contain a minimum harmonisation clause does not necessarily say anything about the article's degree of harmonisation. Whereas

---

[116]   Another widely voiced fear is that of declines in property value due to CO_2 storage in geological reservoirs located under residential areas (see e.g. the cancelled CCS demonstration project in Barendrecht).

such clauses frequently appeared in environmental Directives two to three decades ago, environmental Directives nowadays usually do not contain minimum harmonisation clauses. Schutze has argued that Article 193 TFEU has taken over the role of minimum harmonisation clauses in environmental Directives.[117]

Second, Article 12 lacks the language typically hinting at minimum harmonisation, such as the term 'at least'. This lack of typical minimum harmonisation language is remarkable considering that the CCS Directive contains a number of provisions in which the words 'at least' are used, hinting at the presence of minimum EU standards and minimum harmonisation.[118] Article 15(3), for instance, provides that routine inspections of all CO_2 storage complexes shall be carried out *at least* once a year, leaving Member States free to require competent authorities to inspect those sites more frequently. When looking at the wording of Article 12, the conclusion must be that there seems to be nothing to suggest that the article constitutes minimum harmonisation.

As for the broad logic of Article 12, at first sight it would not seem to oppose stricter national purity criteria. Stricter purity criteria would most likely contain either absolute limit values (hereafter impurity-limit levels) for, or bans on, impure substances, or minimum standards for CO_2 content, along the lines of the stricter criteria proposed by Parliament during the negotiations on the CCS Directive.[119] Such criteria would normally appear to contribute to achieving the objective of Article 12 by reducing the risk of captured CO_2 leaking into the atmosphere and possibly harming the environment and human health. Stricter national purity criteria would arguably be likely to improve the safety of CO_2 *transport.*[120]

Nevertheless, it is not certain whether the same would hold true for the safety of CO_2 *storage.* At the moment, it appears to be unclear what the net effect of the presence of impurities in the CO_2 stream on storage safety will be. A review of the literature suggests that the presence of impurities could both increase as well

---

[117] Schutze (n 72) 175.
[118] These are Articles 7, 9, 14, 15, 18 and 20. Considering the character of these articles which, for instance, deal with reporting requirements, it is not surprising that these provisions set minimum standards. However, the subject matter of Article 12 as such would also seem suitable for minimum standards.
[119] It is, after all, hard to conceive how further national requirements on the purity of the CO_2 stream could be stricter than the criteria contained in Article 12 of the CCS Directive without setting absolute limit-values for impurity substances, introducing a minimum standard for CO_2 content or banning certain impure substances.
[120] On the possible negative consequences of CO_2 stream impurity for CO_2 transport (e.g. pipeline corrosion and hydrate formation), see also Commission, 'Guidance Document 2' (n 33) 68–72.

as decrease the risks of CO_2 storage.[121] Conversely, stricter purity criteria could both increase as well as decrease the safety of CO_2 storage. It is therefore not certain whether the broad logic of Article 12 would, on the whole, allow for stricter national purity criteria.

Yet, even if the broad logic of the provision would allow for stricter national purity criteria, it would probably restrict the way in which Member States can impose such stricter criteria. This is due to the case-by-case approach underlying Article 12. This approach may be inferred from the use of the term 'overwhelmingly', which was, as I have indicated, deliberately chosen by the London Protocol's Scientific Group to prevent setting arbitrary levels for CO_2 stream components and allows for a case-by-case assessment of levels of impurity. The EU legislature adopted the same concept and required the CO_2 stream to consist overwhelmingly of CO_2, despite alternative views taken by some Member States during the negotiations on the CCS Directive. During these negotiations, certain Member States expressed their preference for defining purity/contaminants values at the EU level, while others indicated that they wanted to retain the possibility of assessing the required purity levels on a case-by-case basis.[122]

This case-by-case approach follows not only from the crucial concept of overwhelmingly. It logically follows from the fact that the (desired) concentration

---

[121]   See Tom Mikunda and Heleen de Coninck (CO2ReMoVe), 'Possible impacts of captured CO_2 stream impurities on transport infrastructure and geological storage formations – current understanding and implications for EU legislation' (2011) <www.ecn.nl/docs/library/ report/2011/o11040.pdf> accessed 21 August 2012 and Carbon Capture Journal, 'CO_2 Capture Project CO2 Impurities Study' (8 July 2012) <www.carboncapturejournal.com/displaynews. php?NewsID=974&PHPSESSID=mmpj7lh3tbba5mjhv5q10fkhc0> accessed 21 August 2012. The CO2ReMoVe project is an EU-funded research project (6th Framework Programme) which aims to create a scientifically based reference for the monitoring and verification of geological storage. For more information on the project see <www.co2remove.eu/> accessed 22 October 2012. Mikunda and de Coninck reviewed the literature on the impact of CO_2 stream impurity and concluded that the presence of Sulphur dioxides (SOx) in the CO_2 stream, and the subsequent formation of sulphuric acids in the injection well, could have an impact on the capacity of storage reservoir rock to retain CO_2. The effect could go two ways: SO_x could accelerate mineralisation of CO_2 reducing the possibility of CO_2 leakage, but it could also negatively affect the cap rock and well closing infrastructure. According to Mikunda and the Coninck, the role of other impurities is even less clear at this moment. A recent study on the effects of CO_2 stream impurity conducted within the framework of the CO_2 capture project (see <www.co2captureproject.org/>), suggests that the presence of impurities in the CO_2 stream could lead to a faster retention of CO_2 in rock pores, in turning decreasing the risks of storage. Yet the same study suggests that, at the same time, the consequence of the presence of impurities could be to enlarge the lateral extent of the CO_2 plume; the injected CO_2 is dispersed more widely. The latter could increase storage risks.

[122]   Tim Dixon, 'Future Direction of CCS Directive toward Defining CO2 Quality' <www. co2captureandstorage.info/docs/oxyfuel/Discussion%20Purity/01%20-%20T.%20Dixon%20 %28IEA%20GHG%29.pdf> accessed 11 March 2011.

of impure substances in a CO$_2$ stream is directly related to a large number of case-specific variables, such as the specific components in the feedstock, the capture process, any post-capture processing[123] and the specific transport infrastructure and storage formation.[124] Impurity is not a standalone concept. Whether certain concentrations of incidental and added substances would adversely affect the integrity of transport and storage infrastructure (Article 12(1)(a)) or pose a significant risk to the environment or human health (Article 12(1)(b)), obviously could only be assessed on a case-by-case basis. In the end, the physical characteristics of transport infrastructure and storage formations determine the levels of impurity required. This is illustrated by Article 12(2), which states that the Commission guidelines are 'to help identify the conditions applicable *on a case by case basis* for respecting the criteria laid down in paragraph 1'.[125]

The case-by-case approach underlying Article 12 limits the scope for Member States to further specify the EU purity criteria by means of national regulation. Article 12 would likely not allow Member States to adopt stricter purity criteria making a case-by-case determination of levels of impurity impossible. Whether stricter purity criteria would indeed make a case-by-case determination of impurity levels impossible largely depends on the legislative instrument and manner by which these criteria would be imposed. Were stricter purity criteria to be imposed in a general manner, through, for instance, a decree regulating all individual cases,[126] such criteria might be inconsistent with the broad logic of Article 12.

Yet, were the same criteria imposed through, for instance, permit authorisations, they would likely not go against Article 12. Such specific regulatory instruments would still allow for a case-by-case-assessment, thus preventing arbitrary limit-values and guaranteeing a large degree of operating flexibility. The impurity-limit levels could be determined in individual cases on the basis of, among other things, the Commission's guidelines on CO$_2$ stream purity, and subsequently be imposed through the permitting process.

Likewise, a decree imposing criteria for various categories of CCS projects could conceivably still leave room for exact impurity-levels to be determined on a case-by-case basis. The sequestration of CO$_2$ by means of, for instance, oxyfuel combustion technology is likely to lead to lower levels of CO$_2$ in the captured CO$_2$ stream than CO$_2$ sequestration based on either pre-combustion or post-

---

123   Commission, 'Guidance Document 2' (n 33) 60.
124   In this regard, see also Det Norske Veritas, 'Design and Operation of CO$_2$ Pipelines' (2010) 14.
125   Emphasis added.
126   Such as the amendments proposed by Parliament during the negotiations on the CCS Directive.

combustion technology.[127] Such differences could be accommodated by devising different criteria for the different categories of capture technology.

Finally, the scope for stricter national CO_2 stream-purity criteria can be assessed in the light of the structure and spirit of the CCS Directive. The character of the CCS Directive has to a large extent been determined by the relatively novel nature of CCS technologies. Even though different parts of the CCS chain have been applied for quite a number of years now, experience with fully-integrated[128] large-scale CCS projects is still limited. Moreover, in relation to the long-term geological storage of CO_2 several uncertainties remain.[129]

Being aware of this, the Commission gave itself a large role in the application of the Directive when drafting its proposal for the CCS Directive. For example, the Commission is to review draft storage permits (Article 10) and draft decisions of approval of the transfer of responsibility for a storage site from the operator to the competent authority (Article 18(4)). In doing so, it is to be assisted by a scientific panel composed of geology experts. According to recitals 25 and 33 and the Commission's Q&A on the CCS Directive, the Commission's central role is to ensure a consistent application of the Directive across the EU and to enhance public confidence in CCS.[130] Also, it is to prevent Member States from implementing safety requirements too casually.[131] Competent authorities might hand out storage permits too easily or take on responsibility for a closed storage site too readily. This could lead to safety risks.

The adoption of stricter CO_2 stream-purity criteria by individual Member States would be likely to lead to an inconsistent application of the CCS Directive due to the development of different sets of criteria across the EU. On the face of it, such inconsistent application would not, however, appear to lead to problems from the point of view of the Directive's overarching objective, i.e. minimising

---

[127] Commission, 'Guidance Document 2' (n 33) 66.

[128] The term 'fully-integrated' refers to CCS projects that include all three parts of the CCS value chain, that is capture, transport and storage. Most of the early CCS projects dealt with parts of the value chain only.

[129] These are related to the (time-)limited experience with the geological storage of CO_2 and the question whether stored CO_2 will safely remain in the underground over a period of thousands of years.

[130] Commission, 'Questions and Answers on the Directive on the Geological Storage of Carbon Dioxide' (2008) <http://europa.eu/rapid/pressReleasesAction.do?reference=MEMO/08/798&language=EN> accessed 7 December 2010.

[131] See e.g. the Commission's impact assessment of the CCS Directive, in which it states that the Commission review of draft storage permits is to ensure sound implementation of the risk management framework in the early phase of storage. See Commission, 'Accompanying Document to the Proposal for a Directive of the European Parliament and of the Council on the Geological Storage of Carbon Dioxide: Impact Assessment' COM (2008) 18 final 3.

environmental and health risks from CCS. As argued in relation to the broad logic of Article 12, stricter purity criteria would normally appear to reduce the risk of captured CO_2 leaking into the atmosphere and possibly harming the environment and human health. Such criteria would, arguably, be likely to improve the safety of CO_2 *transport*. Yet, in relation to *storage* safety, the net effects of the absence or decreased presence of certain CO_2 stream impurities are unclear; there could be negative as well as positive effects. Considering the uncertainties surrounding the effects of stricter purity criteria on storage safety, it does not seem possible to say whether, on the whole, the structure/spirit of the CCS Directive would allow for such criteria.

However, an inconsistent implementation of the Directive's CO_2 stream-purity criteria could have other negative consequences. As argued earlier, the existence of different CO_2 stream-purity criteria across the EU could hinder the cross-border transport and storage of captured CO_2, since it would likely increase costs along the CCS value chain. This would make the development of a cross-border EU market for CCS more difficult.

Nevertheless, even though the CCS Directive recognises the possibility of the cross-border application of CCS, the development of a cross-border EU market for CCS is not one of its declared objectives.[132] There is nothing in the CCS Directive to indicate that it aims at fostering the development of a cross-border EU market for CCS. Furthermore, it is highly unlikely that the CCS Directive would have been endorsed by the Council had that been the case. During the negotiations on the CCS Directive, several Member States strongly voiced their concerns over the safety of CCS technology, particularly in relation to the long-term geological storage of CO_2. For those Member States, facilitating the deployment of CCS technology at the EU level already represented a far-reaching concession, never mind going as far as incentivising the cross-border application of CCS.[133]

---

[132] See e.g., Article 24 of the Directive on transboundary cooperation in cases of transboundary transport of CO_2 and transboundary storage sites or transboundary storage complexes.

[133] See e.g., Council of the European Union, 'Proposal for a Directive of the European Parliament and of the Council on the geological storage of carbon dioxide and amending Council Directives 85/337/EEC, 96/61/EC, Directives 2000/60/EC, 2001/80/EC, 2004/35/EC, 2006/12/EC and Regulation (EC) No 1013/2006' 14532/08 footnotes 44–49 <http://register.consilium.europa.eu/pdf/en/08/st14/st14532.en08.pdf> accessed 3 November 2011. The document reveals the worries Member States like Greece, Malta and Ireland had over the environmental safety of the geological storage of CO_2. Greece even proposed (see footnote 44) to restrict the scope of application the CCS Directive to the demonstration phase and thereafter evaluate whether the scope should be widened.

The CCS Directive intends to ensure that those Member States that wish to deploy CCS[134] do so in an environmentally safe manner. It does not, however, aim to facilitate the development of a cross-border EU CCS market. In this regard, it could therefore be argued that an inconsistent implementation of the Directive's CO_2 stream-purity criteria would not conflict with the structure/spirit of the CCS Directive.

3.6.3 THE CONFORMITY OF STRICTER CO_2 STREAM-PURITY CRITERIA WITH PRIMARY EU LAW

Some might argue that an assessment of the scope for EU Member States to adopt stricter CO_2 stream-purity criteria is incomplete without a review of such measures' compatibility with primary EU law. Even in cases where a matter has been exhaustively regulated by EU environmental legislation, stricter national measures must be compatible with the provisions of the Treaty on European Union (TEU) and TFEU, including those on the free movement of goods. This is a direct consequence of the wording of Article 193 TFEU, which, as we have seen earlier, provides that more stringent protective measure 'must be compatible with the Treaties'.

I have argued above that it is not clear that stricter CO_2 stream-purity criteria would be compatible with the TFEU free-movement provisions (Articles 34–36 TFEU), since they would likely increase costs along the CCS value chain. A CO_2 capturer looking to export captured CO_2 for storage to another Member State, the law of which contains stricter requirements on CO_2 stream purity than those in the capturer's national law, will have to further purify the captured CO_2 stream. This will require extra energy and accordingly make the export of captured CO_2 for storage more costly.

When reviewing the compatibility of national measures with primary EU law, the Courts will normally first assess whether the allegedly infringing measure indeed falls within the scope of the applicable provision of primary law, in this case Article 34 TFEU. As a second step, the Courts will determine whether the infringing measure can perhaps be justified, for example on one of the grounds contained in Article 36 TFEU. In so doing, they will review the proportionality of the relevant national measure. The Courts will first inquire whether the measure is suitable for achieving the objective pursued ('suitability test').

---

[134] Article 4(1) explicitly and, considering Member States' sovereignty over their own subsoil, needlessly provides that Member States have the right not to allow for any storage in parts or in the whole of their territory. This clearly reflects the political sensitivity of CO_2 storage among the Member States. On the sovereignty of Member States over their own subsoil and the right not to allow for CO_2 storage in their territory, see further Chapter VIII.

Second, they will inquire whether the national measure goes beyond what is necessary to achieve the objective pursued ('necessity test'), that is, whether a suitable measure exists that is less onerous.[135] A measure must pass both sub-tests to be compatible with the general EU law principle of proportionality. As stricter national CO_2 stream-purity criteria might hinder cross-border trade of captured CO_2 for storage, it could be argued that it is useful to assess how the Courts would likely apply the proportionality principle when reviewing the compatibility of such criteria with the TFEU provisions on the free movement of goods.

However, I doubt the usefulness of such assessment. In practice, the ECJ normally does not conduct such a review when a matter has been exhaustively regulated by EU legislation. The cases referred to in section 3.4.1, above, show that the ECJ, as soon as it finds a certain matter to have been exhaustively regulated by EU law, does not review the compatibility of the relevant national measure with primary EU law, such as Article 34 TFEU.[136] There are two aspects to this. In most of the cases referred to in section 3.4.1, the ECJ decided not to review the compatibility of the relevant national measure with primary EU law only after it had found the measure to be *incompatible* with the applicable Directive/Regulation. However, in *DaimlerChrysler* the ECJ indicated that a review of a national measure's compatibility with primary EU law is also not necessary when the relevant rule *does meet* the requirements of the applicable secondary EU legislation.[137]

The logic of the ECJ's line of reasoning in this regard is compelling. It would arguably be strange for a Member State to be allowed to adopt a national measure on the basis of Article 36 TFEU, while the relevant secondary EU legislation precludes the same measure.[138] Such a situation would come down to the EU

---

[135]   Some authors contend that the Courts also apply a third element in their proportionality test ('proportionality stricto sensu'). See Gráinne de Búrca, 'The Principle of Proportionality and its Application in EC Law' in Ami Barav, Derrick A Wyatt QC and Joan Wyatt (eds), *Yearbook of European Law 1993* (Oxford University Press 1993) 105, 133; Jan H Jans, 'Proportionality Revisited' (2000) 27 Legal Issues of Economic Integration 239, 240; and Jurgen Schwarze, *European Administrative Law* (2nd edn, Sweet & Maxwell 2006) 859. This alleged sub-test requires that even if there are no less restrictive means, it must be established that the measure does not have an excessive impact on the applicant's interests. See Takis Tridimas, 'Proportionality in Community Law: Searching for the Appropriate Standard of Scrutiny' in Evelyn Ellis (ed), *The Principle of Proportionality in the Laws of Europe* (Hart Publishing, 1999) 65, 68. Yet, other authors dispute that the Courts generally apply it as a separate element of their proportionality test. See Tridimas, 'Proportionality in Community Law' 68, and Paul Craig, *EU Administrative Law* (Oxford University Press 2006) 670.

[136]   For a similar kind of reasoning see Ludwig Krämer, *EU Environmental Law* (7th edn, Sweet & Maxwell 2011) 118.

[137]   Case C-324/99 *DaimlerChrysler* (n 61) paras 44–46.

[138]   Even when considering the primacy of primary EU law over secondary EU law.

legislator never being able to prohibit certain national measures if these measures could (perhaps) be justified under Article 36 TFEU. Likewise, it seems unnecessary to review a national measure's compatibility with primary EU law if that measure is compatible with the applicable Directive/Regulation, since the latter may be considered to be in line with the provisions of the Treaties. If not, the specific Directive/Regulation would be unlawful.

The case of *Deutscher Apothekerverband* provides an exception to the above rule.[139] In this case, the ECJ confirmed the principle that a national measure in a sphere which has been the subject of exhaustive harmonisation at the EU level must be assessed in the light of the provisions of the harmonising measure and not those of the Treaty.[140] However, it also provided that when the applicable EU legislation contains a minimum harmonisation clause expressly stating that the power conferred on the Member States must be exercised 'with due regard for the Treaty', 'such a provision does not, therefore, obviate the need to ascertain whether the national prohibition at issue in the main proceedings is compatible with Articles 28 to 30 EC (34 to 36 TFEU)'.[141]

In section 3.6.1, I argued why, in my opinion, the purity of the CO_2 stream has been exhaustively regulated by Article 12, even in relation to the public interest of appeasing societal concerns over CCS. Assuming that the argumentation is correct, the Courts would not be likely to review the compatibility of stricter CO_2 stream-purity criteria with primary EU law, but would instead assess such national measures in the light of Article 12, as I have done above. Article 12 does not contain a minimum harmonisation clause expressly obliging Member States to take due regard of the Treaty when devising stricter national rules.

It could be argued that such a duty instead follows directly from Article 193 TFEU, and therefore that when EU legislation which exhaustively regulates a matter is based on Article 192(1) TFEU, an assessment must be made of a national measure's conformity with primary law, even when the relevant EU legislation does not directly refer to the Treaties. Nevertheless, the case of *DaimlerChrysler*, which involved EU legislation based on Article 192(1) TFEU (Regulation 259/93),[142] seems to suggest that this is not the case. It can therefore be questioned whether an assessment of the proportionality of stricter national CO_2 stream-purity criteria would be useful here.

---

[139]   Case C-322/01 (n 61) *Deutscher Apothekerverband*.
[140]   Ibid para 64.
[141]   Ibid para 65.
[142]   Council Regulation (EEC) 259/93 of 1 February 1993 on the supervision and control of shipments of waste within, into and out of the European Community [1993] OJ L30 (Shipments of Waste Regulation).

3.7 LESSONS FOR MEMBER STATES SEEKING TO ADOPT STRICTER CO$_2$ STREAM-PURITY CRITERIA

When drafting Article 12 of the CCS Directive, the EU legislature had to balance the private interest of sufficient operating flexibility with the public interest of minimising the risks to the environment and human health from contamination of the CO$_2$ stream. Even though a Member State adopting stricter CO$_2$ stream-purity criteria to appease public concerns over CCS would not balance the exact same interests, the purity of the CO$_2$ stream for storage seems to have been exhaustively regulated at EU level, even in relation to this public interest. Therefore, the sole framework for assessing the scope for Member States to adopt stricter CO$_2$ stream-purity criteria (for these reasons) is the relevant harmonising legislation, the CCS Directive.

An analysis of the wording of Article 12 reveals that the EU CO$_2$ stream-purity criteria do not appear to constitute minimum harmonisation. Furthermore, stricter national CO$_2$ stream-purity criteria would seem to be in line with the broad logic of Article 12 only if such criteria would allow for a case-by-case assessment of levels of impurity. General impurity levels applying to all individual cases would seem to collide with the broad logic of Article 12, for such criteria would not respect the case-by-case approach underlying the article. Finally, it is not certain whether the structure/spirit of the CCS Directive would allow for stricter national purity criteria, as the consequences of such criteria for the safety of CO$_2$ storage are unclear.

On the basis of these findings, the scope for EU Member States to adopt stricter CO$_2$ stream-purity criteria under EU law appears to be narrow. The wording of Article 12 does not seem to indicate that the EU CO$_2$ stream-purity criteria are minimum standards. What is more, it is uncertain whether the structure/spirit of the CCS Directive and the broad logic of Article 12 would allow for stricter purity criteria. Even if they would, the latter would, in any case, seem to rule out impurity-levels that apply in all individual cases.

Based on the above, Member States seeking to adopt stricter CO$_2$ stream-purity criteria are advised to design these criteria in such a way that impurity levels can be determined on a case-by-case basis, in relation to the characteristics of each individual project. They would be wise not to adopt general impurity levels that apply in all individual cases, such as the criteria proposed by Parliament during the negotiations on the CCS Directive. Yet, Member States would appear to be allowed to impose stricter CO$_2$ stream-purity criteria through permit authorisations. Likewise a decree imposing ranges of impurity levels for various categories of capture technology would seem permitted as it would conceivably still leave room for exact impurity-levels to be determined on a case-by-case basis.

CHAPTER IV

STORAGE SITE STEWARDSHIP FINANCING AND THE CROSS-BORDER STORAGE OF CO$_2$

> In order not to infringe Article 110 TFEU, Member States are advised to:
>
> - Avoid differentiating the level of the financial security/mechanism charge based on the geographical origin of the CO$_2$ stream
>
> - Make sure that a differentiation of the level of the financial security/ mechanism charge on the basis of objective criteria does not lead to imported CO$_2$ streams primarily falling in the highest tax scales
>
> - Refrain from using flat rates for imposing a financial security/mechanism charge on imported CO$_2$ streams
>
> - Refrain from using fixed scales and abstract criteria for determining the financial security/mechanism charge levied on CO$_2$ streams imported for storage
>
> - Publish the criteria on the basis of which the charge is determined, when using fixed scales and abstract criteria for determining the financial security/mechanism charge levied on CO$_2$ streams imported for storage

4.1 INTRODUCTION

Articles 19 and 20 of the CCS Directive address the financing of the (long-term) stewardship of CO$_2$ storage sites. Under Article 19 (financial security), Member States are to require applicants for a CO$_2$ storage permit to prove that they have the financial means to meet all future obligations, including closure and post-closure requirements. By virtue of Article 20 (financial mechanism), Member States are to require the storage operator to make a financial contribution available to the competent authority before responsibility for the storage site is transferred to the latter.[1] The financial contribution is to cover the costs incurred by the competent authority to ensure that the CO_2 is completely and permanently contained in the storage site. Articles 19 and 20 leave Member States free to decide the arrangements by which to fulfil these obligations.

In its guidance document on Articles 19 and 20 the Commission suggests that the same instrument could be used for both sets of obligations.[2] One of the instruments often mentioned in relation to the financing of (long-term) storage site stewardship is that of a fund.[3] In its report on the Commission's initial proposal for the CCS Directive, Parliament proposed to add to then Article 19 (financial mechanism) a provision requiring Member States that would allow for the geological storage of CO_2 in their territory to establish a fund, to be financed through annual contributions paid by the storage site operator.[4] Likewise, in US literature on the long-term stewardship of CO_2 storage sites a trust fund financed through, for instance, a fee for each tonne of CO_2 injected has frequently been suggested.[5]

Member States can implement the financial requirements of Articles 19 and 20 in many different ways, for instance by imposing the abovementioned fee per tonne of CO_2 stored or by requiring a bank guarantee. In the remainder of this chapter I use a catch-all phrase to cover all different possible instruments for implementing Articles 19 and 20: the financial security/mechanism charge levied by the competent authority. This phrase does not refer to the implementation of Articles 19 and 20 by means of the levy of a charge only. The important thing is that whichever instrument(s) Member States choose, the CO_2

---

[1]   Article 18 of the CCS Directive provides that the storage operator can, under certain circumstances, transfer responsibility for the CO_2 storage site to the competent authority.
[2]   Commission, 'Implementation of Directive 2009/31/EC on the Geological Storage of Carbon Dioxide – Guidance Document 4: Article 19 Financial Security and Article 20 Financial Mechanism' (2011) 41 <http://ec.europa.eu/clima/policies/lowcarbon/ccs/implementation/docs/gd4_en.pdf> accessed 7 May 2012.
[3]   Ibid 2.
[4]   European Parliament, 'Report on the proposal for a directive of the European Parliament and of the Council on the geological storage of carbon dioxide and amending Council Directives 85/337/EEC, 96/61/EC, Directives 2000/60/EC, 2001/80/EC, 2004/35/EC, 2006/12/EC and Regulation (EC) No 1013/2006' (COM(2008)0018 – C6–0040/2008 – 2008/0015 (COD) amendment 110.
[5]   See e.g., California Carbon Capture and Storage Review Panel, 'Technical Advisory Committee Report: Long-Term Stewardship and Long-Term Liability in the Sequestration of CO_2,' (2008) 6 <www.climatechange.ca.gov/carbon_capture_review_panel/meetings/2010–08–18/white_papers/Long-Term_Stewardship_and_Long-Term_Liability.pdf> accessed 7 May 2012; Alexandra B Klass and Elizabeth J Wilson, 'Climate Change and Carbon Sequestration: Assessing a Liability Regime for Long-Term Storage of Carbon Dioxide' (2008) 58 Emory Law Journal 103, 174 and CCSReg Project, 'Carbon Capture and Sequestration: Framing the Issues for Regulation – An Interim Report from the CCSReg Project' (2009) 96 and 113–14 <www.ccsreg.org/pdf/CCSReg_3_9.pdf> accessed 7 May 2012.

storage operator will likely pass on to his customer – the capturer seeking to have his captured CO_2 stored – the costs of the charge levied by the competent authority. In the case of the relevant CO_2 stream coming from another Member State – it was either captured there or captured in a third/non-EU state and brought into circulation there – a financial security/mechanism charge levied by the competent authority might therefore constitute a hindrance to the free movement of goods.[6]

As Member States are unlikely to impose such a charge *by reason of* the captured CO_2 crossing their border (incoming from another Member State) and on 'foreign' CO_2 streams *only*,[7] Article 110 of the Treaty on the Functioning of the EU (TFEU) – and not Article 30 TFEU[8] – is likely to come into play.[9] Article 110 prohibits Member States from imposing on imported products any internal taxation that is in excess of that imposed on similar domestic products or protects competing domestic products. In the *Stadtgemeinde Frohnleiten* case, the ECJ indicated that the term taxation includes a charge on a specific activity in connection with products, such as a levy imposed on an operator of a waste disposal site.[10] I further address the relevance of this case for the financial security/mechanism charge levied under Articles 19 and 20 in section 4.3 below, but there obviously is a strong analogy between such a levy and the financial requirements imposed on a storage operator through the storage permit.

Even though a charge imposed by virtue of Articles 19 and 20 of the CCS Directive would probably not *directly*[11] discriminate against CO_2 streams from

---

[6] In section 4.3 below, I explain why captured CO_2 will probably qualify as a good within the meaning of the free movement of goods.

[7] Most likely, the system will rather be a general one, as a consequence of which a charge is levied on the storage permit applicant/storage operator, irrespective of the (foreign) origin of the captured CO_2.

[8] Article 30 TFEU provides that customs duties on imports and exports and charges with equivalent effect shall be prohibited between Member States.

[9] In the remainder of this work, I will, for the sake of clarity and consistency, refer to Article 110 TFEU, even though the case law or literature cited refers to either one of Article 110's predecessors (Article 90 TEC or Article 95 of the Treaty Establishing the European Economic Community (TEEC)). Lonbay has long ago noted that the impact of Article 110 on Member State tax policies, and thus their economic and social policies, is shown to be enormous. See Julian Lonbay, 'A Review of Recent Tax Cases – Wine, Gambling, Fast Cars and Bananas' (1989) 14 European Law Review 48, 56.

[10] Catherine Barnard, *The Substantive Law of the EU: The Four Freedoms* (3rd edn, Oxford University Press 2010) 55. See Case C-221/06 *Stadtgemeinde Frohnleiten* [2007] ECR I-09643, paras 43 and 47. In this regard, see also Paul Craig and Gráinne de Búrca, *EU Law: Text, Cases and Materials* (4th edn, Oxford University Press 2008) 664. Craig and de Búrca argue that in relation to duties and taxation 'the Court's jurisprudence has consistently looked behind the form of the disputed measure to its substance, and the ECJ has interpreted the relevant Articles in the manner best designed to ensure that the Treaty objectives are achieved'.

[11] The term direct discrimination can be said to refer to discrimination that is explicitly or obviously based on a prohibited ground. By contrast, the concept of indirect discrimination

other Member States (on grounds of nationality),[12] it may nonetheless have an *indirectly* discriminatory effect in that it could make it more difficult for 'foreign' CO_2 streams to be stored than for 'domestic' CO_2 streams.[13] This raises questions as to the compatibility of Member States' financial security/mechanism charges with Article 110 TFEU.

In the following sections, I answer the question of in what way Member States should design the financial security/mechanism charge for it not to breach Article 110 TFEU. To this end, section 4.2 briefly discusses Articles 19 and 20 of the CCS Directive and the financing of storage site stewardship. Section 4.3 returns to the matter of classifying captured CO_2 for geological storage under EU free movement law. I already addressed this issue in the third chapter of this thesis, but will further explore it in this section. Section 4.4 deals with Article 110 in the light of a financial security/mechanism charge being imposed on storage operators. Section 4.5 discusses several Article 110 cases that are of particular relevance in this regard. Finally, section 4.6 gives an overview of the requirements a Member State financial security/mechanism instrument has to meet for it to be compatible with Article 110 TFEU and relates this to the cross-border storage of CO_2.

4.2 ARTICLES 19 AND 20 OF THE CCS DIRECTIVE AND THE FINANCING OF STORAGE SITE STEWARDSHIP

Article 19(1) of the CCS Directive provides that Member States have to ensure that the applicant for a storage permit proves that he can meet all financial

---

deals with seemingly neutral differentiation criteria with a disproportionate impact or effect on a group (or object) that is protected by an explicit prohibition of discrimination. See Christa Tobler, *Indirect Discrimination: A Case Study into the Development of the Legal Concept of Indirect Discrimination under EC Law* (Intersentia 2005) 56–7. For an illustration of the difference between the two concepts, see, for instance, Case C-379/87 *Groener* [1989] ECR 03967, Opinion of AG Darmon, para 5, and Case 109/88 *Danfoss* [1989] ECR 03199, Opinion of AG Lenz, paras 26–29.

12   See (n 7). See also Martin Hedemann-Robinson, 'Indirect Discrimination: Article 95(1) EC Back to Front and Inside Out?' (1995) 1 European Public Law 439, 442 and Stephen Weatherill, *Cases and Materials on EU Law* (9th edn, Oxford University Press 2010) 330. Hedemann-Robinson argues that direct discriminatory taxation is unlikely to occur in practice as such a tax measure would be too obvious a violation of the principle of equality enshrined in Article 110(1) TFEU. Weatherill notes that Member States are commonly more devious than to develop a tax system that lacks origin neutrality; it is possible to avoid direct discrimination on grounds of nationality, but to achieve a similar result by instead basing a taxation system on criteria which indirectly prejudice the imported product.

13   In a somewhat similar vein, see e.g. Case C-313/05 *Brzeziński* [2007] ECR I-00513, Opinion of AG Sharpston, para 13. This case dealt with the taxation of new passenger cars.

obligations under the storage permit (closure and post-closure requirements), as well as any obligations under the EU ETS Directive. As indicated above, Member States are free to decide the arrangements by which to meet this requirement, but the financial security mechanism must be valid and effective before the storage operator starts injecting CO_2. Article 19(2) requires the financial security mechanism to be periodically adjusted in order to take into account possible changes in risks (of leakage) and estimated costs. By virtue of Article 19(3), the financial security instrument shall remain effective until the responsibility for the storage site is transferred to the competent authority and after the withdrawal of a storage permit, until a new permit is issued or the responsibility for the site is transferred to the competent authority.

Whereas Article 19 sees to the financing of storage site stewardship before the transfer of responsibility, Article 20 addresses the same issue in the (long-term) period thereafter. Under Article 20(1), Member States are to ensure that the storage operator makes a financial contribution to the competent authority before responsibility for the site is transferred to the latter. This financial mechanism/contribution must cover at least the anticipated cost of monitoring for a period of thirty years. The contribution seems to generally be intended to guarantee that the competent authority can cover the costs of keeping the injected CO_2 completely and permanently contained in the long run. Again, Member States are free to choose the particular form of the required financial mechanism. Article 20(2) gives the Commission the possibility to issue guidelines in this regard.

As Articles 19 and 20 are not particularly clear and determinate, it does not come as a surprise that the Commission has made use of the possibility to issue guidelines on the implementation of both provisions.[14] Guidance document 4 gives Member States information on and options for the implementation of Articles 19 and 20. The guidelines, among other things, suggest ways to determine the amount of the financial security/mechanism and list the obligations to be covered by both instruments. These obligations include storage site monitoring, surrender of EU ETS emission allowances in case of CO_2 leakage and preventive and remedial action under the Environmental Liability Directive in case of (an imminent threat of) environmental damage occurring.[15] Several

---

14   Commission, 'Guidance Document 4' (n 2).
15   Ibid 9 and 42–43. The latter two obligations follow from Article 1(30) of European Parliament and Council Directive 29/2009/EC of 23 April 2009 amending Directive 2003/87/EC so as to improve and extend the greenhouse gas emission allowance trading scheme of the Community [2009] OJ L140/63 and Article 34 of the CCS Directive. Article 1(30) of Directive 29/2009/EC has included the geological storage (as well as the capture and transport per pipeline) of CO_2 in Annex I to European Parliament and Council Directive 2003/87/EC of 13 October 2003 establishing a scheme for greenhouse gas emission allowance trading within

financial *security* instruments are mentioned, such as funds, financial institution guarantees and insurances.[16] In relation to the options for a financial *mechanism*, the guidelines state that the two articles are linked by similar intent and similar options available to the Member States for their implementation.[17] As indicated above, it is explicitly suggested that the Member States could use the same instrument for meeting the obligations under Articles 19 and 20.[18]

The (predominantly US) literature on storage site stewardship has mainly focused on long-term stewardship and long-term (post-closure) liability for the stored CO_2. As indicated above, the option of a (trust) fund financed through a fee per tonne of CO_2 injected has regularly been suggested.[19] On occasion, it has been proposed to apply a risk-weighted per-tonne fee.[20] Such a fee would likely take into account the chemical composition and geophysical characteristics of the relevant CO_2 stream and storage reservoir respectively. Both variables can vary to a great extent. The chemical composition of the CO_2 stream, for instance,

---

the Community and amending Council Directive 96/61/EC [2003] OJ L275/32 (the EU ETS Directive). Annex I lists the categories of activities to which the EU ETS Directive applies. As a consequence, storage operators need to surrender EU ETS emission allowances for any leaked CO_2. Article 34 of the CCS Directive has added the operation of CO_2 storage sites to the activities listed in Annex III to European Parliament and Council Directive 2004/35/EC of 21 April 2004 on environmental liability with regard to the prevention and remedying of environmental damage [2004] OJ L143 (Environmental Liability Directive). As a consequence, the Environmental Liability Directive's 'strict liability' regime applies to CO_2 storage activities. This means that the operator of a storage site may be held financially liable for environmental damage caused by CO_2 storage activities, even if he is not at fault. In case of (an imminent threat of) environmental damage occurring, he is required to take (preventive and) remedial measures pursuant to Articles 5(1) and 6(1) of the Environmental Liability Directive. By virtue of Article 18(1) of the CCS Directive, the transfer of responsibility for the storage site to the competent authority includes the transfer of both types of liability ('climate' and 'environmental' liability), as well as all legal obligations relating to monitoring and corrective measures.

16 Commission, 'Guidance Document 4' (n 2) 2. See pages 4–6 of the guidance document for a more elaborate discussion of the possible instruments.
17 Ibid 41.
18 Ibid.
19 See United States Environmental Protection Agency (EPA), 'Approaches to Geologic Sequestration Site Stewardship after Site Closure' (2008) 8 <www.epa.gov/> accessed 8 May 2012; Klass and Wilson (n 5) 174; Chiara Trabucchi and Lindene Patton, 'Storing Carbon: Options for Liability Risk Management, Financial Responsibility' in *World Climate Change Report* (2008) 1, 16; CCSReg Project, 'Policy Brief: Compensation, Liability and Long-term Stewardship for CCS' (2009) 7 <www.ccsreg.org/pdf/LongTermLiability_07132009.pdf> accessed 9 May 2012; and Elizabeth J Wilson, Alexandra B Klass and Sara Bergan, 'Assessing a Liability Regime for Carbon Capture and Storage' (2009) 1 Energy Procedia 4575, 4580. For a European context, see, for instance, Gøril Tjetland and others, 'Incentives and Regulatory Frameworks Influence on CCS Chain Establishment' <www.sintef.no/project/ecco/Publications/Conference%2014%20June%202011/10%20-%20Regulatory%20framework%20recommendations%20-%20Goeril%20Tjetland.pdf> accessed 9 May 2012.
20 Klass and Wilson (n 5) 174 and CCSReg Project (n 19) 7.

will vary depending on the (capture) source and the technical requirements for transport, and has the potential to affect the chemistry in the storage reservoir.[21]

As the characteristics of CO_2 storage reservoirs and CO_2 streams will differ from reservoir to reservoir and from stream to stream and these characteristics are highly relevant for determining the risks related to individual CO_2 storage projects, Member States might want to differentiate the financial security/ mechanism charges imposed on storage fee applicants/storage operators. It is, for instance, conceivable that the amount of the financial security/mechanism charge imposed will be higher in the case of CO_2 storage in saline aquifers, as the geophysical characteristics of such formations are generally less well known than those of depleted gas and oil fields.[22] Likewise, a storage operator seeking to inject a CO_2 stream the composition of which has a higher risk profile, is likely to be faced with a higher financial security/mechanism duty than an operator injecting a lower-risk CO_2 stream. In the following, I will explore the possibilities for Member States in designing financial security/mechanism instruments in the light of Article 110 TFEU. Before turning to Article 110 itself, I will first return to a matter already discussed in the third chapter of this thesis: the qualification of captured CO_2 for storage under EU internal market provisions.

4.3 CAPTURED CO_2 FOR STORAGE UNDER EU INTERNAL MARKET PROVISIONS

In section 3.3 of the previous chapter, I briefly discussed the status of captured CO_2 for storage under EU free movement provisions. I argued why, as the law stands after the *Walloon Waste* case,[23] all captured CO_2 is likely to classify as a good within the meaning of the TFEU provisions on the free movement of goods. In Chapter III, it was not opportune to discuss the classification of captured CO_2 for storage under EU free movement provisions any further, since such discussion would not have changed the answer to the question posed.

---

[21] Stijn Santer and others (CO_2Europipe), 'Making CO_2 Transport Feasible: The German Case Rhine/Ruhr Area (D) – Hamburg (D) – North Sea (D, DK, NL)' (2011) 19 <www.co2europipe. eu/Publications/D4.2.2%20-%20Making%20CO2%20transport%20feasible%20-%20the%20 German%20case.pdf> accessed 9 May 2012, and Tom Mikunda and Heleen de Coninck (CO_2ReMoVe [n 121 Chapter III]), 'Possible Impacts of Captured CO2 Stream Impurities on Transport Infrastructures and Geological Storage Formations: Current Understanding and Implications for EU Legislation' (2011) 5 <www.ecn.nl/docs/library/report/2011/o11040.pdf> accessed 9 May 2012.

[22] See e.g. Karen Kirk, 'Safety of CO_2 Storage in the North Sea' <www.zeroco2.no/6-bgs-karen-kirk-storage-in-the-north-sea.pdf> accessed 9 May 2012, and Andreas Bielinski, 'Numerical Simulation of CO_2 Sequestration in Geological Formations' (PhD thesis, University of Stuttgart 2007) 5 <http://d-nb.info/996784012/34> accessed 9 May 2012.

[23] Case C-2/90 *Commission v. Belgium* [1992] ECR I-04431.

However, in this section I will examine the issue more closely. To this end, the European Court of Justice's (ECJ) ruling in *Walloon Waste* will first be discussed in more detail. Subsequently, two relevant post-*Walloon Waste* cases are addressed.

In *Walloon Waste*, the ECJ had to answer the question of whether waste constitutes a good within the meaning of the TFEU provisions on the free movement of goods. The Belgian government had argued that non-recyclable and non-reusable waste could not be characterised as a good within the meaning of the TFEU provisions on the free movement of goods.[24] According to it, waste for disposal had no intrinsic commercial value and could thus not be the subject of a sale. Therefore, operations for the disposal or tipping of such waste had to be covered by the TFEU provisions on the free movement of services.[25] This line of reasoning does not seem to be without (economic) merit.[26] The ECJ, however, thought differently. It indicated that 'objects' which are transported cross-border 'in order to give rise to a commercial transaction' are covered by the TFEU provisions on the free movement of goods, 'irrespective of the nature of those transactions'.[27] The question of whether the object can give rise to a commercial transaction was considered to be decisive by the ECJ.[28]

This criterion stems from the first case in which the ECJ gave a definition of a good within the meaning of the TFEU provisions on the free movement of goods, i.e. *Commission v. Italy*.[29] In this case, the ECJ provided that a good is a product which (a) is capable, as such, of forming the subject of a commercial transaction and (b) can be valued in money.[30] It is precisely the latter criterion

---

[24] Case C-2/90 *Commission v. Belgium* [1992] ECR I-04431 para 25.
[25] This is a distinction which is also commonly made in literature. See L Krämer, 'Environmental Protection and Article 30 EEC Treaty' (1993) 30 Common Market Law Review 111, 115.
[26] See also Jan Jans and Hans Vedder, *European Environmental Law* (3rd edn, Europa Law Publishing 2008) 233.
[27] Case C-2/90 *Commission v. Belgium* (n 23) para 26.
[28] The 'secondary' character of the TFEU provisions on the free movement of services might also have influenced the ECJ's ruling. Article 57 TFEU determines that 'services shall be considered to be "services" within the meaning of the Treaties where they are normally provided for remuneration, *in so far as they are not governed by the provisions relating to freedom of movement for goods, capital and persons*' (emphasis added). However, in Case C-452/04 *Fidium Finanz* [2006] ECR I-0952 the ECJ stated that the phrase 'are not governed by the provisions relating to freedom of movement for goods, capital and persons' relates to the definition of the notion of services and 'does not establish any order of priority between the freedom to provide services and the other fundamental freedoms' (recital 32). According to the ECJ, 'the notion of services covers services which are not governed by other freedoms, in order to ensure that all economic activity falls within the scope of the fundamental freedoms' (recital 32). This seems to suggest that the ECJ does not consider the free movement of services to be secondary to the other fundamental freedoms.
[29] Case 7-68 *Commission v. Italy* [1968] ECR 00423.
[30] Case 7-68 *Commission v. Italy* (n 29).

which was used by the Belgian government in Walloon waste to argue that waste for disposal does not constitute a good, but falls under the TFEU provisions on the free movement of services. In *Walloon Waste*, the ECJ, in relation to recyclable and reusable waste, referred to this condition as a requirement of 'intrinsic commercial value'.

This indicates that the question of whether a product can be valued in money, in essence asks whether the product in question has (positive) intrinsic commercial value. A product which can be negatively valued in money, i.e. represents a cost to the holder of it, would accordingly not constitute a good within the meaning of the TFEU provisions on the free movement of goods. With this in mind, it is difficult to understand why the ECJ in *Walloon Waste* rejected the Belgian government's argument that waste for disposal should not be considered to constitute a good.

The distinction between waste for disposal on the one hand and recyclable and reusable waste on the other, is one that could also be made for captured CO_2. Captured CO_2 *for storage* could be seen as the equivalent of waste for disposal. It is transported and subsequently stored with a view to permanent disposal. Captured CO_2 for storage has a negative intrinsic commercial value, i.e. it represents a cost. The capturer will have to pay both the transporter and the storage operator a service fee for handling the captured CO_2.

On the other hand, captured CO_2 *for reuse* in e.g. enhanced hydrocarbon recovery activities[31] or in industrial or horticultural processes (production of sodas, use in greenhouses) is similar to recyclable or reusable waste. It has a positive intrinsic value – even though that value may in practice be minimal – and can be sold to third parties which use the captured CO_2 in their production processes.[32] CO_2 for reuse would, as long as the applicable safety precautions are respected, seem to meet the criteria in Article 6(1) of Directive 2008/98/EC (new Waste Directive) for no longer being qualified as disposable waste. Accordingly, there is a strong argument for having CO_2 *for reuse* fall under the TFEU provisions on the free movement of goods, whereas CO_2 *for storage* should fall under the provisions on the free movement of services. Yet, on the basis of *Walloon Waste*, all captured CO_2 is likely to classify as a good within the meaning of the TFEU provisions on the free movement of goods. This appears to directly follow from the ECJ's central argument that objects which are transported cross-border in order to give rise to a commercial transaction are

---

31 (n 56) Chapter I.
32 According to some, CO_2 in general lacks inherent value. See Varun Rai, David G Victor and Mark C Thurber, 'Carbon Capture and Storage at Scale: Lessons from the Growth of Analogous Energy Technologies' (2009) 38 Energy Policy 4089, 4090.

covered by the TFEU provisions on the free movement of goods, irrespective of the nature of those transactions.[33]

Nevertheless, in a case following *Walloon Waste*, the ECJ seemed to somewhat relinquish the line that objects which are subject to a cross-border commercial transaction are covered by the free movement of goods provisions, irrespective of the nature of those transactions.[34] In *Jägerskiöld*, the ECJ was asked whether fishing rights (and the related fishing permits) constituted goods, in accordance with its earlier judgment in *Commission v. Italy*.[35, 36] Referring to, inter alia, the *Schindler* case (on the organisation of lotteries),[37] the ECJ in essence held that even if the relevant activity is related to something which may be characterised as a product/good (e.g, a lottery ticket or fishing permit), it will be seen as a service if it allows the receiving entities to engage in a certain activity. The product (the ticket or permit) has value only because it constitutes a prerequisite to engage in a certain activity (participate in the lottery, fish). The ancillary character of the good/product is essential to the activity falling under the free movement of services.

In this regard, it could be questioned whether the ECJ's ruling in the *Jägerskiöld* case is of consequence for the classification of captured CO_2 for storage under EU free movement provisions. The commercial transaction at hand – the capturer paying the storage operator to store the captured CO_2 – does not involve the granting to the paying entity (the capturer) of a right to engage in a certain activity. Rather, the capturer pays the storage operator to perform the service of storing the captured CO_2. The underlying facts of the *Jägerskiöld* case are too different from the scenario in which a capturer pays a storage operator to have his captured CO_2 stored for the point of *Walloon Waste* to no longer hold.

A more recent case on waste disposal seems to confirm the relevance of *Walloon Waste* for the classification of captured CO_2 for storage under EU free movement provisions. In *Stadtgemeinde Frohnleiten*, the Austrian government argued that an Austrian tax on waste disposal fell outside the scope of Article 110 TFEU on

---

[33] Case C-2/90 *Commission v. Belgium* (n 23) para 26.
[34] In *Walloon Waste*, the ECJ in fact referred to 'objects which are shipped across a frontier for the purposes of commercial transactions' (para 26) and not to 'objects subject to a cross-border commercial transaction'. Nevertheless, there does not seem to be a substantial difference between the two formulations. In analogy with the free movement of services, it arguably makes no difference whether the relevant good crosses the border between two Member States or whether the buyer of the good does so; the decisive criterion for the free movement of goods to apply is the cross-border element; the purely internal sale of goods is not covered by the TFEU provisions on the free movement of goods.
[35] Case 7-68 *Commission v. Italy* (n 29).
[36] Case C-97/98 *Jägerskiöld* [1999] ECR I-07319, para 18.
[37] Case C-275/92 *Schindler* [1994] ECR I-01039.

the ground that it did not constitute a levy imposed on products within the meaning of that provision.[38] Analogous to the reasoning of the Belgian government in *Walloon Waste*, the Austrian government held that waste deposited at waste disposal sites had no pecuniary value and that the relevant tax, therefore, was not imposed on goods in trade, but on the supply of *services* rendered by the operator of the waste disposal site.[39]

The ECJ, referring to its judgment in *Walloon Waste*, held that waste, recyclable or not, is to be characterised as goods.[40] Waste for disposal, even if it has no intrinsic value, may give rise to commercial transactions in relation to the disposal or deposit thereof.[41] Furthermore, the ECJ reasoned that Article 110 TFEU has to be interpreted widely and that it applies to internal taxation imposed on the use of imported products where those products are essentially intended for such use and have been imported solely for that purpose.[42] It considered these requirements to have been met in the case at hand.[43]

Finally, the ECJ stated that it followed from the case law (Cases 20/76 *Schöttle* and C-90/94 *Haahr Petroleum*)[44] that a charge imposed not on a product as such, but on a specific activity related to that product (and calculated according to, inter alia, the weight of the product) falls within the scope of Article 110 TFEU.[45] In this regard, it noted that the relevant case law had been developed in the context of goods with a market value offered for sale in the importing state,

---

[38]   Case C-221/06 *Stadtgemeinde Frohnleiten* (n 10) para 34.
[39]   Ibid paras 34–35.
[40]   Ibid para 37.
[41]   Ibid para 38.
[42]   Ibid para 41.
[43]   Ibid para 42.
[44]   In *Schöttle*, the German government had maintained that a German tax on the transport of goods by road was not caught by Article 110 TFEU, since it did not affect the transported product as such. The ECJ, however, held that a tax imposed on the international transport of goods by road according to the distance covered on the national territory and the weight of the goods in question had to be regarded as taxation indirectly imposed on products. According to it, such a tax had an immediate effect on the cost of the national and imported product and, by virtue of Article 110 TFEU, had to be applied in a manner which was not discriminatory to imported products. See Case C-20/76 *Schöttle* [1977] ECR 00247, paras 12–15. See also Case C-90/94 *Haahr Petroleum* [1997] ECR I-04085, paras 38–40. In this regard, see also Case C-221/06 *Stadtgemeinde Frohnleiten* [2007] ECR I-09643, Opinion of AG Sharpston, paras 32–34 and Case C-206/06 *Essent Netwerk Noord BV* [2008] ECR I-05497, Opinion of AG Mengozzi, paras 45–50.
[45]   Para 43. This does, however, seem to contrast with another (earlier) ruling of the ECJ. In Joined Cases C-393/04 and C-41/05 *Air Liquide Industries*, the ECJ held that Article 110 TFEU did not apply since 'the tax is not imposed specifically on exported or imported products or in such a way as to differentiate them, given that it applies to economic activities carried out by industrial, commercial, financial or agricultural undertakings *and not to products as such*' (emphasis added). See Joined Cases C-393/04 and C-41/05 *Air Liquide Industries* [2006] ECR I-05293, para 57.

in which case the effect on the cost of the product is the appropriate criterion for finding a hindrance to the free movement of goods.[46] By contrast, the only commercial transactions to which waste for disposal may give rise are those relating to the disposal or deposit thereof.[47] In the case of waste for disposal, the proper criterion for finding a hindrance to the free movement of goods is the effect of the levy on the price paid for disposal of the waste.[48] Concluding, the ECJ held that Article 110 applied to a waste tax such as that levied by Austria since:[49]

1) the management of waste disposal clearly was an activity related to the waste deposited there;
2) the rate of the relevant levy depended, among other things, on the weight and the nature of the waste; and
3) the operator of the waste disposal site would pass on to the foreign party seeking to have his waste disposed the cost of the levy paid.

In *Stadtgemeinde Frohnleiten*, the ECJ clearly confirmed its earlier ruling in *Walloon Waste*. Waste for disposal does not fall under the free movement of services, but under the free movement of goods, provided that a cross-border element is present. Furthermore, a tax levied on the disposal of such waste in principle falls within the scope of Article 110 TFEU. This finding is not affected by the tax being levied on the disposal of waste and not on the waste itself. By its ruling in *Stadtgemeinde Frohnleiten*, the ECJ upheld the line of *Walloon Waste* that objects which are transported cross-border in order to give rise to a commercial transaction are covered by the TFEU provisions on the free movement of goods, irrespective of the nature of those transactions.

In section 3.3 of the previous chapter, I argued why *Walloon Waste* is of relevance for the classification of captured CO_2 for storage under EU free movement provisions, despite the fact that captured CO_2 for storage no longer classifies as waste under EU law.[50] The cases of *Jägerskiöld* and *Stadtgemeinde Frohnleiten* show that the relevance of *Walloon Waste* has not diminished. Captured CO_2 – whether for storage or not – is most likely to classify as a good within the meaning of the EU provisions on the free movement of goods. Furthermore, as long as it is not charged *by reason of* the captured CO_2 crossing a Member State's border as well as to the exclusion of domestically produced CO_2 streams for storage, any charge levied on the storage of captured CO_2 will likely fall within

---

[46] Case C-221/06 *Stadtgemeinde Frohnleiten* (n 10) para 44.
[47] Ibid para 45.
[48] Ibid.
[49] Ibid paras 46–47.
[50] This is a consequence of Articles 35 and 36 of the CCS Directive.

the scope of Article 110 TFEU.[51] The management of CO_2 storage (sites) clearly is an activity related to the CO_2 stored there, the rate of the relevant financial security/mechanism charge could very well – at least partly – depend on the nature (composition) of the captured CO_2 for storage and the storage operator will probably pass on to the foreign capturer the cost of the levy paid to the competent authority. In the following sections, we will look at the consequences this classification could have in practice.

4.4   ARTICLE 110 TFEU

Article 110 TFEU provides that:

> [N]o Member State shall impose, directly or indirectly, on the products of other Member States any internal taxation of any kind in excess of that imposed directly or indirectly on similar domestic products. Furthermore, no Member State shall impose on the products of other Member States any internal taxation of such a nature as to afford indirect protection to other products.

The first paragraph of Article 110 prohibits Member States from taxing imported products more heavily than 'similar' domestic products. Article 110(1) does not specify when domestic products can be considered to be 'similar' to imported products. However, as we will see below when discussing Article 110(1), the ECJ has in several cases clarified what is to be understood by similar domestic products. Crucial in the second paragraph of Article 110, is the term 'other products'. Like the notion of similarity, the term is not further specified in Article 110. In a number of cases, the ECJ has, however, given guidance on the meaning of this phrase. It basically refers to domestic products which are not similar to the relevant imported products, but do compete with those products. We will get back to the meaning of the concept of 'other products' when discussing Article 110(2), below.

4.4.1   PURPOSE

The purpose of Article 110 is to prevent Member States from circumventing the prohibitions in Articles 28–30 TFEU[52] by imposing discriminatory internal

---

51   The boundaries between Articles 30 and 110 TFEU will, briefly, be further discussed in section 4.4.2 below.

52   Articles 28–30 TFEU prohibit the Member States from imposing customs duties on imports from and exports to other Member States and on products in free circulation in the Member States.

taxation.[53] The provision's role is to further the free movement of goods within the EU.[54] Articles 28–30 TFEU would be of little use if Member States could disadvantage imported and exported products by levying discriminatory taxes on all products *inside* their territory.[55] Article 110 can be seen a specific formulation of the general prohibition of discrimination laid down in Article 18 TFEU.[56]

4.4.2 SCOPE AND CONTENT

Article 110 applies to 'indirect taxes' (taxes on products) as opposed to 'direct taxes' (taxes imposed on income and property).[57] The provision was inspired by corresponding provisions of the United Nations General Agreement on Tariffs and Trade (GATT).[58] Both Article III:2 GATT[59] and Article 110 TFEU prohibit discriminatory internal taxes.[60] They establish a level playing field for imports (and exports – see below) and domestic produce; the prohibition of discrimination ensures that products enjoy equality of conditions in the market – whatever their origin.[61]

Given its importance to the attainment of the single market, the ECJ has interpreted Article 110 widely so as to cover all procedures which, directly or indirectly, undermine the equal treatment of domestic products and imported products.[62] In section 4.3, we saw that this includes taxes levied on a specific *activity* related to a good subject to a cross-border commercial transaction (waste disposal), instead of on a good itself (waste).[63] It also includes taxes on the *use* of

---

[53] See Case 252/86 *Bergandi* [1988] ECR 01343, para 24.

[54] See Case C-166/98 *Socridis* [1999] ECR I-03791, para 16, and Case C-402/09 *Ioan Tatu* [2011] ECR I-02711, Opinion of AG Sharpston, para 30.

[55] See, for instance, also Case C-402/09 *Ioan Tatu* [2011] ECR I-02711, para 53.

[56] See Case C-265/99 *Commission v. France* [2001] ECR I-02305, Opinion of AG Alber, para 36.

[57] See Marco M Slotboom, 'Do Different Treaty Purposes Matter for Treaty Interpretation? The Elimination of Discriminatory Internal Taxes in EC and WTO Law' (2001) 4 Journal of International Economic Law 557, 564, and Barnard (n 10) 52.

[58] See George A Bermann and others, *Cases and Materials on European Union Law* (3rd edn, West Publishing Co 2011), 427, and Slotboom (n 57) 558.

[59] Article III:2 GATT provides that 'the products of the territory of any contracting party imported into the territory of any other contracting party shall not be subject, directly or indirectly, to internal taxes or other internal charges of any kind in excess of those applied, directly or indirectly, to like domestic products. Moreover, no contracting party shall otherwise apply internal taxes or other internal charges to imported or domestic products in a manner contrary to the principles set forth in paragraph 1'.

[60] Slotboom (n 57) 558–59.

[61] Adrian Emch, 'Same Same but Different? Fiscal Discrimination in WTO Law and EU Law: What Are 'Like' Products?' (2005) 32 Legal Issues of Economic Integration 369, 372.

[62] Barnard (n 10) 53. Barnard refers to Case C-221/06 *Stadtgemeinde Frohnleiten* (n 10) para 40.

[63] In relation to the indirect taxation of a product, see also Case C-387/01 *Weigel* [2004] ECR I-04981, para 68. In *Weigel*, the ECJ held that a taxation does not have to be directly paid by

imported products.[64] What is more, for the purposes of the application of Article 110 (and Article 30), it is of little account that the financial charge is not levied by the state (but by a private entity), if the levy of the charge is required under national law.[65] Also, the identity of the person liable for the payment of the charge seems to be of little account if the charge relates to a product or a necessary activity in connection with a product.[66] Even though it is not expressly provided for in Article 110, the provision applies to the discriminatory taxation of both imported and *exported* products.[67]

Articles 30 and 110 TFEU are mutually exclusive.[68] The ECJ has consistently held that both provisions cannot be applied together and that the same charge cannot belong to both categories at the same time: it is either a charge having equivalent effect or an internal taxation.[69] Article 30 applies to all charges exacted at the time of or by reason of importation which are imposed specifically on an imported product to the exclusion of the similar domestic product.[70] Article 110, on the other hand, applies to financial charges levied within a general system of internal taxation applying systematically to domestic and imported goods.[71] In the *Denkavit Loire* case, the ECJ held that a charge must

---

    the purchaser of the product. According to the ECJ, both imported and locally bought used vehicles were taxed equally, since the price paid for a domestic used car included a residual portion of the tax levied at first registration of a car in Austria, which reduced in line with the vehicle's depreciation through use.

[64]   Barnard (n 10) 54. Barnard refers to Case 252/86 *Bergandi* (n 52) para 27.

[65]   See Case C-206/06 *Essent Netwerk Noord BV* [2008] ECR I-05497, paras 45–46. In the *Essent Netwerk Noord* case, the ECJ held that the fact that the relevant price surcharge was levied by an electricity distribution network operator (and not by a state entity) was irrelevant, since the price surcharge was required under national law.

[66]   Ibid paras 48–49.

[67]   See e.g. Case 51/74 *Van der Hulst's zonen* [1975] ECR 00079, paras 34 and 35, Case 46/76 *Bauhuis* [1977] ECR 00005, para 25, and Case 142/77 *Larsen and Kjerulff* [1978] ECR 01543, paras 20–27.

[68]   In most cases, it will probably not be too difficult to determine whether the charge at hand falls under Article 30 or under Article 110. For an overview of more difficult situations, see Craig and de Búrca (n 10) 661–64. See also, Barnard (n 10) 66–69. See also Case C-517/04 *Koornstra* [2006] ECR I-05015, Opinion of AG Stix-Hack, para 35. According to the AG, the ECJ's case law illustrates the difficulties arising in relation to the essential distinction between the two provisions.

[69]   See Case C-101/00 *Tulliasiamies and Siilin* [2002] ECR I-07487, para 115; Case C-234/99 *Nygård* [2002] ECR I-03657 para 17; Case C-383/01 *De Danske Bilimportører* [2003] ECR I-06065, para 33, and Case C-387/01 *Weigel* (n 63) para 63. See also Peter J Oliver, *Oliver on Free Movement of Goods in the European Union* (5th edn, Hart Publishing 2010) 101.

[70]   Alan Dashwood and others, *Wyatt and Dashwood's European Union Law* (6th edn, Hart Publishing 2011) 401.

[71]   Ibid. See also Joined Cases C-34/01 to C-38/01 *Enirisorse* [2003] ECR 14243, Opinion of AG Stix-Hackl, para 182. Referring to Case C-212/96 *Chevassus-Marche* [1998] ECR I-00743 (para 20), Stix-Hackl states that the ECJ has held that the essential feature of a charge having an effect equivalent to a customs duty which distinguishes it from an internal tax resides in the fact that the former is borne solely by an imported product as such whilst the latter is

impose the same duty on national products and identical imported products at the same marketing stage and as a consequence of the same chargeable event, for it to fall within the scope of Article 110.[72] A charge in breach of Article 30 has to be abolished.[73] By contrast, Article 110 only requires the elimination of the discriminatory or protective element (and not of the tax itself).[74]

One of the situations in which there can be boundary-line problems between Articles 30 and 110 is that in which the importing Member State does not make the imported product (or has a very low domestic production of the particular good), but imposes a tax on it nonetheless.[75] In *Co-frutta*, the ECJ held that any charge levied on imported products in such situation does not constitute a charge having equivalent effect under Article 30 but internal taxation within the meaning of Article 110, provided the tax fits within the general scheme of taxation applied systematically to categories of products in accordance with objective criteria irrespective of the origin of the products.[76] The ECJ stated that the compatibility of such a charge with EU law is to be assessed on the basis of Article 110.[77] In the *Co-frutta* case, the latter was possible as there *was* domestic production, if only very little.

Nevertheless, if there is no domestic production whatsoever, it seems impossible to assess the conformity of the relevant tax with Article 110, since there are no similar or competing domestic products the taxation of which can be compared with that of cross-border products. This point was raised in *Commission v. Denmark*.[78] This case centred around a Danish registration duty levied on new cars. As Denmark did not produce any cars, the tax de facto was levied on imported cars only. The Danish government agreed with the Commission's classification of the Danish registration duty as an internal tax within the

---

borne both by imported and domestic products, applying systematically to categories of products in accordance with objective criteria irrespective of the origin of the products.

[72] Case 132/78 *Denkavit Loire* [1979] ECR 01923, para 8. In *Nygård*, the ECJ refined its judgment in *Denkavit Loire* and held that the moment of taxation of domestic and cross-border products may be different as long as, in real economic terms, the different moments of taxation correspond to the same marketing stage. See Case C-234/99 *Nygård* (n 69) para 30.

[73] See Case C-383/01 *De Danske Bilimportører* [2003] ECR I-06065, Opinion of AG Jacobs, paras 26–28.

[74] Ibid.

[75] See Craig and de Búrca (n 10) 662 and Barnard (n 10) 66–7. A similar – but not identical – situation is that in which the detailed rules governing the levying of the charge are such that it is imposed solely on imported products to the exclusion of exported products. In that case, the charge is to be regarded as a charge having equivalent effect under Article 30. See Case 32/80 *Kortmann* [1981] ECR 00251, para 18.

[76] Case 193/85 *Co-frutta* [1987] ECR 02085, para 10. See Barnard (n 10) 66–7. See also Case 90/79 *Commission v. France* [1981] ECR 00283, para 14.

[77] Case 193/85 *Co-frutta* (n 76) para 13.

[78] Case C-47/88 *Commission v. Denmark* [1990] ECR I-04509.

meaning of Article 110, but (inconsistently) considered that the latter provision could not apply since there were no similar or competing domestic products.[79]

The ECJ held that Article 110 could not be invoked against internal taxation imposed on internal products where there is no similar or competing domestic production.[80] In particular, it stated, the provision did not provide a basis for censuring the excessiveness of the level of taxation adopted by Member States for particular products, in the absence of any discriminatory or protective effect.[81] As there was no domestic production of cars or competing products in Denmark, the ECJ concluded that the Danish registration duty did not infringe Article 110.[82] The ECJ did, however, leave open the possibility that such tax could be assessed under Article 30.[83]

As I have argued before, a financial security/mechanism charge levied on storage permit applicants/storage operators is likely to fall within the scope of Article 110 (and not of Article 30).[84] Such levy will probably not be charged at the time of, or on account of, the captured CO_2 crossing the border. Furthermore, it will likely not specifically be imposed on CO_2 streams imported for storage to the exclusion of CO_2 streams produced domestically for storage. Instead, the financial security/mechanism charge will probably be imposed through a general system applicable to all storage permit applicants/storage operators within the territory of the Member State concerned, irrespective of the origin of the relevant captured CO_2 for storage.[85] The financial security/mechanism will likely be required as part of the national storage permitting process and lead to a taxation of the service related to the captured CO_2, the storage of CO_2.

In section 4.3, we have seen that the taxation of an activity/service related to a product subject to a cross-border commercial transaction does not prevent Article 110 from applying. As any financial security/mechanism charge will, in this way, be indirectly imposed on and borne by both imported and domestic CO_2 streams for storage, Article 110 will likely apply.[86] The case of *Essent*

---

[79]   Case C-47/88 *Commission v. Denmark* (n 78) para 6.
[80]   Ibid para 10.
[81]   Ibid.
[82]   Ibid para 11.
[83]   Ibid para 13. In this regard, see also Case C-383/01 *De Danske Bilimportører* (n 69) paras 40–42.
[84]   See section 4.3, above.
[85]   The word 'origin' should be read to refer to the *geographical* origin of the CO_2 stream. As argued in section 4.2, the *source* of the captured CO_2 directly affects the composition of the CO_2 stream, a factor which Member State authorities could very well take into account when setting the level of the financial security/mechanism charge.
[86]   *Stadtgemeinde Frohnleiten* and *Essent Netwerk Noord BV* indicate that it is irrelevant that the storage permit applicant/storage operator (and not the foreign capturer) will formally be liable for the payment of the financial security/mechanism charge.

Netwerk Noord BV shows that it will, in this regard, be of no relevance whether the financial security/mechanism charge is levied by the state or by a private entity; as a consequence of the mandatory implementation of Articles 19 and 20 of the CCS Directive, the levy of the financial security/mechanism charge will be required under national law and will thus fall within the scope of Article 110. In the cases of *Lütticke*[87] and *Firma Fink-Frucht*,[88] the ECJ held that both paragraphs of Article 110 have direct effect; they can be invoked by private parties to challenge a Member State's internal tax system.[89]

4.4.3 ARTICLE 110(1) TFEU

Article 110(1) prohibits Member States from taxing imported and exported products more heavily than 'similar' domestic products. A heavier taxation of cross-border products only constitutes discrimination prohibited under Article 110 if it leads to *similar* cross-border and domestic products being treated *differently*. As indicated earlier, Article 110(1) does not specify when domestic products can be considered to be 'similar' to imported products. Nevertheless, in a number of cases, the ECJ has given guidance on the concept of similarity of cross-border and domestic products.

In several early cases under Article 110, the ECJ applied a formal test, examining whether the relevant products normally came within the same fiscal, customs, or statistical classification.[90] In later cases, the ECJ adopted a broader test which combined a factual comparison of the products with an economic analysis of their use.[91] In order to determine whether two products are similar the ECJ will have regard to whether the products have similar characteristics and whether they meet the same needs from the point of view of the consumer.[92] This approach is well-illustrated by the case of *Commission v. Denmark*, in which the ECJ held in relation to the similarity of wine made from grapes and fruit wine that:

> [I]t is necessary first to consider certain objective characteristics of both categories of beverages, such as their origin, the method of manufacture and their organoleptic properties, in particular taste and alcohol content, and secondly to consider whether

---

[87] Case 57/65 *Lütticke* [1966] ECR 00205, 210.

[88] Case 27/67 *Firma Fink-Frucht* [1968] ECR 00223, 232.

[89] Bermann and others (58) 427.

[90] Craig and de Búrca (n 10) 654; Barnard (n 10) 55–56 and Margot Horspool and Matthew Humphreys, *European Union Law* (6th edn, Oxford University Press 2010) 307. Barnard refers to Case 27/67 *Firma Fink-Frucht* (n 88) 232 and Case 45/75 *Rewe v. HZA Landau* [1976] ECR 00181, para 12.

[91] Barnard (n 10) 56.

[92] Dashwood and others (n 70) 397.

or not both categories of beverages are capable of meeting the same need from the point of view of the consumers.[93]

The Court has variously considered methods of manufacture, properties, origin, meeting the same needs of consumers (both current and future), process of creation, organoleptic properties and objective characteristics in trying to determine similarity.[94] The ECJ's approach in determining similarity has been criticised for its perceived subjectivity. Emch, for instance, has noted that any assessment of likeness should examine the economic relationship between products, viewed from the consumer's perspective; this is best expressed through the degree of the relevant products' substitutability as perceived by consumers (only), rather than through physical or objective characteristics that at times lead to arbitrary generalisations and groupings.[95] According to Emch, the examination of physical characteristics should only be held as valid evidence if it can reliably indicate existing consumer behaviour.[96] Nevertheless, as the law stands, the ECJ appears to apply a combined test under Article 110(1), assessing both the characteristics of the products as well as their use.[97]

As argued before, discrimination against cross-border products can be either direct or indirect.[98] Direct discrimination can take several forms under Article 110, such as only the imported product being subject to the tax (e.g. Case 57/65 *Lütticke*), the imported product being taxed at a different (usually higher) rate (e.g. Cases C-90/94 *Haahr Petroleum* and 21/79 *Commission v. Italy*)[99] and domestic producers having more time to pay the tax than foreign producers (e.g. Case 55/79 *Commission v. Ireland*).[100, 101] Since there are no express defences to Article 110, any directly discriminatory tax breaches Article 110 and cannot be saved.[102]

By contrast, indirect discrimination can be objectively justified.[103] The case law contains many examples of Article 110 cases on indirectly discriminatory tax measures. In *Commission v. France*, the ECJ found a French rule taxing light-tobacco cigarettes more heavily than dark-tobacco cigarettes to be incompatible with Article 110(1) as the latter were almost exclusively domestically produced

---

93   Case 106/84 *Commission v. Denmark* [1986] ECR 00833, para 12.
94   Horspool and Humphreys (n 90) 307.
95   Emch (n 61) 401 and 415.
96   Ibid 406.
97   Barnard (n 10) 56.
98   For the difference between the two forms of discrimination, see (n 11).
99   Case 21/79 *Commission v. Italy* [1980] ECR 00001.
100   Case 55/79 *Commission v. Ireland* [1980] ECR 00481.
101   Barnard (n 10) 57–58, Craig and de Búrca (n 10) 649 and Dashwood and others (n 70) 397–98.
102   Barnard (n 10) 58.
103   Dashwood and others (n 70) 398.

while almost all light-tobacco cigarettes were imported from other Member States.[104] In *Humblot*, a French car tax was held to be indirectly discriminatory since in practice only imported cars fell in the highest tax category and were, thus, taxed significantly higher than domestically produced cars.[105] *Commission v. Greece* dealt with a similar Greek car taxation scheme.[106] In this case, the ECJ ruled that a system of taxation cannot be regarded as discriminatory solely because only imported products come within the most heavily taxed category.[107] As the Commission had failed to show that the Greek scheme could have the effect of favouring the sales of Greek cars, the tax scheme was not contrary to Article 110.[108] Barnard has argued that *Commission v. Greece* demonstrates that where indirect discrimination is alleged under Article 110(1), the ECJ requires the complainant to demonstrate actual (as opposed to potential) disparate impact on the imported goods or actual benefit to the domestic producer.[109]

In the scenario of a financial security/mechanism charge indirectly being imposed on both domestic and cross-border CO_2 streams for storage (a general scheme), both types of CO_2 stream, in principle, would arguably constitute similar products within the meaning of Article 110(1). As for the (physical) characteristics of the two categories of products, both domestic and foreign CO_2 streams will principally consist of CO_2.[110] As a consequence of differing CO_2 sources and technical requirements for transport, the presence of impurities can vary from CO_2 stream to CO_2 stream. This does not, however, mean that different CO_2 streams with varying (degrees of) impurities are not – from the point of view of the characteristics of the products – to be considered similar products within the meaning of Article 110(1). The essential feature of all CO_2 streams for storage is that they principally consist of CO_2, which (obviously) is to be stored underground. From the point of view of the characteristics of the products, any differences in (degrees of) incidental and added substances do not, in my opinion, invalidate the similarity of differently composed CO_2 streams for storage.

---

[104]   See Case C-302/00 *Commission v. France* [2002] ECR I-02055, para 30.
[105]   Case 112/84 *Humblot* [1985] ECR 01367, para 14.
[106]   Case C-132/88 *Commission v. Greece* [1990] ECR-01567.
[107]   Ibid para 18.
[108]   Ibid para 20.
[109]   Barnard (n 10) 60. Barnard also refers to Case 252/86 *Bergandi* (n 53), Case 196/85 *Commission v. France* [1987] ECR 01597 and Case 140/79 *Chemial Farmaceutici* [1981] ECR 00001.
[110]   As indicated in Chapter III, a CO_2 stream for storage must consist 'overwhelmingly' of carbon dioxide (Article 12). It may contain incidental associated substances from the source, capture or injection process and trace substances added to assist in monitoring and verifying CO_2 migration. Concentrations of all incidental and added substances shall, however, be below levels that (a) adversely affect the integrity of the storage site or the relevant transport infrastructure, (b) pose a significant risk to the environment or human health or (c) breach the requirements of applicable EU legislation.

In principle, different CO_2 streams for storage will, therefore, also constitute similar/substitutable products from the point of view of the 'consumer', the storage operator. The storage operator stores and monitors the CO_2 stream. Differences in impurities present in different CO_2 streams do not fundamentally alter the use of the captured CO_2 for storage. The only exception is perhaps the situation in which a certain storage reservoir is suitable for the storage of CO_2 streams with a specific composition (range) only. In this situation, domestic CO_2 streams that would not meet the specific requirements of the relevant storage reservoir arguably should not be considered similar to imported CO_2 streams that do meet these requirements; the two will not be substitutable from the point of view of the storage operator/consumer.

In *Stadtgemeinde Frohnleiten* the ECJ ruled that 'categories of waste similarly intended for disposal by means of long-term depositing *are clearly similar products*'.[111] The ECJ did, however, specifically link this finding to the competent authority drawing no distinction as to the hazardous nature or other characteristics of the waste deposited.[112] This seems to suggest that if Member State authorities differentiate according to the composition of the CO_2 stream – for instance for storage safety reasons – the various CO_2 streams should not necessarily be seen as similar products.[113]

Should Member States differentiate the level of financial security/mechanism charge imposed, they will likely do so according to individual storage projects' differing levels of storage risk. As argued in section 4.2, the composition of the captured CO_2 stream is of great relevance in this regard. CO_2 stream composition, however, is not related the geographic origin of the captured CO_2, but directly follows from the capture source and the technical requirements for transport. The (geographic) origin of the captured CO_2 stream would, therefore, normally be irrelevant for setting the level of the financial security/mechanism charge.[114] As a consequence, Member States are more likely to discriminate against cross-border CO_2 streams for storage in an indirect than in a direct manner, even more so since any financial security/mechanism charge lacking

---

111   Case C-221/06 *Stadtgemeinde Frohnleiten* (n 10) para 59 (emphasis added). Even though CO_2 captured and transported for storage does not classify as a waste under the Waste Directive and the Regulation on shipments of waste, there is arguably a strong analogy: both non-recyclable waste and CO_2 captured for storage are disposed of. To this extent, see section 3.3 of Chapter III.

112   Case C-221/06 *Stadtgemeinde Frohnleiten* (n 10) para 59.

113   In this regard, see also Case C-209/98 *Sydhavnens* [2000] ECR I-03743, para 61 (definition of product market for competition law purposes).

114   Unless certain Member States/regions export a particular uniform CO_2 stream composition. In that case, however, it still arguably is the CO_2 stream composition and not the (geographic) origin *as such* that is decisive.

(geographic) origin neutrality would arguably constitute too obvious a violation of Article 110.

Based on the above, I would argue that when a financial security/mechanism charge is indirectly imposed on both domestic and cross-border CO_2 streams for storage, the two categories of products would, *in principle*, constitute similar products within the meaning of Article 110(1). Domestic and cross-border CO_2 streams for storage could be considered not to be similar within the meaning of Article 110(1) only when there are, in relation to a specific storage reservoir, reasons for differentiating between different CO_2 streams according to their composition. Furthermore, should a financial security/mechanism charge discriminate against cross-border CO_2 streams for storage, then such taxation will likely be indirectly – as opposed to directly – discriminatory. In the next section, we will briefly discuss Article 110(2), which is of relevance for those situations in which domestic and cross-border CO_2 streams are not considered to be similar products within the meaning of Article 110(1).

4.4.4 ARTICLE 110(2) TFEU

When cross-border and domestic products are not similar within the meaning of the first paragraph of Article 110, they can still be in competition with each other under the second paragraph of the provision. The ECJ is said to apply a two-stage methodology under Article 110(2).[115] At the first stage, it examines whether there is some competitive relationship between the two products in order to render Article 110(2) applicable at all. At the second stage, the ECJ considers whether the tax system is protective of the domestic product.

For the determination of a competitive relationship between products, the degree of substitution between two products is one of the most important factors.[116] In *Commission v. UK*,[117] the ECJ held that Article 110(2) concerns products which are, even only potentially, in competition with each other.[118] The ECJ argued that in order to determine the existence of a competitive relationship under Article 110(2), it is necessary to consider not only the state of the market at a particular moment but also the further potential for the substitution of products for one another in case of intensification of trade.[119]

---

[115] Craig and de Búrca (n 10) 658. Craig and de Búrca refer to Case 170/78 *Commission v. United Kingdom* [1983] ECR 02265.
[116] Dashwood and others (n 70) 399.
[117] Case 170/78 *Commission v. UK* [1980] ECR 00417.
[118] Dashwood and others (n 70) 399.
[119] Case 170/78 *Commission v. UK* (n 117) para 6.

In order to prove the protectionist effects of a tax scheme, it does not suffice to demonstrate that there is a different rate of taxation; the effect on price must be such as to afford protection to domestic products.[120] For instance, while in *Commission v. UK*[121] the tax on cheap (imported) wine was so high as to afford protection to beer (mainly domestically produced), in *Commission v. Sweden*[122] the ECJ found that the difference in price between beer and wine pre-tax and after tax was very similar: even though taxes for wine were higher, THIS difference was not capable of influencing consumer behaviour in the sector concerned.[123]

It is not always easy to determine whether products are similar or in competition.[124] This is partly caused by the fact that the tests performed by the ECJ under both paragraphs of Article 110 overlap to a certain degree. As we have seen, the test performed under Article 110(1) is a mixture of an analysis of the characteristics of the cross-border and domestic products and of the question whether the two products meet the same needs from the point of view of the consumer.[125] Under Article 110(2), the main question is whether the cross-border and domestically produced products are substitutable from a consumer point of view. What is more, as part of the test performed under the second paragraph of Article 110, the ECJ also takes into account factors such as the manufacture and composition of the product.[126] These are aspects typically analysed under Article 110(1).

In a number of early Article 110 cases, the ECJ was not overly concerned with the question whether the relevant tax scheme fell within the scope of Article 110(1) or (2).[127] According to Craig and de Búrca, and Barnard, the ECJ subsequently became more rigorous in distinguishing between Article 110(1) and (2).[128] This is

---

[120] Dashwood and others (n 70) 399.

[121] Case 170/78 *Commission v. UK* (n 117).

[122] Case C-167/05 *Commission v. Sweden* [2008] ECR I-02127.

[123] Dashwood and others (n 70) 399–400.

[124] Ibid 400.

[125] The latter constitutes what is referred to as demand-side substitution in EU competition law. Contrary to demand-side substitution, supply-side substitution refers to suppliers being able to switch production to the relevant products and to market them in the short term without incurring significant additional costs or risks in response to small and permanent changes in relative prices. For the definition of the relevant market, demand-side substitutability is generally considered to constitute the most immediate and effective disciplinary force on the suppliers of a given product. See Commission, 'Commission Notice on the Definition of the Relevant Market for the Purposes of Community Competition Law' [1997] OJ C372.

[126] Barnard (n 10) 61.

[127] Craig and de Búrca (n 10) 656.

[128] Ibid 656 and Barnard (n 10) 64. See, however, Gareth Davies, 'Process and Production Method-Based Trade Restrictions in the EU' (2008) 20 <http://papers.ssrn.com/sol3/papers.cfm?abstract_id=1118709> accessed 13 June 2012. Davies refers to para 56 of the ECJ's ruling in Case C-221/06 *Stadtgemeinde Frohnleiten* to argue that the ECJ globalises Article 110,

to be welcomed as both paragraphs have different remedies.[129] A breach of Article 110(1) means that the offending Member State has to equalise the taxes on domestic and imported goods.[130] Breach of Article 110(2) requires the Member State to remove the protective effect, but this does not necessarily mean equalising the tax burden on the imported goods.[131] What is more, under Article 110(1), there is the possibility for an indirectly discriminatory tax to be objectively justified, provided that it meets the proportionality test.[132] This option does not exist for a protective tax under Article 110(2).

As argued above, a financial security/mechanism charge indirectly imposed on both domestic and cross-border CO_2 streams for storage, would, in principle, likely fall within the scope of Article 110(1) and thus not under Article 110(2). In general, domestic and cross-border CO_2 streams for storage are to be considered as similar products within the meaning of the first paragraph of Article 110. The only exception likely is the situation in which a certain storage reservoir is suitable for the storage of CO_2 streams with a specific composition (range) only. In that case, domestic CO_2 streams that do not meet these specific requirements are not, from the point of view of the storage operator, substitutable for imported CO_2 streams that do meet these requirements; the two do not compete for the same storage reservoir. As a consequence, a financial security/mechanism charge indirectly imposed on imported CO_2 streams only does not appear to fall within the scope of Article 110(1). What is more, since the substitutability test is likewise applied under Article 110(2), such a charge also does not appear to fall within the scope of the second paragraph of Article 110. It seems to fall outside the scope of Article 110 in its entirety.

Nevertheless, by virtue of the *Co-frutta* and *Commission v. Denmark* cases discussed above, a financial security/mechanism charge indirectly imposed on imported CO_2 streams only (by reason of there being no domestic products) will in principle fall within the scope of Article 110, even though it cannot infringe either Article 110(1) or 110(2). A financial security/mechanism charge imposed on the operator of a storage reservoir can – logically – infringe Article 110 only if imported and domestic CO_2 streams for storage are able to compete for that particular reservoir. Should the financial security/mechanism charge solely imposed on imported CO_2 streams be of such an amount that the free movement of goods within the internal market would be impeded, then it may fall within the scope of Article 30.

---

providing a single conceptual framework for similar products and non-similar products that are in competition.
[129] Barnard (n 10) 64.
[130] Craig and de Búrca (n 10) 656.
[131] Ibid and Barnard (n 10) 65.
[132] On the proportionality test, see section 3.6.3 of Chapter III.

4.4.5 THE (EXEMPTION) SYSTEM APPLIED BY THE ECJ

At first sight, the text of Article 110 appears to be rigid.[133] As argued before, it does not seem to allow for any (de minimis) exceptions to the rules that imported products must not be taxed more heavily than similar domestic products and taxation of imported products may not lead to non-similar competing domestic products being protected.[134] There is nothing in the wording of Article 110 to suggest that it is possible to justify discriminatory or protective internal taxes. However, the ECJ has traditionally applied Article 110 in a manner allowing the Member States a considerable discretion to determine the content of their own taxation policy (system and level of taxation).[135] The formula traditionally used by the ECJ in this regard is that:

> Community law at its present stage of development does not restrict the freedom of each Member State to lay down tax arrangements which differentiate between certain products on the basis of objective criteria, such as the nature of the raw materials used or the production process employed. Such differentiation is compatible with Community law if it pursues objectives of economic policy which are themselves compatible with the requirements of the Treaty and its secondary legislation, and if the detailed rules are such as to avoid any form of discrimination, direct or indirect, in regard to imports from other Member States or any form of protection of competing domestic products.[136]

Hedemann-Robinson has labelled the ECJ's approach in this regard as an 'inverted approach'.[137] According to him, the 'model' approach would be to first determine possible indirect discrimination and then to assess potential objective justifications.[138] Instead, the ECJ has often first investigated whether the tax measure is justifiable (on the basis of the above formula) and has then required the tax regime to be devoid of discriminatory effect.[139] If the latter requirement is not satisfied then the ECJ has, in the main, assumed that the tax is automatically unlawful under Article 110(1).[140] Hedemann-Robinson has argued

---

133   Geert van Calster, 'Greening the EC's State Aid and Tax Regimes' (2000) 21 European Competition Law Review 294, 310.

134   See Case C-383/01 *De Danske Bilimportører* [2003] ECR I-06065, Opinion of AG Jacobs, para 28 and Van Calster (n 133) 310.

135   See Case C-402/09 *Ioan Tatu*, Opinion of AG Sharpston (n 54) paras 32 and 54. See also Craig and de Búrca (n 10) 651; Horspool and Humphreys (n 90) 305; Barnard (n 10) 52; Dashwood and others (n 70) 396, and Ben JM Terra and Peter J Wattel, *European Tax Law* (2nd edn, Kluwer 2012) 13.

136   Case 243/84 *Johnnie Walker* [1986] ECR 00875, para 22.

137   Hedemann-Robinson (n 12).

138   Ibid 447–48.

139   Ibid 449.

140   Ibid. Hedemann-Robinson's article focuses exclusively on indirect discrimination under Article 110(1).

that the primary reason for the ECJ to often follow such approach is the sovereignty in the area of the taxation of goods still retained by the Member States.[141] In support of his claim that the ECJ often applies an inverted approach in its Article 110 rulings, he has referred to a number of cases.

The case of *Hansen* centred around German tax rules favouring the production of certain types of spirits by means of a reduced rate of taxation.[142] A German trader had imported spirits which he deemed to be entitled to these tax advantages, but which were taxed at the ordinary rate. Subsequently, the trader had brought proceedings before a German court, which referred a number of questions to the ECJ. One of the questions the German court asked was whether Article 110 had to be interpreted as not allowing for imported products to be taxed more heavily than similar home-produced goods (taxed at the lowest rate), even if the lowest rate of charge was applicable only to a small proportion of domestic production and only for special social reasons.[143] The ECJ held that at the (then) present stage of development and in the absence of any harmonisation of the relevant provisions, EU law did not prohibit Member States from granting tax advantages, in the form of exemption from or reduction of duties, to certain types of spirits or to certain classes of producers.[144]

The ECJ acknowledged that tax advantages like the German rules could serve legitimate economic or social purposes such as the use of certain raw materials by the distilling industry and the continued production of particular spirits of high quality.[145] It did, however, state that such tax advantages had to be extended without discrimination to spirits coming from other Member States (meeting the same criteria).[146] Hedemann-Robinson has stressed that the ECJ did not consider whether the German tax rule could have been upheld for objective reasons notwithstanding its discriminatory impact against imported spirits, but simply insisted on the tax conferring equality of fiscal treatment to imports.[147]

The parallel cases of *Vinal* and *Chemial Farmaceutici* dealt with Italian tax rules which imposed heavier charges on synthetic alcohol than on alcohol obtained by fermentation, on the basis of the raw materials and the manufacturing processes employed.[148] In both cases, the ECJ had been asked by an Italian court whether the Italian tax arrangements were compatible with Article 110. In its answer, the

---

[141] Hedemann-Robinson (n 12) 458–59.
[142] Case 148/77 *Hansen* [1978] ECR 01787.
[143] Ibid para 4.
[144] Ibid para 16.
[145] Ibid.
[146] Ibid paras 17 and 20.
[147] Hedemann-Robinson (n 12) 459.
[148] Case 46/80 *Vinal* [1981] ECR 00077 and Case 140/79 *Chemial Farmaceutici* (n 109).

ECJ reiterated the *Johnnie Walker*[149] formula.[150] According to it, the Italian tax rules met these requirements.[151] It pursued an objective of legitimate industrial policy: the promotion of the production of alcohol obtained by fermentation in order to reserve the raw material of synthetic alcohol for other more important economic uses.[152] Moreover, the Italian rules could not be considered as discriminatory.[153] On the one hand, the ECJ noted, imports from other Member States of alcohol obtained by fermentation qualified for the same tax treatment as Italian alcohol produced by fermentation.[154] On the other hand, even though the rate of tax prescribed for synthetic alcohol hindered the import of synthetic alcohol from other Member States, it likewise made the development of (non-existing) Italian production of synthetic alcohol more difficult.[155]

According to Hedemann-Robinson, the ECJ should have weighed the considerations related to the supply of finite resources (the raw material of synthetic alcohol) in the light of having established the discriminatory impact of the tax system on imported denatured alcohol, rather than considering them at an initial stage in its analysis.[156]

Another case cited by Hedemann-Robinson is *Bergandi*.[157] *Bergandi* concerned a French tax on automatic game machines. The central question was whether Article 110 allowed Member States to impose on automatic games machines principally of foreign origin a tax three times higher than that imposed on machines principally of domestic manufacture.[158] Referring to the case of *Commission v. Denmark*,[159] the ECJ stated that a national system of taxation, despite making no formal distinction according to the origin of the products, undeniably contains discriminatory or protective characteristics (is indirectly discriminatory) if it has been adjusted so that the bulk of domestic production comes within the most favourable tax category, whereas almost all imported products come within the most heavily taxed category.[160]

The ECJ held that it had nevertheless stated that, at the present stage of development and in the absence of harmonisation of the relevant provisions, EU

---

149 Case 243/84 *Johnnie Walker* (n 136).
150 Case 46/80 *Vinal* (148) para 13 and Case 140/79 *Chemial Farmaceutici* (n 109) para 14.
151 Case 46/80 *Vinal* (n 148) para 14 and Case 140/79 *Chemial Farmaceutici* (n 109) para 15.
152 Ibid.
153 Case 46/80 *Vinal* (n 148) para 15 and Case 140/79 *Chemial Farmaceutici* (n 109) para 16.
154 Ibid.
155 Ibid.
156 Hedemann-Robinson (n 12) 460.
157 Case 252/86 *Bergandi* (n 53).
158 Ibid para 22.
159 Case 171/78 *Commission v. Denmark* [1980] ECR 00447.
160 Case 252/86 *Bergandi* (n 53) para 28.

law does not prohibit the Member States from establishing a system of taxation differentiated according to various categories of products provided that the tax benefits granted serve legitimate economic or social purposes.[161] It recognised that the desire to encourage the use of particular categories of game machines and to discourage the use of others could be a legitimate social purpose.[162] The ECJ ruled that a system of taxation graduated according to the various categories of automatic games machines, which is intended to achieve legitimate social objectives and which procures no fiscal advantage for domestic products to the detriment of similar or competing products, is not incompatible with Article 110.[163]

Hedemann-Robinson has noted that even though the ECJ initially conceded that Article 110 allowed Member States to differentiate between various categories of product provided that the tax difference served legitimate economic or social purposes, it added the caveat that such differentiation could not be compatible with Article 110 where the tax system bore a discriminatory or protective effect against imports.[164] According to him, the ECJ, through the application of its inverted standard formula, effectively prevented the national court from entering into an examination of whether the tax could be objectively justified.[165]

The latest case cited by Hedemann-Robinson dates from 1990.[166] The ECJ, however, has used the *Johnnie Walker* formula in a number of cases since.[167] These cases indicate that the ECJ still applied the inverted approach as recently as 2006.[168]

Partly referring to the same cases cited by Hedemann-Robinson,[169] Barnard has argued that any scheme of taxation which makes a distinction between products based on objective criteria (unrelated to origin) is compatible with Article 110.[170] In this regard, Barnard has cited the *Johnnie Walker* formula, while leaving out

---

[161]   Case 252/86 *Bergandi* (n 53) para 29.
[162]   Ibid para 30.
[163]   Ibid para 32.
[164]   Hedemann-Robinson (n 12) 461.
[165]   Ibid 461.
[166]   Case 132/88 *Commission v. Greece* (n 106). See section 4.4.3 above.
[167]   See, for instance, Case C-90/94 *Haahr Petroleum* (n 44), para 29; Case C-213/96 *Outokumpu* [1998] ECR I-01777, para 30; Joined Cases C-290/05 and C-333/05 *Nádasdi* [2006] ECR I-10115, para 51; and Case C-221/06 *Stadtgemeinde Frohnleiten* (n 10) para 56.
[168]   In three of the four cases, the ECJ can be said to have applied the inverted approach, first examining whether the relevant tax could be justified and then examining whether it was discriminatory or not. See Case C-213/96 *Outokumpu* (n 167) paras 30–35; Joined Cases C-290/05 and C-333/05 *Nádasdi* (n 167) paras 51–57; and Case C-221/06 *Stadtgemeinde Frohnleiten* (n 10) paras 56–57.
[169]   For instance, Case 252/86 *Bergandi* (n 53) and Case 140/79 *Chemial Farmaceutici* (n 109).
[170]   Barnard (n 10) 52.

the requirement that the 'rules are such as to avoid any form of discrimination, direct or indirect, in regard to imports from other Member States or any form of protection of competing domestic products'.[171] According to Barnard, the first question to ask in relation to Article 110 is whether there are any objective criteria – unrelated to origin – for distinguishing between products. If this question can be answered in the positive, the relevant tax measure falls outside the scope of Article 110.[172] The question whether it places a greater burden on products coming from or going to other Member States will then not be tackled.

It might be a question of semantics, but the problem with Barnard's approach seems to be that it rules out the possibility of a distinction between products based on objective criteria (unrelated to origin) leading to indirect discrimination. In *Johnnie Walker*, the ECJ clearly stated that Member States are free to differentiate between products, *provided* that such differentiation does not lead to imported products being discriminated against, either directly *or indirectly*. Objective criteria that are not related to origin can nonetheless lead to a greater burden being placed on cross-border products than on domestic products.[173] If that is the case, it is necessary to assess whether the indirectly discriminatory tax measure can be objectively justified. So, if there are objective criteria (unrelated to origin) for distinguishing between domestic and cross-border products, the next question is whether these criteria lead to a greater burden being placed on cross-border products.[174] Only if this question can be answered in the negative will the relevant tax measure fall outside the scope of Article 110. Cases like *Vinal* and *Chemial Farmaceutici* indicate that the ECJ is not always consistent in this regard; sometimes it is willing to accept certain objective criteria, even though they (clearly) have an indirectly discriminatory effect.

In sum, even though Article 110 does not appear to allow for any exceptions to the general non-discrimination principle enshrined in the provision, the ECJ has in a number of Article 110 cases taken Member State tax autonomy as a point of departure. Member States are free to differentiate between (categories of) products, provided that the differentiation is based on objective criteria, pursues objectives compatible with EU law and does not lead to either direct or indirect

---

[171] Barnard (n 10) 52.

[172] Ibid 54.

[173] In this regard, see also Case 74/76 *Ianelli* [1977] ECR 00557, para 19. In this case, the ECJ held that the fact that a tax or levy is a special charge or appropriated for a specific purpose cannot prevent its falling within the field of application of Article 110. In other words, the differentiation being based on objective criteria does not suffice for the relevant tax scheme to fall outside the scope of Article 110.

[174] In this regard, see also Case C-221/06 *Stadtgemeinde Frohnleiten*, Opinion of AG Sharpston (n 44) paras 44–46.

discrimination of cross-border products. If these criteria are met, the differentiating tax provisions appear to fall outside the scope of Article 110. Directly and indirectly discriminatory tax measures fall within the scope of Article 110. The former are prohibited per se. Indirectly discriminatory taxes are compatible with Article 110 only if they can be objectively justified and are proportionate. The ECJ has in case like *Vinal* and *Chemial Farmaceutici* indicated that it might, under certain circumstances, be willing not to test whether indirectly discriminatory tax measures can be objectively justified; it will then simply deem those measures not to be discriminatory, meaning that they will fall outside the scope of Article 110.

I doubt, however, whether these cases point to the general rule (apparently perceived by Barnard) that any scheme of taxation which makes a distinction between products based on objective criteria (unrelated to origin) is compatible with Article 110.[175] The inverted approach does not boil down to the ECJ always accepting objective criteria for differentiating between products without assessing possible discriminatory effects; it will still require differentiating tax measures not to be discriminatory (see e.g., *Hansen* and *Bergandi*). The safest bet is probably to expect tax rules that do not differentiate according to origin, but can lead to a heavier burden being placed on cross-border products, to have to be objectively justified. One of the possible objective justifications is the protection of the environment.[176] In the next section, we will look at a number of cases that are of particular relevance for the requirements imposed by Article 110 on a financial security/mechanism instrument.

4.5 SELECTED ARTICLE 110 TFEU CASES

In the following sections, I will discuss three cases that might be of particular relevance for the requirements imposed by Article 110 on a Member State financial security/mechanism. These cases set the scope for differentiation in environmental standards (*Outokumpu*), indicate when imported waste and domestically produced waste are considered to be similar (*Stadtgemeinde Frohnleiten*) and define the way in which internal taxation of imported products can be determined (*Commission v. Greece*).

---

[175] See also Van Calster (n 133) 310. Van Calster has noted that the ECJ's case law on the possibility for Member States to introduce tax differentiation is unclear.
[176] Barnard (n 10) 60.

4.5.1   CASE C-213/96 *OUTOKUMPU*

A first case which might be of particular relevance is *Outokumpu*. The case of *Outokumpu* dealt with a Finnish tax on electricity. For electricity produced in Finland, the amount of the duty depended on the method of production of the electricity and the raw materials used for production. By contrast, the duty levied on imported electricity was a flat rate higher than the lowest but lower than the highest rate applicable to domestically produced electricity. Finnish metallurgic company Outokumpu had imported electricity from Sweden and had been charged the levy applicable to imported electricity. Subsequently, Outokumpu instituted national proceedings against the tax decision, claiming, among other things, that the discriminatory tax infringed Article 110. The relevant Finnish court decided to stay proceedings and refer to the ECJ, inter alia, the question of whether the Finnish electricity tax indeed was discriminatory within the meaning of Article 110.

The ECJ first reiterated the *Johnnie Walker* formula, stating that Member States are free to establish a tax system that differentiates between products, provided that such differentiation is based on objective criteria, pursues objectives compatible with EU law and does not discriminate (directly or indirectly) against cross-border products.[177] The ECJ then argued that Article 110 does not preclude the rate of an internal tax on electricity from varying according to the manner in which the electricity is produced and the raw materials used for its production, in so far as that differentiation is based on environmental considerations.[178] Subsequently, the ECJ stated that it had consistently held that Article 110 is infringed where the taxation on the imported product and that on the similar domestic product are calculated in a different manner on the basis of different criteria which lead, if only in certain cases, to higher taxation being imposed on the imported product.[179] According to the ECJ, the latter was the case; imported electricity, whatever its method of production, was subject to a flat-rate duty higher than the lowest duty charged on domestically produced electricity.[180] The ECJ held that it was immaterial in this regard that domestically produced electricity in some cases was taxed more heavily than imported electricity.[181] In order to ascertain whether the Finnish tax system was compatible with Article 110, the tax burden imposed on imported electricity had to be compared with the lowest tax burden imposed on electricity of domestic origin.[182]

---

[177]   Case C-213/96 *Outokumpu* (n 167) para 30.
[178]   Ibid para 31.
[179]   Ibid para 34.
[180]   Ibid para 35.
[181]   Ibid para 36.
[182]   Ibid.

The Finnish government had argued that in view of the characteristics of electricity, the differential rates applicable to domestically produced electricity could not be applied to imported electricity; once electricity enters the distribution network, the origin and consequently the method of production cannot be determined.[183] In such circumstances, it considered the application of a flat rate the only logical way of treating imported electricity in an equitable manner.[184] In response, the ECJ stated that it had held before that practical difficulties cannot justify the application of internal taxation which discriminates against products from other Member States.[185] The ECJ acknowledged that it is indeed extremely difficult to determine precisely the method of production of imported electricity, but pointed out that the Finnish tax rules did not even give the importer the opportunity of demonstrating that the electricity imported by him had been produced by a particular method in order to qualify for the rate applicable to electricity of domestic origin produced by the same method.[186] It stressed that objectively justified tax differences must be abolished if such abolition is the only way of avoiding direct or indirect discrimination against imported products.[187]

Outokumpu appears to be a clear example of directly discriminatory taxes; imported electricity was charged a flat rate tax, while the tariffs for domestically produced electricity varied according to the method of production. The case shows that Member States, in principle, are free to differentiate the taxation of complex products such as electricity on the basis of, for instance, environmental criteria related to the way in which the product was produced. However, in doing so, they should either charge imported products the lowest tariff applicable to similar domestic products or give the importer the opportunity to demonstrate that the imported product was produced in such a way as to qualify for the rate applicable to domestic products produced in the same way. The fact that it might in practice be difficult to determine the way in which an imported product was produced does not mean that Member States are allowed to charge a tariff other than the lowest tariff applicable to domestic products. Practical difficulties are not an excuse for taxing imported products more heavily than similar domestic products.

In relation to the imposition of a financial security/mechanism charge, the case of *Outokumpu* seems to suggest that Member States can differentiate the charge based on objective criteria related to, for instance, the composition of the CO_2 stream. Such differentiation would arguably have the ultimate objective of guaranteeing the environmental safety of CO_2 storage, particularly when riskier

---

[183] Case C-213/96 *Outokumpu* (n 167) para 37.
[184] Ibid para 37.
[185] Ibid para 38.
[186] Ibid para 39.
[187] Ibid para 40.

CO_2 streams are being stored. Both the financial security and the financial mechanism are to provide the financial means for ensuring that the CO_2 is completely and permanently contained. Even though Member States will likely have the freedom to differentiate the financial security/mechanism charge according to CO_2 stream composition, they will be under the obligation to allow the party importing captured CO_2 for storage to demonstrate that the relevant CO_2 stream meets the criteria for being charged at a particular rate. In practice, this will probably come down to the storage operator being given the possibility to show that by storing the relevant imported CO_2 stream, he meets the criteria for being charged a particular financial security/mechanism tariff. Should it in practice be difficult to determine the composition of the CO_2 stream (to be) imported then the storage operator cannot be charged a flat rate duty. He must then be charged the lowest tariff applicable to domestic CO_2 streams for storage.

The practical problem raised in *Outokumpu* might likewise occur in relation to the transport of captured CO_2 for storage. Like electricity, the different types of CO_2 stream will blend if transported through the same pipeline. Initially, this might not be a widespread phenomenon. Even though the cross-border deployment of CCS could take place earlier than initially expected,[188] it will probably take a while before more complicated, (highly) meshed CO_2 transport networks develop.[189] Yet should captured CO_2 be transported through meshed networks, then practical difficulties will not, by virtue of *Outokumpu*, be an excuse for discriminating against 'similar' CO_2 streams imported for storage. As argued before, the level of the financial security/mechanism charge could be related to the composition of the CO_2 stream that is to be stored. Should Member States decide to use a single instrument to implement the obligations under Articles 19 and 20 of the CCS Directive, they will likely determine the level of the financial security/mechanism charge as part of the storage permitting process, possibly on the basis of the (expected) characteristics of the CO_2 stream produced by the relevant capturer.

Nevertheless, in the case of a (highly) meshed CO_2 transport network, these might not be the characteristics of the CO_2 stream actually injected, due to the particular CO_2 stream having blended with other (differently composed) CO_2 streams. This raises questions in relation to the imposition of the financial security/mechanism charge. What, for instance, would happen when the risk profile of the actually injected CO_2 stream appears to be higher than that of the captured CO_2 stream on which the financial security/mechanism charge was originally based? It does not seem likely that it will then be possible to impose an

---

[188]   See the Introduction to this thesis.
[189]   It would make sense to not expect such networks developing before the large-scale commercial deployment of CCS takes off. Most likely, this will not be before 2030.

additional financial security/mechanism charge on the party responsible for the presence of the substances causing the higher risk profile as it would perhaps not even be possible to identify that entity.

Potential problems related to the blending of different CO_2 streams in a meshed transport network could be solved by introducing a certain (minimum) quality required for feeding CO_2 streams into a Member State's CO_2 transport network. This network could be managed by a single transport system operator, as is often the case with the gas transmission network, who charges the capturers for feeding into the network. In that way, individual storage operators will have the guarantee that they receive CO_2 streams with a (fairly) standardised composition. The storage operators could then pass on to the transport system operator the costs of the financial security/mechanism charge. In turn, the transport system operator can pay the storage operator from the proceeds of the transport tariffs charged to the capturers.

Another interesting possibility is that of a Member State choosing to become a host to large-scale storage of captured CO_2 streams, including streams from other Member States. In that case, this Member State is likely to be faced with higher risks related to CO_2 storage and associated long-term liability. This could impact the level of financial security/mechanism charges levied on storage operators. As more captured CO_2 is stored in the same storage reservoirs, financial security/mechanism charges could rise. If a majority of the CO_2 stored in the Member State consists of CO_2 streams originating in other Member States, this could de facto lead to 'domestic' capturers seeking to have their captured CO_2 stored indirectly having to pay higher financial security/mechanism charges than they would have otherwise.

However, it is not likely that this will lead to problems with Article 110 TFEU. First, the higher financial security/mechanism charges will probably be charged to foreign and domestic capturers alike. Both domestic and foreign capturers will probably be confronted with higher financial security/mechanism charges if they want to have their CO_2 stored in a field in which CO_2 storage is taking or has already taken place. Second, even if domestic capturers could, in such a situation, be said to be discriminated against, there is nothing in EU law that prevents Member States from treating domestic products less favourably than imported products ('reverse discrimination').[190]

---

[190] Barnard (n 10) 58.

4.5.2   CASE C-221/06 *STADTGEMEINDE FROHNLEITEN*

Of all Article 110 cases, the facts in *Stadtgemeinde Frohnleiten* probably come closest to those of the case in which a financial security/mechanism duty is indirectly imposed on a CO_2 stream imported for storage. *Stadtgemeinde Frohnleiten* centred around an Austrian levy on the depositing of waste. Under the Austrian Law on the rehabilitation of disused hazardous sites, inter alia, the long-term depositing of waste – including the depositing of waste in geological structures – was subject to a disused hazardous site levy. Yet, the depositing of waste for the sake of the safeguarding or rehabilitating of either suspected contaminated sites or disused hazardous sites (both entered in national registers) was exempted from the disused hazardous site levy. The levy was calculated on the basis of the weight of the waste and was to be paid by the operator of the waste disposal site.

In 2001 and 2002, Gemeindebetriebe Frohnleiten, the municipal waste company for the town of Frohnleiten had deposited several tonnes of waste from an Italian waste disposal site which was being rehabilitated in accordance with the Italian plan for the rehabilitation of contaminated sites. As the Italian waste came from a contaminated site, Gemeindebetriebe Frohnleiten had applied for an exemption of the disused hazardous site levy. The application for exemption was ultimately denied by the Federal Minister for Environment, Youth and Family, since the waste did not come from a site entered in the Austrian registers of suspected contaminated sites and disused hazardous sites. Thereupon, Gemeindebetriebe Frohnleiten challenged the Minister's decision before the Austrian administrative court, claiming, in essence, that the disused hazardous site levy fell within the scope of Article 110 and that the latter provision would be breached if a higher levy were imposed on imported products than on similar domestic products. The administrative court decided to stay proceedings and refer to the ECJ the question of whether the levy (and its exemption scheme), as applied by the Minister, was compatible with, inter alia, Article 110.[191]

The ECJ started out by noting that, according to settled case law, Article 110(1) is infringed where the tax charged on the imported product and that charged on the similar domestic product are calculated in a different manner on the basis of different criteria which lead, if only in certain cases, to a higher taxation being imposed on the imported product.[192] A system of taxation is compatible with Article 110 only if it excludes any possibility of imported products being taxed more heavily than domestic products and cannot in any event have

---

[191]   Case C-221/06 *Stadtgemeinde Frohnleiten* (n 10) para 24.
[192]   Ibid para 49.

discriminatory effect.[193] The ECJ noted that a national provision reserving the benefit of exemption from internal taxation to certain domestic products, to the exclusion of imported products, was liable to lead to higher taxation being imposed on imported products than on domestic products.[194] The Austrian government had disputed the discriminatory effect of the levy (and its exemption scheme), arguing that it did not establish any inequality of treatment between similar situations.[195] It basically held that imported contaminated waste was not the same as domestic contaminated waste since it did not come from an Austrian contaminated/hazardous site.[196] These arguments were rejected by the ECJ.

First, reiterating the *Johnnie Walker* formula, the ECJ held that even if the objective of rehabilitating domestic sites contaminated by waste was compatible with the requirements of EU law, a national provision leading to higher taxation being imposed on imported products than on domestic products could not itself be compatible with Article 110.[197] Second, the ECJ argued, waste imported from other Member States and waste produced in Austria had to be regarded as similar products.[198] According to it, categories of waste similarly intended for disposal by means of long-term depositing clearly were similar products, even more so since the relevant Austrian exemption rules drew no distinction according to the hazardous nature or other characteristics of the waste deposited.[199] This was not altered by the difference of origin between domestic waste and waste imported from other Member States.[200] The ECJ acknowledged that it had held in *Walloon Waste*[201] that national rules treating domestic and imported waste differently were not discriminatory in view of the environmental legal principle that environmental damage should be rectified at the source (consistent with the international environmental legal principles of proximity and self-sufficiency).[202, 203]

Nevertheless, that finding concerned a possible justification of national measures hindering the free movement of goods under Article 34 TFEU (and not

---

[193] Case C-221/06 *Stadtgemeinde Frohnleiten* (n 10) para 50.
[194] Ibid para 52.
[195] Ibid para 53.
[196] Ibid paras 53–54.
[197] Ibid para 57.
[198] Ibid para 58.
[199] Ibid para 59.
[200] Ibid para 60.
[201] Case C-2/90 *Commission v. Belgium* (n 23).
[202] The international environmental legal principle of proximity determines that waste must, in principle, be disposed of as close as possible to the point of generation. The self-sufficiency principle refers to the necessity for each country to be self-sufficient in waste management. See Elli Louka, *International Environmental Law: Fairness, Effectiveness, and World Order* (3rd edn, Cambridge University Press 2012) 92–3.
[203] Case C-221/06 *Stadtgemeinde Frohnleiten* (n 10) para 61.

Article 110 TFEU).[204] Moreover, it related to non-hazardous waste, which fell outside the scope of then Directive 84/631/EEC[205] (Shipments of Waste Directive).[206] Since *Walloon Waste*, the ECJ argued, Directive 84/631/EEC had been replaced by Regulation 259/93,[207] the latter in principle covering all waste (hazardous and non-hazardous) and taking account of the principles of proximity and self-sufficiency.[208] The ECJ noted that Regulation 259/93 gave Member States the possibility to take measures to prohibit or to object to shipments of waste and to raise reasoned objections to planned shipments of waste, in order to implement the principles of proximity and self-sufficiency at EU and national levels.[209] Therefore, the ECJ argued, once a Member State had refrained from using this possibility, it could not pose restrictions on the free movement of shipped waste in its territory on the basis of the principles of proximity and self-sufficiency at EU and national levels.[210] As the Austrian authorities had authorised the shipment of the Italian waste to Austria under Regulation 259/93, they could not restrict the free movement of the shipped waste in Austria on the basis of the principles of proximity and self-sufficiency.[211]

Finally, referring to its judgment in *Outokumpu*, the ECJ noted that it could indeed be extremely difficult for the Austrian authorities to ensure that sites located in other Member States satisfied the requirements laid down in Austrian legislation for being categorised as disused hazardous sites or contaminated sites, but that that legislation did not even give the importer the opportunity of proving that the imported waste qualified for the exemption applicable to waste from disused hazardous sites or suspected contaminated sites in Austria.[212] Moreover, again referring to *Outokumpu*, the ECJ stressed that objectively justified tax differences must be abolished if such abolition is the only way of avoiding direct or indirect discrimination against imported products.[213]

---

204   Case C-221/06 *Stadtgemeinde Frohnleiten* (n 10) para 69.
205   Council Directive 84/631/EEC of 6 December 1984 on the supervision and control within the European Community of the transfrontier shipment of hazardous waste [1984] OJ L326 (Hazardous Waste Directive).
206   Case C-221/06 *Stadtgemeinde Frohnleiten* (n 10) para 62.
207   Council Regulation EEC 259/93 of 1 February 1993 on the supervision and control of shipments of waste within, into and out of the European Community [1993] OJ L30 (Shipments of Waste Regulation). In 2007, Regulation 259/93 was replaced by European Parliament and Council Regulation EC 1013/2006 of 14 June 2006 on shipments of waste, OJ 2006 L190.
208   Case C-221/06 *Stadtgemeinde Frohnleiten* (n 10) paras 63–64.
209   Ibid para 66.
210   Ibid para 67.
211   Ibid para 68.
212   Ibid para 71.
213   Ibid para 72.

As for the compatibility of the Austrian levy (and its exemption scheme) with Article 110, the importance of the ECJ's ruling in *Stadtgemeinde Frohnleiten* seems to lie in its analysis of the similarity of imported waste and domestically produced waste. As argued in section 4.4.3, the ECJ seems to clearly regard categories of waste similarly intended for disposal by means of long-term depositing as similar products. The only exception would possibly be the situation in which Member State authorities differentiate according to the characteristics of the waste; if the relevant tax rules distinguish between different types of waste, these different types of waste might not be seen as similar products. Even though domestically produced waste and imported waste have different origins, this cannot invalidate their similarity within the meaning of Article 110.

First, the finding of *Walloon Waste* that national measures treating domestic and imported waste differently on the basis of the source principle did not discriminate was related to the specific context of Article 34 TFEU.[214] Second, in the situation of the shipment of waste between Member States, the principles of self-sufficiency and proximity (similar to the source principle) were covered under Regulation 259/93 and are still covered under its successor, Regulation 1013/2006.[215] Member States have the option to prohibit or object to shipments of waste for disposal on the ground that the principles of self-sufficiency and proximity are not respected.[216] As the field of hindering the import of waste on these grounds has thus been exhaustively regulated, Member States cannot hinder the import of waste on these grounds if they have not used the possibility under Regulation 1013/2006. Importantly, the ruling in *Stadtgemeinde Frohnleiten* confirms the judgment in *Outokumpu*, reiterating that practical difficulties cannot be an excuse to treat domestic and imported products differently; there should at least be the opportunity to prove that imported products qualify for the same tax treatment as similar domestic products.

---

[214] In its ruling in *Stadtgemeinde Frohnleiten*, the ECJ did not say so, but it probably referred to Article 110 in essence being more of a non-discrimination provision than Article 34. Yet in the early days of European integration, there was considerable academic discussion over the precise meaning and scope of then Article 34 TEEC, with some authors taking the view that Article 34 was little more than a specific application of the general principle of non-discrimination on grounds of nationality. In this regard, see, for instance, Laurence W Gormley, *Prohibiting Restrictions on Trade Within the EC* (T.M.C. Asser Instituut 1985) 8–19. In this regard, see sections 4.4.1 and 4.4.2 above.

[215] Recital 5 of the preamble to Regulation 1013/2006 provides that 'in the case of shipments of waste for disposal, Member States should take into account the principles of proximity, priority for recovery and self-sufficiency at Community and national levels, in accordance with Directive 2006/12/EC of the European Parliament and of the Council of 5 April 2006 on waste, by taking measures in accordance with the Treaty to prohibit generally or partially or to object systematically to such shipments'.

[216] To this extent, see Article 11 of Regulation 1013/2006.

With regard to a financial security/mechanism charge being levied on imported CO_2 streams, the ruling in *Stadtgemeinde Frohnleiten* allows for several interesting points to be made. First, if Member State authorities differentiate the level of the financial security/mechanism charge according to the composition of the CO_2 stream, there is a chance that the differently composed CO_2 streams with not be regarded as similar products by the ECJ. If the latter is the case, such differentiation will also not be discriminatory; different products may be taxed differently, unless they are in competition for the same storage reservoir and the tax differentiation leads to domestic CO_2 streams being protected. In that case, the financial security/mechanism charge will be in breach of Article 110(2). This line of reasoning seems to be supported by *Outokumpu*, in which the ECJ essentially ruled that differentiation between products is allowed when the relevant products are, on the basis of environmental considerations, not to be regarded as similar products.

Second, at first sight the ECJ's reasoning in relation to *Walloon Waste* and the Regulation on the shipment of waste seems to suggest that Member States might perhaps be allowed to try and discourage the importation of CO_2 streams for storage from other Member States by means of tax differentiation, on the grounds of self-sufficiency and proximity. As a consequence of Articles 35 and 36 of the CCS Directive, captured and transported CO_2 for storage has been removed from the scope of the Waste Directive (Directive 2006/12/EC)[217] and the Regulation on the shipments of waste. As captured CO_2 for storage does not formally qualify as a waste under EU law, Member States do not have the possibility to prohibit or object to (to be) imported CO_2 streams.[218] Inverting the ECJ's line of reasoning, this appears to suggest that Member States might perhaps be allowed to, on grounds of self-sufficiency and proximity, try and discourage such imports by taxing them differently.

This conclusion should, however, be put in the perspective of the different character of Articles 110 and 34 TFEU. Article 110 essentially being a non-discrimination provision, the analogy between *Walloon Waste* and this provision should perhaps not be drawn. The point of departure of Article 110 is that the discriminatory taxation of cross-border products is not allowed. Only indirectly discriminatory tax measures may be objectively justified. Differentiation between domestic and imported products allowed under Article 34 will, therefore, not be allowed under Article 110 if it leads to direct discrimination against cross-border products. The differentiating measures in *Walloon Waste*

---

[217] European Parliament and Council Directive 2006/12/EC of 5 April 2006 on waste [2006] OJ L114. This Directive has now been replaced by European Parliament and Council Directive 2008/98/EC of 19 November 2008 on waste [2008] OJ L312/3.

[218] The only option they would have is to completely ban the geological storage of CO_2 in their territory.

arguably were directly discriminatory.[219] What is more, it could be argued that the EU legislator, by removing captured CO_2 for storage from the scope of the Waste Directive and the Regulation on the shipments of waste (Articles 35 and 36 of the CCS Directive) intended this option not to be available for Member States.

Finally, *Stadtgemeinde Frohnleiten* underlines the point that practical difficulties are never an excuse for taxing domestic and imported CO_2 streams for storage differently. There must be an opportunity for the storage operator to prove that the CO_2 stream (to be) imported meets the requirements for being levied the same financial security/mechanism charge as similar domestic CO_2 streams. As indicated by AG Sharpston in her Opinion in *Stadtgemeinde Frohnleiten*, the importing Member State may in that case require that evidence be adduced that removes the risk of tax evasion.[220]

4.5.3 CASE C-74/06 *COMMISSION V. GREECE*

Commission v. Greece dealt with a Greek tax on the registration of imported second-hand cars. The case gives a good overview of the long line of case law dealing with registration taxes imposed on second-hand imported cars and the freedom for Member States to determine the taxable value of imported products by means of general and abstract criteria. The case followed Case C-375/95 *Commission v. Greece*,[221] in which the ECJ found Greece to have failed to fulfil its obligations under Article 110 TFEU. The Greek authorities had applied a special consumer tax and a flat-rate added special duty on private cars imported into or assembled in Greece. The taxable value of these cars was determined by reducing the price of equivalent new cars by 5% for each year of age of the cars concerned, subject, as a rule, to a maximum reduction of 20%. In order to comply with the judgment of the ECJ in Case C-375/95, Greece had reformed its tax scheme by replacing the special consumer tax and flat-rate added special duty with a registration tax and by revising the method of taxation of the taxable value of the cars concerned.

The Commission, however, was concerned that the new Greek scheme still was contrary to Article 110 and started infraction proceedings under ex Article 226 TEC (now Article 258 TFEU).[222] According to it, the Greek legislation applied a scale which was based on a single criterion of vehicle depreciation (how long the

---

[219] Under the relevant Belgian law, the dumping of waste from outside of the Walloon provinces in Belgium was in principle prohibited.
[220] See Case C-221/06 *Stadtgemeinde Frohnleiten*, Opinion of AG Sharpston (n 44) para 50.
[221] Case C-375/95 *Commission v. Greece* [1997] ECR I-05981.
[222] Case C-74/06 *Commission v. Greece* [2007] ECR I-07585, paras 7–8.

vehicle had been in use at the taxable date).[223] The Commission argued that this did not reflect the actual depreciation with sufficient precision to avoid any risk of imported second-hand vehicles being subject to a tax which exceeded the amount of the residual tax incorporated in the value of similar second-hand vehicles already registered in Greece.[224]

Referring to *Commission v. Denmark*,[225] the ECJ held that a Member State may not tax imported second-hand cars based on a value higher than the real value of the car with the result that imported cars are taxed more heavily than similar second-hand cars on the domestic market.[226] By virtue of the *Weigel* case,[227] the value attributed to the imported second-hand car by the authorities should faithfully reflect the value of a similar second-hand car already registered on the domestic market.[228] The ECJ noted that it had held before that the taking into account of the actual depreciation of cars need not necessarily involve an assessment or expert examination of each of them.[229] To avoid the administrative burden inherent in such a system, a Member State may establish, by means of fixed scales calculated on the basis of objective criteria, a value for second-hand cars which, as a general rule, would be very close to their actual value.[230]

The ECJ further argued that, in order for a system of taxation of imported second-hand cars on the basis of general criteria to be compatible with Article 110, it had to be structured in such a way, making allowance for the reasonable approximations inherent in any system of that type, as to exclude any discriminatory effect.[231] Assessing the Greek taxation scheme, the ECJ then held that the taxable value of imported cars was, in essence, based on a single criterion, namely the age of the car.[232] According to the ECJ, the failure to take into account the mileage of a car prevented the scale used in the Greek tax scheme from leading to a 'reasonable approximation' of the actual value of

---

[223] Case C-74/06 *Commission v. Greece* (n 223) para 7.
[224] Ibid.
[225] Case C-47/88 *Commission v. Denmark* (n 78).
[226] Case C-74/06 *Commission v. Greece* (n 223) para 28.
[227] Case C-387/01 *Weigel* (n 63).
[228] Case C-74/06 *Commission v. Greece* (n 223) para 28.
[229] Ibid para 29.
[230] Ibid.
[231] Ibid para 31. In a case following *Commission v. Greece*, *Ioan Tatu*, the ECJ held that by introducing into the calculation of the tax the car's age and actual annual average kilometrage, and by adding to the use of those criteria the option of taking into account, at not excessive cost, the condition of the vehicle and its equipment by means of an inspection by the competent motor vehicle registration authority, the relevant Romanian legislation ensured that the tax was reduced in accordance with a reasonable approximation of the actual value of the vehicle. See Case C-402/09 *Ioan Tatu* (n 55) para 44.
[232] Case C-74/06 *Commission v. Greece* (n 223) para 39.

imported second-hand vehicles.[233] The ECJ reiterated that a fixed scale may be used, but that such a scale should nevertheless enable a value to be established that will, as a general rule, be very close to the actual value.[234]

Finally, the ECJ stated that for a tax scheme such as the Greek one to be compatible with Article 110(1), the criteria on which the fixed method of calculating the depreciation of vehicles is based must be made known to the public.[235] What is more, the ECJ held that the compatibility with Article 110 of taxation of imported second-hand vehicles on the basis of general criteria presupposes that the owner of such a vehicle is able to challenge the application of a fixed method of calculation to the vehicle in order to demonstrate that it leads to taxation exceeding the amount of the residual tax incorporated in the value of similar used vehicles already registered in the national territory.[236] The ECJ stated that if a fee is due for use of a complaints procedure, such a fee is likely to deter a car owner from appealing to a complaints commission only if it represents a significant proportion of the amount of the registration tax challenged.[237]

In *Commission v. Greece* the ECJ gave considerable guidance as to the scope for Member States under Article 110(1) to differentiate taxes imposed on similar domestic and imported products. The bottom line is that Member States are, for administrative reasons, allowed to determine the taxable value/basis of imported products on the basis of abstract and general criteria. The use of such criteria, however, should lead to a taxable value/basis that is very close to the actual value/content of the imported product and thus not to a taxation that is higher than appropriate on the basis of the taxable value/basis of the product. As long as this is the case, not every imported product has to be individually assessed. Furthermore, Member States will have to make public the criteria on which the determination of the taxable value/content is based and there will have to be a possibility for the importer to challenge the tax charged on this basis on the ground that it exceeds the tax imposed on similar domestic products.

Even though the line of cases summarised in *Commission v. Greece* relates to the imposition of a registration tax on imported second-hand cars, there appear to be some interesting potential lessons for the (indirect) imposition of a financial security/mechanism charge on CO_2 streams imported for storage. First, *Commission v. Greece* seems to suggest that Member States would, in principle, be allowed to use fixed scales calculated on the basis of objective criteria for

---

[233]   Case C-74/06 *Commission v. Greece* (n 223) para 43.
[234]   Ibid para 42.
[235]   Ibid para 46.
[236]   Ibid para 59.
[237]   Ibid para 53.

determining the level of financial security/mechanism charge (indirectly) imposed on imported CO_2 streams. The objective criteria used in this regard could, for instance, as suggested in section 4.2 above, be related to the risks associated with the composition of the CO_2 stream. Nevertheless, the use of abstract and general criteria should lead to a taxation of imported CO_2 streams that closely resembles the actual composition and corresponding risks of the relevant imported CO_2 stream. The use of fixed scales may not lead to domestic CO_2 streams with a similar composition being taxed more favourably due to, for instance, a different system being applied for determining the level of taxation of those CO_2 streams. Finally, the criteria on the basis of which CO_2 streams for storage will (indirectly) be taxed will have to be made public and there will have to be a possibility for the entity importing the CO_2 stream to challenge the tax charged on the ground that it exceeds the tax imposed on similar domestic CO_2 streams for storage.

Whether Member State authorities will indeed use fixed scales based on general criteria for determining the level of the financial security/mechanism charge is uncertain. First, the methodology of using such fixed scales seems to primarily be appropriate for the taxation of mass consumption goods, such as second-hand cars, that have the potential to be individually imported on a large scale. The character of CO_2 streams for storage obviously is different, possibly resulting in smaller-scale importation. It can therefore be questioned whether such as scheme would really be necessary for the imposition of a financial security/mechanism charge.

What is more, Article 12(3)(a) of the CCS Directive obliges Member States to ensure that the storage operator accepts and injects CO_2 streams only if an analysis of the composition of the streams and a risk assessment has been carried out. For each individual storage project, there will thus be an assessment of the composition of the CO_2 stream and the risks related to the storage of that stream. That assessment could serve as a basis for determining the level of the financial security/mechanism charge. Finally, the risks related to the storage of individual CO_2 streams are not only dependent on the composition of the relevant CO_2 stream, but also follow from the geochemical characteristics of the storage reservoir concerned. The latter might perhaps be difficult to capture in fixed scales. *Commission v. Greece*, however, indicates that Member States would have a certain discretion, should they decide to determine the level of the financial security/mechanism charge by means of fixed scales based on general and abstract criteria.[238]

---

[238] The main findings of *Commission v. Greece* were confirmed by the ECJ in Case C-402/09 *Ioan Tatu* (n 55).

4.6 LESSONS FOR THE DESIGN OF FINANCIAL SECURITY/MECHANISM CHARGES

The above analysis shows that the taxation of the service of storing captured CO_2 will likely fall within the scope of Article 110. Within Article 110, the imposition of a financial security/mechanism charge will probably be captured by 110(1) and not by 110(2). When a storage reservoir is suitable for the storage of CO_2 streams with a specific composition (range) only and there are no domestic CO_2 streams that meet the relevant requirements, the imposition of a financial security/mechanism charge would in principle fall under the scope of Article 110, even though it cannot infringe either Article 110(1) or 110(2). However, should the financial security/mechanism charge indirectly imposed on imported CO_2 streams only be of such an amount that the free movement of goods within the internal market would be impeded, then the Courts may consider it under Article 30 TFEU. There does not appear to be a general rule that tax rules which differentiate between products on the basis of objective criteria (unrelated to origin), but could still lead to a heavier burden being placed on cross-border products, are per se compatible with Article 110.

Outokumpu suggests that Member States would be allowed to differentiate the financial security/mechanism charge on the basis of objective criteria related to, for instance, the composition of the CO_2 stream. Even though Member States will likely have this freedom, they will, by virtue of *Outokumpu* and *Stadtgemeinde Frohnleiten*, be under the obligation to allow the party importing captured CO_2 for storage to demonstrate that the relevant CO_2 stream meets the criteria for being charged at a particular rate. In practice, this will probably come down to the storage operator being given the change to show that by storing the relevant imported CO_2 stream, he meets the criteria for being charged a particular financial security/mechanism tariff. Should it be difficult to determine the composition of the CO_2 stream (to be) imported, then the storage operator cannot be charged a flat rate duty. He must then be charged the lowest tariff applicable to domestic CO_2 streams for storage. Different treatment may not lead to discrimination against imported CO_2 streams for storage. In addition, *Outokumpu* indicates that practical difficulties will not be an excuse for discriminating against 'similar' imported CO_2 streams for storage.

On the basis of *Commission v. Greece*, Member States would, in principle, appear to be allowed to use fixed scales calculated on the basis of objective criteria for determining the level of the financial security/mechanism charge (indirectly) imposed on imported CO_2 streams. Nevertheless, the use of abstract and general criteria should lead to a taxation of imported CO_2 streams that closely resembles the actual composition and corresponding risks of the relevant imported CO_2

stream. The use of fixed scales may not lead to domestic CO_2 streams with a similar composition being taxed more favourably due to, for instance, a different system being applied for determining the level of taxation of those CO_2 streams. Finally, the criteria on the basis of which CO_2 streams for storage are (indirectly) taxed will have to be made public and there will have to be a possibility for the entity importing the CO_2 stream to challenge the tax charged on the ground that it exceeds the tax imposed on similar domestic CO_2 streams for storage.

Should a (highly) meshed CO_2 transport network develop, then it could, by virtue of *Commission v. Greece*, be problematic to impose a financial security/ mechanism charge on the basis of (inter alia) the characteristics of the CO_2 stream actually stored. If the stored CO_2 stream has a risk profile higher than that of the CO_2 stream contractually delivered by a foreign capturer (and equal to or lower than that contractually delivered by a domestic capturer), that capturer will arguably be (indirectly) charged too high a financial security/ mechanism charge. The charge will, in that case, not closely resemble the actual composition and risks of that particular CO_2 stream and likely discriminate against the CO_2 stream imported for storage.

Considering the direct effect of Article 110, the foreign capturer could then challenge the legality of the national rule imposing the financial security/ mechanism charge on the ground of an alleged incompatibility with Article 110 TFEU (see *Outokumpu*). In the event of the blending of domestic and imported CO_2 streams for storage, Member States would be wise to make sure that the imported CO_2 streams for storage are charged in accordance with their actual composition to avoid their being discriminated against. In the absence of a centrally managed network, for instance in a transitory phase from a point-to-point/hub and spoke network to a (highly) meshed network, Member State authorities should be particularly aware of the risk of discriminatory taxation of imported CO_2 streams for storage.

Based on the analysis in this chapter, a number of concrete recommendations for the design of the financial security/mechanism charge can be made.

First, Member States should avoid differentiating the level of the financial security/mechanism charge based on the (geographical) origin of the CO_2 stream for storage. Direct discriminatory taxation is prohibited per se under Article 110. Second, Member States can, in principle, safely differentiate the level of the financial security/mechanism charge on the basis of objective criteria related to, for instance, the composition of the CO_2 stream or the geochemical characteristics of the storage reservoir, but would be wise to make sure that such differentiation does not lead to imported CO_2 streams primarily falling in the highest tax scales.

Third, Member States are advised not to use flat rates to impose a financial security/ mechanism charge on imported CO_2 streams for storage. Flat rates will arguably only be allowed under Article 110 if they are equal to the lowest level of charge imposed on domestic CO_2 streams for storage. Fourth, even though Member States would probably be allowed to use fixed scales and abstract criteria for determining the level of the financial security/mechanism charge imposed on imported CO_2 streams, they should be very careful in doing so. Particularly in the transitory phase from a point-to-point/hub and spoke network to a (highly) meshed network (risk of blending of streams), Member States should be careful to charge any imported CO_2 stream in accordance with its actual composition and corresponding risks. They are therefore not advised to base the level of the financial security/mechanism charge on the composition of the CO_2 stream *actually stored.*

Finally, when using fixed scales and abstract criteria for determining the financial security/mechanism charge levied on CO_2 streams imported for storage, Member States are advised to publish the criteria on the basis of which the charge is determined. Also, they will have to make sure that there is a possibility for the entity importing the relevant CO_2 stream to challenge the financial security/mechanism imposed on the relevant CO_2 stream. In view of the above, the scientific uncertainties still surrounding the (long-term) effects of the geological storage of CO_2 and the fact that storage operators will, in any case, conduct an analysis of the relevant CO_2 stream for storage as well as a risk assessment, it may be wise not to use fixed scales and abstract criteria for determining the financial security/ mechanism charge levied on CO_2 streams imported for storage. Should new scientific knowledge/developments lead to interim changes to financial security/mechanism charges imposed, then the above lessons likewise apply.

CHAPTER V

REFUSING ACCESS TO CCS INFRASTRUCTURE AND THE GENERAL EU LAW PRINCIPLE OF LOYALTY

- Article 21(2)(b) CCS Directive does not appear to infringe Article 4(3) TEU in conjunction with Article 194(1)(c) TFEU

- Individuals would not seem able to invoke Articles 4(3) and 194(1)(c) before a national court to challenge the validity of Article 21(2)(b)

- Even though Articles 4(3) and 194(1)(c) do not appear to narrow their scope for implementing Article 21(2)(b), Member States would be wise to require CO_2 transport or storage operators refusing access on the ground mentioned in Article 21(2)(b) to expand their capacity when it is economically sensible to do so or when the potential customer is willing to pay for the expansion

5.1   INTRODUCTION

Article 21 of the CCS Directive deals with the access of parties who do not control the relevant infrastructure (third-party access) to the CO_2 transport network and storage sites. Article 21(2) provides that Member States are free to determine the manner in which third-party access to CO_2 transport and storage infrastructure is provided, as long as such access is provided in a fair, open and non-discriminatory manner and a number of considerations are taken into account. By virtue of Article 21(2)(b), each Member State shall take into account 'the proportion of its CO_2 reduction obligations pursuant to international legal instruments and to Community legislation that it intends to meet through capture and geological storage of CO_2'.

The wording of Article 21(2)(b) is not very precise, but the provision seems to require Member States to take into account their own (anticipated) need for national transport and storage capacity when developing a regime for third-party access to CO_2 transport and storage infrastructure. Accordingly, it appears to give Member States the possibility to require the operators of CO_2 transport and storage infrastructure to refuse to grant access to the relevant infrastructure when the capacity concerned is needed to meet part of the Member State's international and EU obligations to reduce greenhouse gas emissions. Think, for instance, of national legislation dedicating certain (or perhaps even all)[1] CO_2 transport and storage infrastructure for internal (national) use only. At least part of the capacity of such infrastructure would then be reserved for the transport and storage of CO_2 captured within the Member State concerned.[2]

The consequent (partial) unavailability to third parties of CO_2 transport and storage capacity in some Member States could possibly lead to problems for other Member States in trying to meet their national targets for the reduction of greenhouse gas emissions.[3] For certain industries, such as the cement industry, CCS might well be the only viable means of achieving deep emission cuts.[4] At the same time, potential CO_2 storage capacity is spread unevenly over Europe and certain Member States have significantly less potential storage capacity than others.[5] The necessity to deploy CCS in order to achieve deep emission cuts in certain sectors in combination with the uneven spread of potential CO_2 storage capacity over Europe means that for some Member States access to CO_2 transport

---

[1] This obviously depends on the amount of capacity needed and the (future) quantity of capacity available.
[2] In relation to the long-term reservation of transport capacity, see also Chapter VII on the development and management of CO_2 transport infrastructure and EU antitrust law.
[3] See also Hans Vedder, 'An Assessment of Carbon Capture and Storage under EC Competition Law' (2008) 29 European Competition Law Review 586, 598. Vedder argues that Article 21(2) (b) 'could effectively foreclose an entire Member State's carbon storage market. A Member State planning to achieve the overwhelmingly majority of its carbon dioxide reduction obligations by means of CCS could invoke this to claim that all carbon storage capacity is reserved'.
[4] United Nations Industrial Development Organization (UNIDO), 'Carbon Capture and Storage in Industrial Applications: Technology Synthesis Report' (2010) 60 <www.unido.org/fileadmin/user_media/Services/Energy_and_Climate_Change/Energy_Efficiency/CCS/synthesis_final.pdf> accessed 18 May 2011.
[5] Thomas Vangkilde-Pedersen and others (EU GeoCapacity), 'Assessing European Capacity for Geological Storage of Carbon Dioxide' (2009) <www.geology.cz/geocapacity/publications/D16%20WP2%20Report%20storage%20capacity-red.pdf> accessed 13 July 2011; Joris Morbee, Joana Serpa and Evangelos Tzimas, 'The Evolution of the Extent and the Investment Requirements of a Trans-European CO_2 Transport Network' (2010) 1 <http://publications.jrc.ec.europa.eu/repository/bitstream/111111111/15100/1/ldna24565enn.pdf> accessed 7 December 2011; and Commission, 'Impact Assessment – Accompanying Document to the Communication from the Commission to the European Parliament, the Council, the European Economic and Social Committee and the Committee of the Regions: Energy Infrastructure Priorities for 2020 and Beyond – A Blueprint for an Integrated European Energy Network' SEC (2010) 1395 final 21 <http://eur-lex.europa.eu/LexUriServ/LexUriServ.do?uri=SEC:2010:1395:FIN:EN:PDF> accessed 2 January 2011.

and storage infrastructure in other Member States might be crucial for achieving deep national emissions reductions.

The above raises questions as to the conformity of Article 21(2)(b) with EU law. In particular, it could, at first sight, be questioned whether Article 21(2)(b) is compatible with one of the general principles of EU law, the principle of EU loyalty/sincere cooperation[6] (principle of loyalty).[7] Article 4(3) of the Treaty on European Union (TEU)[8] provides that pursuant to the principle of loyalty, the EU and the Member States shall, in full mutual respect, assist each other in carrying out tasks which flow from the Treaties.[9]

Furthermore, by virtue of the new Article 194(1)(c) of the Treaty on the Functioning of the European Union (TFEU),[10] 'EU policy on energy shall aim, *in a spirit of solidarity between Member States*, to (…) promote energy efficiency and energy saving and the development of new and renewable forms of energy'.[11] Article 194(1)(c) seems to require secondary EU climate and energy legislation, like the CCS Directive, to be drafted 'in a spirit of solidarity between Member States'.[12] At first sight, it is questionable whether Article 21(2)(b) meets that requirement.

---

[6]   Article 4(3) TEU uses the words 'principle of sincere cooperation'. However, for the sake of recognisability, I will hereafter use an abbreviation of the principle's 'old' name (principle of EU loyalty), i.e. the principle of loyalty.

[7]   There are also other questions that could be asked, such as the provision's conformity with the free movement of goods provisions in Articles 34–36 TFEU. I will here, however, focus on the conformity of Article 21(2)(b) with the general EU law principle of loyalty as the provision's emphasis on Member States' self-interest seems to stand in sharp contrast with the cooperative conduct prescribed by this principle.

[8]   For reasons of consistency, I will refer to Article 4(3) TEU in the remainder of this writing, even though the case law or literature cited referred to ex Article 10 of the Treaty establishing the European Community (TEC) or ex Article 5 of the Treaty establishing the European Economic Community (TEEC).

[9]   As indicated by AG Mischo in his Opinion in the *Booker* case, Directives, as well as other secondary EU legislation, must conform to the general principles of EU law See Joined Cases C-20/00 and C-64/00 *Booker* [2003] ECR I-07411, Opinion of AG Mischo, paras 57 and 58. The term 'Treaties' refers to the Treaty on European Union (TEU) and the Treaty on the Functioning of the European Union (TFEU).

[10] The article was introduced by the Treaty of Lisbon, which amended the EC Treaty and entered into force on 1 December 2009.

[11] Emphasis added.

[12] The notion of solidarity is pervasive in EU law. References to an EU principle of solidarity between the Member States can, for instance, also be found in Articles 3(3) TEU and 191(2) and (3) TFEU. Article 3(3) TEU provides that the EU shall promote solidarity among Member States. By virtue of Articles 191(2) and (3) 'Union policy on the environment shall aim at a high level of protection *taking into account the diversity of situations in the various regions of the Union*' (emphasis added) and 'in preparing its policy on the environment, the Union shall take account of (…) the economic and social development of the Union as a whole and the balanced development of its regions'.

In addition to the validity of Article 21(2)(b), both provisions might also affect the Member States' scope for implementing the provision in their national legal order. National measures implementing secondary EU legislation must respect the general principles of EU law as well as the Treaty provisions.[13] On the face of it, Article 4(3) TEU in particular appears to be capable of narrowing the Member States' scope for implementing Article 21(2)(b). As the principle of loyalty allegedly includes the obligation for Member States to mutually assist each other,[14] it appears doubtful whether the Member States would comply with their duties under Article 4(3) when implementing Article 21(2)(b) of the CCS Directive.

In the following sections, I answer the questions of to what extent Article 21(2)(b) of the CCS Directive is compatible with Article 4(3) TEU in conjunction with Article 194(1)(c) TFEU and in what way both provisions could narrow the Member States' scope for implementing Article 21(2)(b). To this end, section 5.2 briefly explores (the context of) Article 21(2)(b). Section 5.3 addresses the concept of general principles of EU law. Section 5.4 examines the general EU law principle of loyalty. Section 5.5 gives a brief outline of the background of Article 194(1) TFEU and its solidarity clause.

Section 5.6 analyses the conformity of Article 21(2)(b) with Articles 4(3) and 194(1)(c). Section 5.7 discusses the possible effects of Articles 4(3) and 194(1)(c) on the Member States' scope for implementing Article 21(2)(b). This includes an analysis of the possibilities for individuals to invoke Articles 4(3) and 194(1)(c) for challenging the validity of national measures implementing Article 21(2)(b) before a national court. Finally, section 5.8 draws a number of lessons for the design of Member States' regimes for third-party access to CO_2 transport and storage infrastructure.

5.2   ARTICLE 21(2)(B) OF THE CCS DIRECTIVE[15]

As indicated earlier, Article 21(2) states that Member States are free to determine the manner in which third-party access to CO_2 transport and storage

---

[13]   Case 106/77 *Simmenthal* [1978] ECR 00629, para 17.

[14]   Cases 235/87 *Mateucci* [1988] ECR 05589, para 19, and C-251/89 *Athanasopoulos* [1991] ECR I-02797, para 57. See Fabian Amtenbrink and Hans Vedder, *Recht van de Europese Unie* (4th edn, Boom Juridische Uitgevers 2010) 159. Ambtenbrink and Vedder state that 'ook tussen de lidstaten onderling heeft het loyaliteitsbeginsel gevolgen. Zo kwam het Hof in de zaken *Mateucci* en *Athanasopoulos* tot de conclusie dat de lidstaten onderlinge bijstand moeten geven'.

[15]   For more information on the concept of third-party access to CO_2 transport and storage infrastructure, see section 6.2 of the next chapter.

infrastructure is provided, as long as access is provided in a fair, open and non-discriminatory manner and a number of considerations are taken into account:

a) the storage capacity which is or can reasonably be made available within the areas determined under Article 4 of the Directive, and the transport capacity which is or can reasonably be made available;
b) the proportion of its CO_2 reduction obligations pursuant to international legal instruments and to Community legislation that a Member State intends to meet through CCS;
c) the need to refuse access where there is an incompatibility of technical specifications which cannot be reasonably overcome; and
d) the need to respect the duly substantiated reasonable needs of the owner or operator of the storage site or of the transport network and the interests of all other users of the storage or the network or relevant processing or handling facilities who may be affected.

In the impact assessment to its 2008 proposal for the CCS Directive, the Commission indicates that there were two options for an enabling legal framework for access to the CO_2 transport network and storage sites.[16] The first option was to follow a basic approach requiring access to networks as well as to storage sites to be granted on a non-discriminatory basis, subject to limitations on access for reasons justified by public interest pursuant to Articles 52 and 62 TFEU.[17] The second option was to impose specific rules for achieving equal access – such as those in the Gas and Electricity Directives[18] (i.e. regulated third-party access) – including, if required, unbundling provisions.[19] The Commission opted for the first option as it considered the market for CCS to be at an early stage and, according to the commission, indications were that there would be

---

[16]   Commission, 'Accompanying Document to the Proposal for a Directive of the European Parliament and of the Council on the Geological Storage of Carbon Dioxide: Impact Assessment' COM (2008) 18 final 49.
[17]   Article 52(1) TFEU contains the express, Treaty-based derogations to the prohibition of restrictions on the freedom of establishment of nationals from another Member State in Article 49 TFEU. It provides that the provisions on the right of establishment do not prejudice the applicability of provisions providing for special treatment of foreign nationals on grounds of public policy, public security or public health. Article 62 TFEU makes the express derogations in Article 52(1) TFEU applicable to the free movement of services.
[18]   European Parliament and Council Directive 2009/73/EC of 13 July 2009 concerning common rules for the internal market in natural gas [2009] L211 and European Parliament and Council Directive 2009/72/EC of 13 July 2009 concerning common rules for the internal market for electricity [2009] L211.
[19]   These are provisions requiring the transport of CO_2 to be separated from the other two activities in the CCS value chain, i.e. capture and storage.

separate operators for the combustion and capture phase on the one hand and transport and storage on the other.[20]

During the negotiations on the CCS Directive, Parliament proposed to add a consideration to the list in then Article 20(2) (now Article 21(2)), which was to counter the various grounds for refusal. Amendment 114 of Parliament's report on the Commission proposal suggested adding 'the need to ensure that adequate provisions are made to establish conditions for cross-border and transit flows of CO_2 in a manner that avoids distortions of competition resulting from the geographical location of potential users within the EU'.[21] The justification for the amendment provided that 'it is important to ensure that the operators in Member States such as Greece are not unduly disadvantaged by their geographical location or potential limitations on development of storage sites'.[22]

Interestingly, recital 38 of the preamble to the CCS Directive provides that

> [A]ccess to CO_2 transport networks and storage sites, *irrespective of the geographical location of potential users within the Union,* could become a condition for entry into or competitive operation within the internal electricity and heat market, depending on the relative prices of carbon and CCS[23]

The phrase 'irrespective of the geographical location of potential users within the Union' was not in the equivalent recital 29 of the original Commission proposal. Amendment 22 of Parliament's report on the proposal for the CCS Directive proposed to amend recital 29 to read: 'transparent and non-discriminatory access to CO_2 transport networks and storage sites, irrespective of the geographical location of potential users within the European Union, *should be* a condition'.[24] The final text of recital 38, in which the words 'should

---

[20] Commission, 'Impact Assessment CCS Directive' (n 16) 45. By contrast, in section 6.3.1.4 of the following chapter, I argue that future EU CCS markets could well be characterised by some form of vertical integration, be it through ownership structures or through long-term (take-or-pay) contracts. In relation to the gas sector take-or-pay contracts are also known as ship-or-pay contracts. Ship-or-pay contracts are long-term contracts requiring payment regardless of whether any gas is actually shipped. See Goldman Sachs, 'Pipeline Financing Discussion' (presentation held at the Wyoming Natural Gas Pipeline Authority, 25 August 2003) <www.wyopipeline.com/information/presentations/2003/Goldman%20Sachs_files/frame.htm> accessed 8 February 2012.

[21] European Parliament, 'Report on the proposal for a directive of the European Parliament and of the Council on the geological storage of carbon dioxide and amending Council Directives 85/337/EEC, 96/61/EC, Directives 2000/60/EC, 2001/80/EC, 2004/35/EC, 2006/12/EC and Regulation (EC) No 1013/2006 (COM(2008)0018 – C6 0040/2008 – 2008/0015(COD))' A6–0414/2008 amendment 114 <www.europarl.europa.eu/oeil/FindByProcnum.do?lang=en&procnum=COD/2008/0015> accessed 26 May 2011.

[22] European Parliament (n 21) amendment 114.

[23] Emphasis added.

[24] Ibid.

be' were replaced by 'could become', seems to suggest that the location of the operator is not relevant, whereas the phrase 'irrespective (...) Union' presumably was chosen to underline the importance of third-party access to transport and storage infrastructure to operators in certain locations (such as Greece).[25]

As argued above, Article 21(2)(b) appears to provide a ground for refusal of third-party access to CO_2 transport and storage infrastructure on the basis of the proportion of its international and EU CO_2 reduction obligations[26] a Member State seeks to meet through CCS.[27] The 'obligations pursuant to international legal instruments' likely refer to the targets for the reduction of greenhouse gases to which the EU-15[28] committed under the Kyoto-protocol as well as any reduction targets the 27 Member States may commit to under a post-Kyoto international climate regime.

As for the obligations pursuant to Community legislation, this is likely to refer to the reduction of greenhouse gases by the installations participating in the EU's greenhouse gas emission trading scheme, the EU ETS. As indicated in Chapter I, the main application of CCS is likely to be at large point sources of CO_2 emissions, such as power plants and industrial plants. These installations are covered by the EU ETS. Since the beginning of 2013, the EU ETS no longer has national caps but a single EU-wide cap. It could therefore be questioned whether it is correct to speak of 'obligations for Member States' in this regard. As we will see later, this is of importance for the compatibility of Article 21(2)(b) with the general EU law principle of loyalty.

---

25   The justification for amendment 22 again provided that 'it is important to ensure that operators in Member States such as Greece are not unduly disadvantaged by their geographic location or potential limitations on development of storage sites'. See European Parliament (n 21) amendment 22.

26   Article 21(2)(b) refers to the CO_2 reduction obligations and not to the *greenhouse gas* reduction obligations of Member States. It is submitted, however, that the phrase CO_2 reduction obligations should be read as *greenhouse gas* reduction obligations as all international (UNFCCC Kyoto Protocol) and EU (European Parliament and Council Decision 406/2009 of 23 April 2009 on the effort of Member States to reduce their greenhouse gas emissions to meet the Community's greenhouse gas emission reduction commitments up to 2020 [Effort Sharing Decision]) reduction obligations are related to *greenhouse gas* emissions, including other gases than CO_2 such as CH_4 (methane) and N_2O (nitrous oxide).

27   On the meaning of Article 21(2)(b), see also Laetitia Birkeland and others, 'Improving the Regulatory Framework, Optimizing Organization of the CCS Value Chain and Financial Incentives for CO_2-EOR in Europe' (2011) 16 <http://bellona.org/ccs/ccs-news-events/publications/article/improving-the-regulatory-framework-optimizing-organization-of-the-ccs-value-chain-and-financial-inc.html> accessed 2 January 2011. Birkeland and others state that if a Member States has already met its reduction obligations through CCS, the Directive suggests that third parties will not have any right to access under fair and open conditions.

28   These are Austria, Belgium, Denmark, Finland, France, Germany, Greece, Ireland, Italy, Luxembourg, The Netherlands, Portugal, Spain, Sweden and the United Kingdom.

5.3 GENERAL PRINCIPLES OF EU LAW[29]

In the literature, general principles of EU law have been given a large number of (slightly) different definitions.[30] Yet, these different definitions have two things in common. First, they all underline the general character of the *general* principles of EU law.[31] Second, they stress the fundamental importance of these principles to the EU legal system.[32] As stated by Tridimas 'the term general principles may be reserved for fundamental propositions of law which underlie a legal system and from which concrete rules or outcomes may be derived'.[33]

General principles of EU law are commonly subdivided into different categories.[34] Generally, the different scholarly classifications make a distinction between those general principles that have more of a fundamental rights character ('constitutional principles') and principles which are characterised as 'institutional' or 'administrative' principles.

In 2010, Rosas and Armati outlined a hierarchy of norms in the EU legal order.[35] On top of that hierarchy are the value foundations of the EU legal order, as laid down in Article 2 TEU. Second come the general principles of EU law (including fundamental rights) and written primary law. Even though Rosas and Armati apparently ranked general principles and primary law on equal footing, they argued that fundamental rights[36] '(...) should now be seen as having a somewhat enhanced status as compared to other parts of primary law'.[37] According to Rosas and Armati, the European Court of Justice's (ECJ's) ruling in *Kadi*[38] has

---

[29] For a comprehensive overview of the concept of general principles of EU law, see Takis Tridimas, *The General Principles of EU law* (2nd edn, Oxford University Press 2006).

[30] See e.g. Bruno de Witte, 'Institutional Principles: A Special Category of General Principles of EC Law' in Bernitz and Nergelius (eds), *General Principles of European Community Law* (Kluwer Law International 2000) 143, 143; Eric Engle, 'General Principles of European Environmental Law' (2009) 17 Penn State Environmental Law Review 215, 217–18 and Tor-Inge Harbo, 'The Function of the Proportionality Principle in EU Law' (2010) 16 European Law Journal 158, 159.

[31] See e.g., Harbo (n 30) 159 and Engle (n 30) 218.

[32] See e.g., Ola Wiklund and Joxerramon Bengoetxea, 'General Constitutional Principles of Community Law' in Ulf Bernitz and Joakim Nergelius (eds), *General Principles of European Community Law* (Kluwer Law International 2000) 119, 120, and Tridimas (n 28) 1.

[33] Tridimas (n 29) 1.

[34] See e.g. Jürgen Schwarze, 'Enlargement, the European Constitution, and Administrative Law' (2004) 53 International and Comparative Law Quarterly 969, 982. Schwarze, makes a distinction between 'constitutional principles' – fundamental rights such as the principle of legal certainty – and general principles 'of an administrative nature'.

[35] Allan Rosas and Lorna Armati, *EU Constitutional Law: An Introduction* (2nd edn, Hart Publishing 2010) 42.

[36] These are part of the first class of general principles of EU law we have identified above (constitutional principles).

[37] Rosas and Armati (n 35) 44.

[38] Joined Cases C-420/05 P and C-415/05 P *Kadi* [2008] ECR I-06351.

confirmed the tendency in earlier case law according to which the EU constitutional order consists of some core principles which may prevail over provisions of the Treaties and thus over written primary law.[39]

For the sake of brevity, I will not further discuss Rosas and Armati's claim that general principles of EU law with a fundamental rights character have a (slightly) higher status than primary EU law. Yet, what can be concluded from various alleged hierarchies of EU norms[40] is that the general principles of EU law 'stand at the pinnacle of the hierarchy of norms established by EU law'[41] and that they are superior to secondary EU law.[42]

As general principles of EU law take precedence over secondary EU law, they may be relied upon as grounds for judicial review of EU acts. Article 263 TFEU mentions the grounds on which an EU act may be annulled by the Courts, among which is the 'infringement of the Treaties or of any rule of law relating to their application'. The phrase 'any law relating to their application' has been used by the Courts as the basis for the doctrine that an EU act may be quashed for infringement of a general principle of EU law.[43] Similarly, general principles of EU law can be invoked under Article 267 TFEU[44] to indirectly challenge the validity of EU acts before national courts.[45] There is widespread agreement that

---

[39]   Rosas and Armati (n 35) 43.

[40]   See for instance, Joakim Nergelius, 'General Principles of Community Law in the Future: Some Remarks on their Scope, Applicability and Legitimacy' in Bernitz and Nergelius (n 32) 223, 229–230; Koen Lenaerts and Marlies Desomer, 'Towards a Hierarchy of Legal Acts in the European Union? Simplification of Legal Instruments and Procedures' (2005) 11 European Law Journal 744, 745; and Tridimas (n 28) 50–1.

[41]   Koen Lenaerts and José A Gutiérrez-Fons, 'The Constitutional Allocation of Powers and General Principles of EU Law' (2010) 47 Common Market Law Review 1629, 1636.

[42]   In the case of *Nold*, the ECJ already held that general principles of EU law with a fundamental rights character are superior to EU secondary law. See Case 4/73 *Nold* [1974] ECR 00491, para 13. In the case of *Woodspring*, the ECJ indicated that the same goes for 'institutional' general principles of EU law, such as the principle of proportionality. On the precedence of general principles of EU law over secondary EU law, see also Lenaerts and Desomer (n 39) 745. General principles of EU law also take precedence over conflicting national law that falls within the scope of EU law. See Lenaerts and Gutiérrez-Fons (n 41) 1629.

[43]   Trevor Hartley, *The Foundations of European Union Law* (7th edn, Oxford University Press 2010) 143. See also Henry G Schermers and Denis F Waelbroeck, *Judicial Protection in the European Union* (Kluwer Law International 2001) 30–31, and Koen Lenaerts and Dirk Arts, *Procedural Law of the European Union* (Sweet & Maxwell 1999) 197.

[44]   Article 276 TFEU contains the preliminary reference procedure, by which the ECJ can give preliminary rulings on the interpretation of the Treaties and the validity and interpretation of acts of the institutions, bodies, offices or agencies of the Union.

[45]   In this regard, see also Xavier Groussot, *General Principles of Community Law* (Europa Law Publishing 2006) 364. On the impact of the general principles of EU law in French public law, Groussot states that 'the use of general principles of Community law before the national courts to indirectly challenge the validity of Community legislation appears to be a real success story'. In section 5.7, I further address the issue of direct effect of general principles of EU law in general and the principle of loyalty in specific.

general principles of EU law are to be applied by the EU institutions and that the Courts may condemn a breach of these principles as they would condemn the breach of any written norm.[46]

Vandamme has argued that the Courts interpret Regulations and Directives as much as possible in conformity with the general principles of EU law and the Treaty.[47] Accordingly, it is only when this 'consistent interpretation' has reached its limits that EU legislation will be annulled for breach of a general principle of EU law.[48] Referring to the *Wachauf* case[49] and the Opinion of AG Slynn in the *Klench* case,[50] Vandamme has claimed that as long as a Directive leaves Member States enough discretion to implement it into national legislation in a way that does not infringe the general principles of EU law, it will not be struck down by the Courts.[51] In other words, the Courts will declare a Directive to infringe a general principle of EU law only when it leaves the Member States insufficient discretion to implement it into national law in a way that does not infringe the relevant general principle.

In *Wachauf*, the ECJ held that the relevant provisions of the Regulation in question[52] left 'the competent national authorities a sufficiently wide margin of appreciation to enable them to apply those rules in a manner consistent with the requirements of the protection of fundamental rights', the latter being 'an integral part of the general principles of the law'.[53] As the Member States had sufficient discretion to apply the relevant provisions without infringing the general principles of EU law, these provisions itself could not, in the ECJ's view, be said to infringe the general principles of EU law.[54] In his Opinion in *Klensch*, AG Slynn (hypothetically) stated that:

> [I]t is clearly the duty of the Council and the Commission when making a regulation within the framework of a common organization of the market to comply with Article 40(3) of the EEC Treaty and to exclude any discrimination between producers or consumers within the Community. If in a particular case it were contended that

---

46 Schwarze (n 34) 970.
47 Thomas A Vandamme, *The Invalid Directive* (Europa Law Publishing 2005) 201.
48 Ibid.
49 Case 5/88 *Wachauf* [1989] ECR 02609.
50 Joined Cases 201 and 202/85 *Klensch*, Opinion of AG Slynn.
51 See Vandamme (n 47) 201.
52 Council Regulation (EEC) 857/84 of 31 March 1984 adopting general rules for the application of the levy referred to in Article 5c of Regulation No 804/68 in the milk and milk products sector [1984] OJ L90.
53 Case 5/88 *Wachauf* (n 49) paras 17 and 22.
54 Ibid para 22.

the choices given *must* lead to discrimination, the proper course would be to challenge the validity of the regulation[55]

Vandamme has argued that, even though the *Wachauf* case and AG Slynn's Opinion in *Klensch* concerned Regulations, this approach can also be taken to Directives.[56] In support, he has referred to the Opinion of AG Mischo in the *Booker*[57] case.[58] In this case, the AG stated that 'it is difficult to see the justification for saying that, when implementing Directives, the Member States are freed from their obligation to respect the fundamental rights enshrined in the Community legal order'.[59]

Yet, it can be doubted whether this quote by AG Mischo should be read as supporting Vandamme's claim that a Directive will not be struck down by the Courts if Member States can implement it without infringing the general principles of EU law. In his opinion, the AG underlines the point that the general principles of EU law apply to national measures implementing EU law.[60] According to AG Mischo, the obligation for Member States to respect the general principles of EU law likewise applies in the implementation of Directives.[61] The AG does not, however, state that the above quoted 'discretion rule' from *Wachauf* similarly applies in the case of a Directive.[62]

Nevertheless, the *Socridis* case seems to lend support to Vandamme's claim.[63] This case was preceded by national proceedings in which the Société Critouridienne de Distribution (Socridis) had challenged the validity of Directives 92/83/EEC and 92/84/EEC, both dealing with the taxation of alcohol.[64] Socridis had argued that the two Directives were incompatible with Article 95(2) of the Treaty Establishing the European Community (TEC – now Article 110(2) TFEU). According to it, they introduced a system of taxation authorising discriminatory and anti-competitive practices that indirectly

---

[55]   Joined Cases 201 and 202/85 *Klensch*, Opinion of AG Slynn, 3500 (emphasis added).
[56]   Vandamme (n 47) 202.
[57]   Case C-20/00 *Booker*, Opinion of AG Mischo (n 9) para 52.
[58]   Ibid.
[59]   Ibid para 52.
[60]   Ibid paras 45–49.
[61]   In this regard, see also Schermers and Waelbroeck (n 43) 35–36.
[62]   The AG refers to para 19 of *Wachauf* and not to the above quoted paras 17 and 22.
[63]   Case C-166/98 *Socridis* [1999] ECR I-03791.
[64]   Council Directive 92/83/EEC of 19 October 1992 on the harmonisation of the structures of excise duties on alcohol and alcoholic beverages [1992] OJ L316 and Council Directive 92/84/EEC of 19 October 1992 on the approximation of the rates of excise duty on alcohol and alcoholic beverages [1992] OJ L316.

favoured wine production to the detriment of beer production.[65] The ECJ, referring to paragraph 22 of its judgment in *Wachauf*, held that:

> [T]he Court has consistently held that directives do not infringe the Treaty if they leave the Member States a sufficiently wide margin of appreciation to enable them to transpose them into national law in a manner consistent with the requirement of the Treaty

According to the ECJ, Member States retained such margin of appreciation since both Directives merely required them to introduce a minimum excise duty on beer.[66]

Even though *Socridis* dealt with the alleged invalidity of a Directive based on its incompatibility with the *Treaty*, the ECJ's statement could, in my view, be extrapolated to *the general principles of EU law*. As we have seen above, general principles of EU law have a status in the EU legal hierarchy which is at least equal to that of primary EU law. As secondary EU law has to conform to both primary EU law and the general principles of EU law, the ECJ's statement would also seem to apply when the validity of (a provision of) secondary EU law is challenged on the ground of its alleged incompatibility with a general principle of EU law. This view arguably is supported by the ECJ's reference to *Wachauf*, a case which centred on the conformity of secondary EU law with the general principles of EU law.

The above cases do not provide much clarity as to what exactly the ECJ considers to be a sufficiently wide margin of appreciation. However, based on the case law as well as on the Courts' tendency to interpret Regulations and Directives as much as possible in conformity with the general principles of EU law and the Treaty, the Courts cannot be expected to easily reach the conclusion that a Member State has insufficient discretion to avoid infringing the general principles of EU law.[67]

---

[65] Case C-166/98 *Socridis* (n 63) para 3. Article 95(2) TEC provided that 'no Member States shall impose on the products of other Member States any internal taxation of such a nature as to afford indirect protection to other products'.

[66] Ibid para 20.

[67] In this regard, Vandamme has argued that in the process of the substantive review of Directives, especially those which are in complex and technical areas, the Courts generally to a large extent respect the discretionary powers of the EU institutions. See Vandamme (n 47) 24–5. A good example in this regard is the review of EU acts based on their alleged incompatibility with the principle of proportionality. See, for instance, Marijn Holwerda, 'Subsidizing Carbon Capture and Storage Demonstration' (2010) 4 Carbon and Climate Law Review 228.

5.4 THE GENERAL EU LAW PRINCIPLE OF LOYALTY

Article 4(3) TEU provides that:

> [P]ursuant to the principle of sincere cooperation, the Union and the Member States shall, in full mutual respect, assist each other in carrying out the tasks which flow from the Treaties. The Member States shall take any appropriate measure, general or particular, to ensure fulfilment of the obligations arising out of the Treaties or resulting from the acts of the institutions of the Union. The Member States shall facilitate the achievement of the Union's tasks and refrain from any measure which could jeopardise the attainment of the Union's objectives

The principle of loyalty as incorporated in Article 4(3) TEU is formulated in terms of a positive and a negative obligation, to which the Member States are subject in their dealings with the EU and as between themselves.[68] In practice, Member States have a whole range of specific obligations under Article 4(3), such as the duties to adopt all necessary measures to ensure that a Directive is fully effective[69] and to refrain from adopting legislation that infringes EU competition rules.[70]

However, their duties under Article 4(3) are not restricted to obligations towards the EU (institutions). In the cases of *Mateucci* and *Athanasopoulos*, the ECJ stated that Member States are under the duty to assist each other in meeting their obligations under EU law.[71] They have the obligation to cooperate loyally and sincerely in the implementation of EU law.[72] Member States, in short, appear to have a mutual assistance obligation under Article 4(3).

---

[68] Koen Lenaerts and Piet Van Nuffel, *European Union Law* (3rd edn, Sweet & Maxwell 2011) 147. See also Case C-412/04 *Commission v. Italy* [2008] ECR I-00619, Opinion of AG Colomer, para 40.

[69] See Cases C-336/97 *Commission v. Italy* [1999] ECR I-03771, para 19, and C-178/05 *Commission v. Greece* [2007] ECR I-04185, para 25.

[70] For a full oversight of Member State obligations under Article 4(3) see, for instance, Schermers and Waelbroeck (n 43) 113–15 and Lenaerts and Van Nuffel (n 68) 149–53. For examples of specific Member State duties under Article 4(3), see, for instance, also Joined Cases 205 to 215/82 *Skimmed milk powder* [1983] ECR 02633, para 42; Case C-511/03 *Ten Kate* [2005] ECR I-08979, para 31; Case C-2/05 *Herbosch Kiere* [2006] I-01079, para 22, and Joined Cases C-231/06 to 233/06 *Jonkman* [2007] I-105149, para 37; Case C-91/08 *Wall AG* [2010] ECR I-02815, para 69 and Opinion 1/09 of the Court (Full Court), delivered on 8 March 2011, para 68.

[71] Cases 235/87 *Mateucci*, para 19 and C-251/89 *Athanasopoulos*, para 57 (n 14). See Amtenbrink and Vedder (n 14) 159.

[72] Schermers and Waelbroeck (n 43) 114. In this regard, see also Case C-73/08 *Bressol* [2010] ECR I-02735, Opinion of AG Sharpston, para 154.

According to Temple Lang, Article 4(3) TEU is relied on most often when a new legal issue or a new situation arises, where there is little or no relevant case law, and arguments are being based on general principles of law.[73] An examination of the case law on Article 4(3) learns that the provision has most often been used by Member States challenging the validity of EU acts under Article 263 TFEU.[74] However, private parties have also relied on Article 4(3) in actions for annulment of EU measures[75] and in preliminary reference procedures concerning national measures.[76] A number of actions for annulment of EU measures suggest that the chances of applicants successfully invoking Article 4(3) for challenging the validity of EU measures are not too great.[77]

Nevertheless, conclusions should not be drawn too easily from the case law on Article 4(3). A complicating factor is that the Courts very often base their judgments on Article 4(3) without expressly referring to the provision.[78] In general, the Courts do not always make express reference to general principles whenever they propound new rules of law.[79] Furthermore, when Article 4(3) is mentioned, it is almost always mentioned briefly, without discussion.[80] Yet, according to Temple Lang, the Courts do cite Article 4(3) when it is necessary to be very clear about the legal basis of an important and controversial conclusion.[81]

---

[73] John Temple Lang, 'The Development by the Court of Justice of the Duties of Cooperation of National Authorities and Community Institutions under Article 10 EC' (2008) 31 Fordham International Law Journal 1483, 1526.

[74] See e.g. Case C-242/97 *Belgium v. Commission* [2000] ECR I-03421, para 29; Case C-263/98 *Belgium v. Commission* [2001] ECR I-06063, para 32; Case C-332/00 *Belgium v. Commission* [2002] ECR I-03609, paras 49–50; Case C-382/99 *Netherlands v. Commission* [2002] ECR I-05163, para 21; Case C-512/99 *Germany v. Commission* [2003] ECR I-00845, para 20; Case C-344/01 *Germany v. Commission* [2004] ECR I-02081, Opinion of AG Léger, para 18; Case C-339/00 *Ireland v. Commission* [2003] ECR I-11757, paras 40 and 64; Case C-46/03 *UK v. Commission* [2005] ECR I-10167, Opinion of AG Stix-Hackl, para 61; and Case C-46/03 *UK v. Commission* [2005] ECR I-10167, para 27. Considering the fact that the principle of loyalty predominantly governs the relation between the Member States and the EU (institutions), this does not come as a surprise.

[75] See e.g. Case T-158/96 *Acciaierie di Bolzano* [1999] ECR II-03927, para 74; Joined Cases T-116/01 and T-118/01 *P&O European Ferries* [2003] ECR II-02957, para 198; Case T-340/04 *France Télécom* [2007] ECR II-00573, para 114 and Case T-60/05 *UFEX* [2007] ECR II-3397, para 165.

[76] See e.g. Case C-147/01 *Weber's Wine World* [2003] ECR I-11365, para 29; Joined Cases C-482/01 and C-493/01 *Orfanopoulos* [2004] ECR I-05257, Opinion of AG Stix-Hackl, para 69; Joined Cases C-392/04 and C-422/04 *i-21 Germany* [2006] ECR I-08559, para 43.

[77] In all four cases cited in footnote 75 the actions of the applicants were dismissed.

[78] John Temple Lang, 'The Duties of National Authorities and Courts under Article 10 EC: Two More Reflections' (2001) 26 European Law Review 84, 85. Temple Lang has argued that it therefore seems unrealistic and no longer useful to estimate the total number of judgments ultimately based on Article 4(3). See Temple Lang, 'The Development' (n 72) 1499.

[79] Hartley (n 43) 143.

[80] Temple Lang, 'The Development' (n 73) 1488.

[81] Temple Lang, 'The Duties' (n 78) 85.

Finally, it is worth mentioning that the principle of loyalty is closely related with – and sometimes equal to[82] – another central notion in the EU legal order, i.e. that of solidarity. The ECJ has in the past characterised the principle of loyalty as an expression of EU solidarity.[83] As argued by Wouters, the principle of loyalty can be seen as the (internal) institutional dimension of solidarity.[84] This notion of solidarity can also be found in the provision which is briefly discussed in the next section, i.e. Article 194(1) TFEU.

5.5   ARTICLE 194(1) TFEU

In the above, I argued that Article 194(1)(c) appears to require secondary EU climate and energy legislation, like the CCS Directive, to be drafted 'in a spirit of solidarity between Member States' and that it is questionable whether Article 21(2)(b) of the CCS Directive meets that requirement. In the following, I briefly explore the (legislative) background of Article 194(1) in order to have a better understanding of the intentions behind the provision. This will improve the analysis in the remainder of this chapter.

Article 194(1) TFEU states that:

> [I]n the context of the establishment and functioning of the internal market and with regard for the need to preserve and improve the environment, Union policy on energy shall aim, in a spirit of solidarity between Member States, to:
> (a) ensure the functioning of the energy market;
> (b) ensure security of supply in the Union;
> (c) promote energy efficiency and energy saving and the development of new and renewable forms of energy; and
> (d) promote the interconnection of networks

---

[82]   National courts sometimes refer to the principle of loyalty as the principle/duty of solidarity. See, for instance, Case C-453/00 *Kuhne & Heitz* [2004] ECR I-00837, Opinion of AG Léger, para 20.

[83]   Lenaerts and Van Nuffel (n 68) 147. Lenaerts and Van Nuffel refer to Joined Cases 6 and 11/69 *Commission v. France* [1969] ECR 00523, para 16, in which the ECJ mentioned 'the solidarity which is at the basis of these obligations as of the whole of the Community system in accordance with the undertaking provided for in Article 5 of the Treaty [now Article 4(3) TEU]'.

[84]   Jan Wouters, 'Constitutional Limits of Differentiation: The Principle of Equality' (2001) University of Leuven Institute for International Law Working Paper No 4 <https://www.law.kuleuven.be/iir/nl/onderzoek/wp/WP04e.pdf> accessed 25 July 2011. See also, Epaminondas A Marias, 'Solidarity as an Objective of the European Union and the European Community' (1994) 2 Legal Issues of European Integration 85, 98–99. See also Schermers and Waelbroeck (n 43) 112–15. Schermers and Waelbroeck, for instance, make a distinction between the duty of loyal and sincere cooperation on the one hand and solidarity on the other, the latter referring to the principle of solidarity between the Member States. This suggests that they see the principle of loyalty as a form of solidarity between the Member States and the EU.

The origin of Article 194(1) lies in the 2004 Treaty establishing a Constitution for Europe (Constitutional Treaty). Article III-256 of the Constitutional Treaty introduced a new provision on energy, the first paragraph of which read that:

> [I]n the context of the establishment and functioning of the internal market and with regard for the need to preserve and improve the environment, Union policy on energy shall aim to:
> (a) ensure the functioning of the energy market;
> (b) ensure security of supply in the Union, and
> (c) promote energy efficiency and energy saving and the development of new and renewable forms of energy

In 2005, as is well known, the Constitutional Treaty was rejected by both French and Dutch voters in national referenda held as part of the ratification process in both Member States. In the subsequent drafting of the Treaty of Lisbon, it was decided to stick to the introduction of a new title on energy in the TEC, but to insert a reference to 'the spirit of solidarity between Member States', as well as a new point (d) on the promotion of the interconnection of energy networks, into the text of Article III-256.[85]

Simultaneously, the decision was taken to amend Article 100 TEC (measures in case of severe difficulties in the supply of certain products) and to insert 'a reference to the spirit of solidarity between Member States and to the particular case of energy as regards difficulties in the supply of certain products'.[86] During the preparatory discussions on the June 2007 European Council conclusions in the General Affairs and External Relations Council (GAERC), a number of delegations had made several suggestions to amend the text of the Treaties in order to reflect more recent developments, including 'the need to address energy security'.[87]

The text of the amended first paragraph of Article III-256 of the Constitutional Treaty was to read that 'in the context of the establishment and functioning of the internal market and with regard for the need to preserve and improve the

---

[85] See point 19(m) of Annex I to the Presidency conclusions of the Brussels European Council (21/22 June 2007). See Council of the European Union, 'Brussels European Council 21/22 June 2007 – Presidency Conclusions' 11177/1/07 REV 1, 21 <http://register.consilium.europa.eu/pdf/en/07/st11/st11177-re01.en07.pdf> accessed 1 September 2011. Annex I to the June 2007 European Council conclusions contained the mandate for the 2007 Intergovernmental Conference (IGC), which was to draft the later Treaty of Lisbon.

[86] See point 19(q) of Annex 1.

[87] Council of the European Union, 'Pursuing the Treaty Reform Process' 10659/07 POLGEN 67, 4 <http://register.consilium.europa.eu/pdf/en/07/st10/st10659.en07.pdf> accessed 1 September 2011.

environment, Union policy on energy shall aim, *in a spirit of solidarity between Member States*, to (…)'.[88]

Likewise, paragraph 1 of Article 100 TEC was to be replaced by the following wording:

[W]ithout prejudice to any other procedures provided for in the Treaties, the Council, on a proposal from the Commission, may decide, *in a spirit of solidarity between Member States*, upon the measures appropriate to the economic situation, in particular if severe difficulties arise in the supply of certain products, *notably in the area of energy*'[89]

The final Treaty of Lisbon introduced a new Article 176 A to the TEC, which was renumbered to Article 194 TFEU. In a declaration to its final act of 17 December 2007, the Intergovernmental Conference that drafted the Treaty of Lisbon stated that 'the conference believes that Article 176 A does not affect the right of Member States to ensure their energy supply under the conditions provided for in Article 297[90] (TEC)'.[91]

As the preparatory documents of Article 194 reveal, the provision's solidarity clause was introduced to address the concerns of a number of Member States about security of energy supply. Even though the documents reviewed show no such evidence, it is likely that the Russian–Ukrainian gas disputes, which reached a climax at the beginning of 2006, played an important role in this regard.[92] In the next section, the conformity of Article 21(2)(b) with the general EU law principle of loyalty and Article 194(1)(c) TFEU is analysed.

---

[88] Council of the European Union, 'Brussels European Council 21/22 June 2007' (n 85) 29 (emphasis added).

[89] Ibid.

[90] Now Article 347 TFEU. Article 297 TEC provided that 'Member States shall consult each other with a view to taking together the steps needed to prevent the functioning of the common market being affected by measures which a Member State may be called upon to take in the event of serious internal disturbances affecting the maintenance of law and order, in the event of war, serious international tension constituting a threat of war, or in order to carry out obligations it has accepted for the purpose of maintaining peace and international security'.

[91] Declaration 35 of the Final Act of the conference of the representatives of the governments of the Member States 2007/C 306/02 <http://eur-lex.europa.eu/LexUriServ/LexUriServ.do?uri=OJ:C:2007:306:0231:0271:EN:PDF> accessed on September 2011.

[92] See also, David Benson and Andrew Jordan, 'A Grand Bargain or an "Incomplete Contract"? European Union Environmental Policy after the Lisbon Treaty' (2008) 17 European Energy and Environmental Law Review 280, 284. In Member States like Poland and the Baltic countries, the issue of energy security of supply and the relationship with Russia is politically sensitive.

5.6 THE CONFORMITY OF ARTICLE 21(2)(B) CCS DIRECTIVE WITH ARTICLES 4(3) TEU AND 194(1)(C) TFEU

In section 5.3, we have seen that the Courts will likely not find (a provision of) a Directive to infringe a general principle of EU law if Member States are left sufficient discretion to implement it in a way that does not infringe the relevant general principle of EU law. As indicated earlier, Article 21(2)(b) provides that a Member State, in designing its regime for access of third parties to CCS infrastructure, takes into account 'the proportion of its CO_2 reduction obligations pursuant to international legal instruments and to Community legislation that it intends to meet through capture and geological storage of CO_2'.

Article 21(2)(b) provides that Member States are free to design their third-party access regime for CO_2 transport and storage infrastructure in any way they see fit, as long as such access is provided in a fair, open and non-discriminatory manner and the proportion of their international and EU greenhouse gas[93] reduction commitments that they intend to meet through CCS is taken into account. As argued before, the wording of Article 21(2)(b) is not very precise. The provision seems to require Member States to take into account their own (anticipated) need for national transport and storage capacity when developing a regime for third-party access to CO_2 transport and storage infrastructure. Yet Article 21(2)(b) does not specify what is to be understood by the 'taking into account' of a Member State's own need of national transport and storage capacity.

Based on the wording of the provision, I would argue that Article 21(2)(b) appears to give Member States the *possibility* to require the operators of CO_2 transport and storage infrastructure to refuse to grant access to the relevant infrastructure when the capacity concerned is needed to meet part of the Member State's international and EU obligations to reduce greenhouse gas emissions.[94] Article 21(2)(b) does not, however, seem to oblige Member States to

---

[93] See n 26.
[94] In relation to access to upstream gas pipeline networks, European Parliament and Council Directive 2009/73/EC concerning common rules for the internal market in natural gas and repealing Directive 2003/55/EC [2009] OJ L211 (Gas Directive) contains a similar provision. Article 34(2)(c) of that Directive determines that Member States, when designing their third-party access regime to upstream gas infrastructure, may take into account 'the need to respect the duly substantiated reasonable needs of the owner or operator of the upstream pipeline network for the transport and processing of gas and the interests of all other users of the upstream pipeline network or relevant processing or handling facilities who may be affected'. An almost identical provision can be found in Article 21(2)(d) of the CCS Directive, which provides that the Member States take into account 'the need to respect the duly substantiated reasonable needs of the owner or operator of the storage site or of the transport

make sure that access to CO_2 transport and storage infrastructure is refused in the case that a Member State needs the capacity concerned for meeting part of its international/EU greenhouse gas reduction obligations. The considerations mentioned in paragraphs (a) to (d) of Article 21(2) instead appear to list a number of possible grounds for refusal of third-party access to CO_2 transport and storage infrastructure to be incorporated in the national third-party access regime.

As Member States have the possibility to require CO_2 transport and storage operators to refuse to grant access to the relevant infrastructure when the capacity concerned is needed to meet part of a Member State's international and EU reduction obligations, Article 21(2)(b) arguably leaves Member States sufficient discretion to implement the provision in a way that would not infringe Article 4(3) TEU. Accordingly, the Courts would likely not find Article 21(2)(b) to infringe Article 4(3) TEU in conjunction with Article 194(1)(c) TFEU.

Yet, this does not mean that the role of the latter two provisions is played out. As I have argued earlier, national implementing measures have to respect the general principles of EU law as well as the Treaty provisions. In the following section, we explore the boundaries set by Articles 4(3) and 194(1)(c) for Member State implementation of Article 21(2)(b).

5.7 ARTICLES 4(3) TEU AND 194(1)(C) TFEU AND MEMBER STATES' IMPLEMENTATION OF ARTICLE 21(2)(B) CCS DIRECTIVE

As stated in the Introduction, one way to implement Article 21(2)(b) is to adopt legislation dedicating certain (or perhaps even all) CO_2 transport and storage infrastructure for internal (national) use only. At least part of the capacity of this infrastructure will then be reserved for the transport and storage of CO_2 captured within the Member State concerned. The consequent (partial) unavailability of CO_2 transport and storage capacity in some Member States could lead to problems for other Member States in trying to meet their national targets for the reduction of greenhouse gas emissions. For certain industries such as the cement industry, CCS might well be the only viable means of achieving deep emissions cuts. As potential CO_2 storage capacity is unevenly spread over Europe and certain Member States have significantly less potential storage capacity than others, access to foreign CO_2 transport and storage infrastructure might be crucial for these Member States to achieve deep national emissions reductions.

---

network and the interests of all other users of the storage or the network or relevant processing or handling facilities who may be affected'.

In the 2009 GeoCapacity[95] report referred to in the Introduction, European CO_2 storage capacity estimates are given.[96] The GeoCapacity report contains only one case study crossing national borders (Estonia – Latvia). Yet, several of the capacity estimates suggest that certain Member States might sooner or later need to have access to other Member States' CO_2 transport and storage infrastructure, should they want to deploy CCS on a large-scale.

Greece's conservatively estimated total potential CO_2 storage capacity, for instance, would be sufficient for storing three and a half years' of national CO_2 emissions from large point sources.[97] Older estimates suggest that Belgium might be in a similar position.[98] The Czech Republic and Slovenia have a conservatively estimated potential storage capacity that suffices for storing 11 and 13 years' of national CO_2 emissions from large point sources respectively. Even a country like Poland has a (conservatively) estimated total potential storage capacity that would suffice to store some 15 years' worth of national CO_2 emissions from large point sources only.

The uneven distribution of potential CO_2 storage capacity in Europe is further illustrated by the finding that of the conservatively estimated total EU CO_2 storage capacity in deep saline aquifers[99] (95724 Mt), more than half is located in three countries (Germany, Spain and Norway).[100] Considering the to date very limited public acceptance of onshore geological storage of CO_2, the necessity for certain Member States to have access to foreign transport and ultimately offshore storage capacity might even be bigger than these numbers suggest. In this respect, the possibility for Member States to require the operators of CO_2 transport and storage infrastructure to refuse to grant access to the relevant infrastructure on the basis of own anticipated use could prove a significant hindrance to the cross-border deployment of CCS in the EU.

5.7.1   ARTICLE 4(3) TEU

As we have seen in section 5.4, the principle of loyalty in Article 4(3) TEU also puts obligations on Member States vis-à-vis each other. In the cases of *Mateucci* and *Athanasopoulos*, the ECJ ruled that Member States are under the duty to assist each other in meeting their obligations under EU law. They have the

---

[95]   The EU GeoCapacity project is an EU-funded research project assessing the European capacity for the geological storage of CO_2.

[96]   Vangkilde-Pedersen and others (n 5).

[97]   Ibid 158.

[98]   Ibid.

[99]   See section 1.4.1.3 of Chapter I.

[100]   Vangkilde-Pedersen and others (n 5) 158.

obligation to cooperate loyally and sincerely in the implementation of EU law. At first sight, the mutual assistance obligation under Article 4(3) seems to significantly narrow the scope of Member States in the implementation of Article 21(2)(b) of the CCS Directive. It could be argued that national implementing measures requiring CO_2 transport and storage operators to refuse to grant access on the ground mentioned in Article 21(2)(b) are incompatible with the mutual assistance obligation under Article 4(3). In a situation where a capturer from another Member State needs access to a Member State's CO_2 transport and/or storage infrastructure, such measures could, at first glance, lead to the latter not meeting its obligations under Article 4(3).

However, a closer look at both cases indicates that this preliminary conclusion may not necessarily be warranted. In *Mateucci*, the ECJ essentially stated that a Member State infringes Article 4(3) TEU where it acts in compliance with a bilateral agreement signed with another Member State and, in doing so, prevents the other Member State from complying with its obligations under EU law.[101] In *Athanasopoulos*, the ECJ held that Article 4(3) TEU obliges a Member State to cooperate in good faith with the institutions of another Member State where those institutions are responsible for complying with obligations arising out of secondary EU law.[102] Both cases show that the mutual assistance obligation requires the Member States to either stop acting in a way that prevents another Member State from complying with its obligations under EU law or to start acting in a way that enables another Member State to meet its EU obligations. The crucial aspect in this regard appears to be the necessity for another Member State to meet its obligations under EU law.

However, this aspect seems to be lacking in the scenario sketched above. When a capturer seeks access to transport and/or storage infrastructure in another Member State for meeting a *domestic* greenhouse gas emission reduction target, no host state *EU* obligations seem to be involved. The only exception would probably be the situation in which the national reduction target contributes towards meeting an EU reduction target. The latter, however, is unlikely to be the case for the part of the national target covering the sectors that will likely deploy CCS.

As argued in section 1.1 of the first chapter, the installations most likely to capture CO_2 are those participating in the EU ETS. Since the beginning of 2013, the EU ETS has a single EU-wide cap. This single EU-wide cap guarantees that the equivalent emission reductions are realised,[103] regardless of any national

---

[101]   Case 235/87 *Mateucci* (n 14) paras 12–23.
[102]   Case C-251/89 *Athanasopoulos* (n 14) para 57.
[103]   The mechanism to ensure that operators do not exceed the emissions cap is incorporated in Article 16(3) of European Parliament and Council Directive 2003/87/EC of 13 October 2003 establishing a scheme for greenhouse gas emission allowance trading within the Community

reduction targets for the sectors covered by the EU ETS.[104] Therefore, Member States arguably had EU emission reduction obligations for those installations which are likely to apply CCS until 2013 only.[105]

Furthermore, Member States are under no EU obligation to deploy CCS activities. On the contrary, the CCS Directive explicitly provides that Member States have the right 'not to allow any storage in parts or on the whole of their territory, or to give priority to any other use of the underground, such as exploration, production and storage of hydrocarbons or geothermal use of aquifers'.[106]

As no EU obligations are involved since the beginning of 2013, a Member State adopting national measures to implement Article 21(2)(b) cannot be said to act in a way that prevents another Member State from meeting its obligations under EU law or, alternatively, to refrain from acting in a way that enables another Member State to meet its obligations. Therefore, it would arguably not breach the requirements in Article 4(3) TEU.

Should EU obligations nonetheless be involved, the mutual assistance obligation under Article 4(3) might arguably oblige the receiving Member State to require the operator refusing access to expand capacity when it is economically sensible to do so or when the potential customer is willing to pay for that expansion. Article 21(4) of the CCS Directive provides that Member States shall take the

---

and amending Council Directive 96/61/EC (EU ETS Directive), which provides that operators who do not surrender sufficient allowances by 30 April of each year to cover their emissions during the preceding year shall be held liable for the payment of an excess emission penalty of EUR 100 for each tonne of CO_2 equivalent emitted for which they have not surrendered allowances.

[104] By contrast, national reduction targets for sectors not covered under the EU ETS, the so-called non-ETS sectors (agriculture, building, transport and waste) *do* contribute towards meeting the EU target for these sectors of a 10% emissions reduction (compared to 2005 levels) by 2020. Article 3(1) of the Effort sharing Decision provides that the Member States must limit their greenhouse gas emissions by at least the percentage set for that Member State in Annex II to the decision. For the non-ETS sectors, Member States clearly do have EU obligations to reduce national greenhouse gas emissions.

[105] These obligations consist of the emission reduction target resulting from the national allocation plans (NAPs), which Member States had to submit in first and second phases of the EU ETS (2005–2008, 2008–2012). Article 9(1) of the 'old' EU ETS Directive (Directive 2003/87/EC prior to having been amended by European Parliament and Council Directive 2009/29/EC of 23 April 2009 amending Directive 2003/87/EC so as to improve and extend the greenhouse gas emission allowance trading scheme of the Community) provided that Member States had to develop a national plan stating the total quantity of allowances (national emissions cap) that they intended to allocate for the relevant period. By virtue of Article 9(3) of the Directive, the Commission could reject the national allocation plan.

[106] See recital 19 of the preamble to the CCS Directive and Article 4(1) of the Directive. On the Member States sovereignty in relation to the storage of CO_2 in their territory, see further Chapter IX, which deals with EU law and the possibilities to force Member States to allow CO_2 storage in their territory.

measures necessary to ensure that the operator refusing access on the grounds of lack of capacity or a lack of connection makes any necessary enhancements as far as it is economically sensible to do so or when a potential customer is willing to pay for them, provided that this would not negatively impact on the environmental security of the transport and geological storage of CO_2.[107]

In relation to CO_2 transport infrastructure, capacity expansion would appear to be relatively straightforward; either existing pipeline and compression infrastructure can be modified or new infrastructure can be constructed. However, with regard to CO_2 storage infrastructure things might be more difficult. As the capacity of any given CO_2 storage reservoir is fixed, capacity enlargement can only take place through the development of a new storage reservoir. Considering the high costs involved with storage site characterisation, it might not automatically be economically sensible to develop a new storage reservoir and it can be questioned whether a potential customer would be willing to pay for such costs.

Article 21(4) refers to Article 21(3) of the CCS Directive, which provides that transport network operators and operators of storage sites may refuse access on the grounds of lack of capacity. Article 21(4) does not refer to the grounds for refusal of access in Article 21(2). However, it could be argued that the mutual assistance obligation under Article 4(3) TEU likewise obliges Member States to ensure that an operator refusing third-party access on the ground mentioned in Article 21(2)(b) expands capacity when it is economically sensible to do so or when the potential customer is willing to pay for that expansion. The ground for refusal of access in Article 21(2)(b) is, after all, based on a lack of capacity.

One way to prevent the situation that third-party access to CO_2 transport and/or storage infrastructure needs to be refused due to lack of capacity is to require project developers to determine the level of customer interest by 'market testing' the demand for new capacity ('open season procedure').[108] Infrastructure can then be developed in such a way as to accommodate expected future demand.[109]

---

[107] A similar provision can be found in Directive 2009/72/EC (Gas Directive – Article 35(2)).

[108] 'Open season' refers to the requirement to publish plans of potential infrastructure projects to allow third parties to express interest in access to new infrastructure. See NERA Economic Consulting, 'Developing a Regulatory Framework for CCS Transportation Infrastructure (Vol. 2 of 2): Case Studies' (2009) 16 <www.decc.gov.uk/assets/decc/What%20we%20do/ UK%20energy%20supply/Energy%20mix/Carbon%20capture%20and%20storage/1_200906 17131350_e_@@_ccsreg2.pdf> accessed 3 October 2011. In this regard, see further Chapter VII on the development and management of CO_2 transport infrastructure and EU antitrust law.

[109] NERA Economic Consulting (n 108) 50. NERA recommend imposing a requirement that all developers of new CO_2 pipelines hold open seasons. According to NERA, the obligation to hold open seasons would facilitate the formation of coalitions to exploit economies of scale in

In that way, the ground for refusal in Article 21(2)(b) does not have to be used. Obviously, the mutual assistance obligation can require project developers to hold an open season procedure through the Member States only; private parties not being addressees of any of the Treaty provisions, Article 4(3) cannot directly require project developers to do so.

5.7.2 ARTICLE 194(1)(C) TFEU

Similarly, it is unlikely that Article 194(1)(c) TFEU narrows Member States' scope when implementing Article 21(2)(b) into national law. Article 194(1)(c) requires EU energy policy to promote energy efficiency and energy saving and the development of new and renewable forms of energy, in a spirit of solidarity between Member States. The provision appears to be directed towards those drafting EU energy policy – the EU legislature – and not towards the Member States.[110] It seems to require EU energy policy to reflect a spirit of solidarity, imposing on those who draft EU energy policy and legislation the obligation to compose texts that are characterised by such a spirit. Accordingly, Article 194(1)(c) does not seem to narrow the Member States' scope for implementing Article 21(2)(b) into national law.

Even if Article 194(1)(c) *can* be read as imposing obligations on the Member States, it is unclear what these obligations entail. It could, for instance, be questioned whether the article's solidarity clause imposes obligations in the area of EU climate and energy policies in the first place. While a literal reading of Article 194(1)(c) suggests that the phrase 'in a spirit of solidarity between Member States' likewise applies to EU energy policy aiming to promote energy efficiency and the development of renewable energy, it is questionable whether such reading accurately reflects the intentions of the EU legislature. As we have seen in section 5.5, the article's solidarity clause was introduced to address several Member States' concerns over energy security of supply. As such, it has a strong security of supply connotation.

Notwithstanding the legislative history of Article 194(1)(c), the concept of solidarity can be said to generally be embedded in EU climate and energy policies/legislation. Recital 8 of the preamble to the Effort sharing decision, for instance, indicates that the different national emission reduction targets imposed

---

the provision of CO_2 pipelines and inform the market of potential pipeline developments and to facilitate the widest possible participation among those willing to commit funds to the project.

[110] Notwithstanding the Member States' being part of the legislature through the Council. Yet what is meant here is that the Member States are not being addressed in their capacity of entities implementing secondary EU legislation.

by it are based 'on *the principle of solidarity between Member* States and the need for sustainable economic growth across the Community, taking into account the relative per capita GDP of Member States'.[111]

Similarly, the new Article 10(2)(b) of the EU ETS Directive (auctioning of allowances)[112] provides that the total quantity of allowances to be auctioned by each Member State shall be composed by having, among other things, '10% of the total quantity of allowances to be auctioned being distributed amongst certain Member States *for the purpose of solidarity* and growth within the Community'.[113] Apparently, the EU legislature does in practice take the concept of solidarity into account when drafting EU climate and energy legislation.

Apart from the strong security of supply background of Article 194(1)(c), it can be questioned whether the phrase 'to promote energy efficiency and energy saving and the development of new and renewable forms of energy' should be understood to cover climate change mitigation instruments such as CCS. On the one hand, it could be argued that energy efficiency and renewable energy are mere examples of the range of possible policy instruments – including CCS – that can be used for tackling climate change. On the other, it could be asked why the drafters of Article 194(1)(c) did not use more general words if they wanted to refer to climate and energy policy instruments in general.

What is more, it seems difficult to maintain that CCS policies fall under the heading of either energy efficiency and energy saving or the development of new and renewable forms of energy. As indicated in section 1.1.5 of the first chapter, CO_2 capture technologies lead to increased energy use instead of energy saving. Furthermore, CCS technologies do not appear to stimulate the development of new and renewable forms of energy. On the contrary, CCS has often been criticised for enabling the prolonged use of 'old' forms of energy, such as electricity production by means of coal combustion.[114]

Finally, as Article 194(1)(c) does not clarify what is meant by the phrase 'in a spirit of solidarity between Member States', it is very difficult, perhaps even impossible, to determine when this requirement is met. The phrase can mean many different things, depending on one's definition of the term solidarity. Andoura, Hancher and van der Woude have argued that without any definition

---

[111]   Emphasis added.
[112]   The provision was introduced by Article 1 of Directive 2009/29/EC.
[113]   Emphasis added.
[114]   See e.g. Greenpeace, 'Greenpeace Q&A on Carbon Capture and Storage (CCS)' <www.greenpeace.org/international/Global/international/planet-2/report/2008/5/q-a-on-carbon-capture-storag.pdf> accessed 15 September 2011.

of the principle of solidarity, or any guidance on how to apply it when developing a new energy policy, it remains unclear whether the solidarity clause in Article 194 TFEU will receive any application in practice, or whether any concrete obligation will derive from it for the EU and the Member States.[115]

The lack of definition of the term in Article 194(1) is exacerbated by solidarity arguably being one of the most elusive concepts of EU law.[116] Accordingly, scholars adopt different usages of solidarity and express considerable variations in confidence as to its value and likely impacts.[117] Borgmann-Prebil and Ross have contended that:

> [S]olidarity as an idea, let alone in any distinctively European version, has proved notoriously difficult to pin down despite being a popular target for inquiry across many disciplines and a recurrent focus for the rhetoric of political struggle or ideology.[118]

Based on the above, it is unclear in what way Article 194(1)(c) could narrow Member State's scope when implementing Article 21(2)(b) into national law. Apart from Articles 4(3) and 194(1)(c) apparently not narrowing Member States' scope for implementing Article 21(2)(b), there also seem to be procedural hurdles for challenging the validity of national measures implementation measures on the ground of an infringement of both provisions. Articles 4(3) and 194(1)(c), arguably, cannot be invoked by individuals to challenge the validity of national measures implementing Article 21(2)(b) before a national court. Before

---

[115] Sami Andoura, Leigh Hancher and Marc van der Woude, 'Towards a European Energy Community: A Policy Proposal' (2010) 98–9 <www.notre-europe.eu/uploads/tx_publication/Etud76-Energy-en.pdf> accessed 5 September 2011. Andoura, Hancher and van der Woude further state that the Lisbon Treaty's new energy policy does not offer prospect of a radical change from the situation under the TEC.

[116] Both Treaties contain many references to solidarity, without clearly defining the concept. See e.g. Article 2 TEU, which mentions the values on which the EU is founded, among which solidarity. See also Article 3(2) TEU (among other things solidarity between generations and between Member States in general), Articles 21 EU (solidarity in EU's external action), Articles 24, 31 and 32 TEU (solidarity between Member States in Common Foreign and Security Policy (CFSP)), Articles 67 and 80 TFEU (solidarity in asylum, immigration and border control-policies) and Article 222 TFEU (solidarity in the event of a terrorist attack or man-made or natural disaster). For other connotations of solidarity in the EU legal order see e.g. Egle Dagilyte, 'Solidarity after the Lisbon Treaty: A new General Principle of EU Law?' 1–2 <www.pravo.hr/_download/repository/Dagilyte_abstract.pdf> accessed on 23 October 2012.

[117] Yuri Borgmann-Prebil and Malcolm Ross, 'Promoting European Solidarity: Between Rhetoric and Reality?' in Malcolm Ross and Yuri Borgmann-Prebil (eds), *Promoting Solidarity in the European Union* (Oxford University Press 2010) 1, 20.

[118] Ibid 1.

indicating why both provisions are likely not to have direct effect,[119] I briefly turn to the concept of the direct effect of Treaty provisions.

5.7.3   THE DIRECT EFFECT OF TREATY PROVISIONS

The doctrine of direct effect allows EU citizens to invoke EU law before a national court.[120] It can be traced back to the ECJ judgment in the *Van Gend en Loos* case.[121] The essential question in *Van Gend en Loos* was whether ex Article 12 of the Treaty Establishing the European Economic Community (TEEC – now Article 30 TFEU)[122] could be invoked by Member State nationals in national legal proceedings. The ECJ held that according to the spirit, general scheme and wording of the Treaty, Article 12 TEEC had to be interpreted as producing direct effects and creating individual rights which national courts must protect.[123] Assessing these three aspects, it developed a number of criteria to be met for a Treaty provision to have direct effect. The provision must be clear, negative ('shall not'), unconditional, contain no reservation on the part of the Member State and not be dependent on any national implementing measure.[124]

In a number of cases following *Van Gend en Loos*, the ECJ is said to have relaxed the criteria for a Treaty provision to have direct effect.[125] In the cases of *Salgoil*, *Van Duyn*, *Reyners* and *Defrenne*, the ECJ relaxed the requirements of clarity, unconditionality and negative phrasing.[126] In the latter two cases, the ECJ did so by identifying and isolating the principle behind the relevant Treaty provision.

---

[119]   The doctrine of direct effect allows Union citizens, when the applicable provision of EU law meets a number of conditions, to invoke that provision in the national legal order.

[120]   The concept of direct effect is often confused with that of 'direct applicability'. The latter refers to the situation in which a provision of EU law penetrates directly into the legal orders of the Member States, without action by national bodies being required to give effect to it. The case law of the Courts establishes that direct applicability is by no means a necessary condition for direct effect. See Anthony Arnull, *The European Union and its Court of Justice* (2nd edn, Oxford University Press 2006) 185–186.

[121]   Case 26/62 *Van Gend en Loos* [1963] ECR 00001.

[122]   The provision prohibited between Member States customs duties on imports and exports and all charges having equivalent effect.

[123]   Case 26/62 *Van Gend en Loos* (n 121) 13.

[124]   Paul Craig and Grainne de Búrca, *EU Law: Text, Cases, and Materials* (4th edn, Oxford University Press 2007) 275.

[125]   Ibid 275–77 and Damian Chalmers, Gareth Davies and Giorgio Monti, *European Union Law* (2nd edn, Oxford University Press 2010) 271–72.

[126]   Case 13/68 *Salgoil* [1968] ECR 00453; Case 41/74 *Van Duyn* [1974] ECR 01337; Case 2/74 *Reyners* [1974] ECR 00631; and Case 43/75 *Defrenne* [1976] ECR 00455. In this regard, see Craig and de Búrca (n 124) 275–77 and Chalmers and others (n 124) 271–72.

Following these cases, the case of *Zaera* demonstrated that the ECJ's willingness to creatively deal with the requirements in *Van Gend en Loos* is not infinite.[127] In *Zaera*, a Spanish national sought to rely on ex Article 2 TEEC, which provided inter alia that 'the Community shall have as its task (...) an accelerated raising of the standard of living', to challenge a rule of national law which prohibited the overlapping of a retirement pension with his employment in the public service.[128] The ECJ stated that:

> [A]rticle 2 of the Treaty describes the task of the European Economic Community. The aims laid down in that provision are concerned with the existence and functioning of the Community; they are to be achieved through the establishment of the Common Market and the progressive approximation of the economic policies of Member States, which are also aims whose implementation is the essential object of the Treaty. With regard to the promotion of an accelerated standard of living, in particular, it should therefore be stated that this was one of the aims which inspired the creation of the European Economic Community and which, *owing to its general terms and its systematic dependence on the establishment of the Common Market and progressive approximation of economic policies, cannot impose legal obligations on Member States or confer rights on individuals.*[129]

This extract from the ECJ's ruling indicates that there were two declared reasons for the ECJ not to render Article 2 TEEC direct effect. First, the promotion of an accelerated standard of living was dependent on and to be realised through further economic integration; the relevant phrase in Article 2 TEEC was not unconditional. Second, the obligation in Article 2 TEEC was worded in general terms and was therefore, not precise enough. Yet, in the above-mentioned cases, the conditionality and the lack of precision did not prevent the ECJ from rendering the relevant provisions direct effect. In my opinion, the underlying reason for the ECJ not to declare Article 2 TEEC to have direct effect was a different one: as one of the declared overarching objectives of the EU, the provision was simply not meant to confer rights on individuals.[130]

Political objectives such as those incorporated in Article 2 TEEC are intended to give political direction, which is illustrated by their usually being placed at the beginning of Treaty texts.[131] As AG Mancini held in his Opinion in Zaera, the

---

[127] Case 126/86 *Zaera* [1987] ECR 03697.
[128] See Paul Craig and G de Búrca, *EU Law: Text, Cases, and Materials* (3rd edn, Oxford University Press 2003) 188–89.
[129] Case 126/86 *Zaera* (n 127) paras 10–11 (emphasis added).
[130] The ECJ acknowledges this in paras 10–11 of its judgment.
[131] In this regard, see also the Opinion of AG Mancini in *Zaera*, who points out that the relevant phrase in Article 2 TEEC repeated the undertaking contained in the third recital of the preamble to the Treaty. See Case 126–86 *Zaera* [1987] ECR 03697, Opinion of AG Mancini, para 10.

provision concerned contained 'expressions of intent, purpose and motive, rather than rules that are of direct operative effect'.[132] In this way, Article 2 TEEC differed from the relevant provisions in *Van Gend en Loos*, *Salgoil*, *Van Duyn*, *Reyners* and *Defrenne*, which were all, except for Article 141 TEC in *Defrenne*,[133] free movement provisions. Taking into mind that the Courts have traditionally been particularly progressive in this area, the outcome of *Van Gend en Loos*, *Salgoil*, *Van Duyn* and *Reyners*, arguably, is not surprising.

Chalmers et al. have argued that the ECJ, since *Defrenne*, has further moved away from using the criteria set out in *Van Gend en Loos*.[134] According to them, several cases[135] indicate that the test has now become whether the substance of the provision is sufficiently precise and unconditional.[136] Yet, as Craig and de Búrca argue, the test for direct effect of Treaty provisions seems to include the requirement that the relevant provision is intended to confer rights on individuals.[137] In the above-mentioned cases, this arguably was the underlying criterion for determining whether the relevant Treaty provision had direct effect. Even though it did not explicitly state so, the ECJ was, in my opinion, willing to ignore some of the *Van Gend en Loos* criteria in *Salgoil*, *Van Duyn*, *Reyners* and *Defrenne*, since it ultimately considered the relevant provisions to confer rights on individuals.

These cases have in common that the ECJ would have risked creating a lacuna in the legal protection of individuals in areas which it has traditionally valued highly (free movement and equal treatment). As it considered the relevant provisions to confer rights on individuals, the latter had to be able to invoke these rights before a national court. I would therefore argue that in order for a Treaty provision to have direct effect, it should be sufficiently precise, unconditional *and* intended to confer rights on individuals. When assessing the possible direct effect of a Treaty provision, it is important to take notice of the ECJ's practice of focussing on the basic principle behind the relevant provision.

---

132 Case 126–86 *Zaera* Opinion of AG Mancini (n 131) para 10.
133 Even though the wording of Article 141 TEC might perhaps not have been that precise, the provision can arguably, much more than Article 2 TEEC, be said to have been intended to create rights for individuals: the right for men and women to receive equal pay for equal work.
134 Chalmers and others (n 125) 272.
135 Case 148/78 *Ratti* [1979] ECR 01629; Joined Cases C-397/01–403/01 *Pfeiffer* [2004] ECR I-08835; and Joined Cases C-152/07–154/07 *Arcor* [2008] ECR I-05959.
136 Chalmers and others (n 125) 272–73. The three cases cited by Chalmers and others concerned the direct effect of a provision in a Directive and not, like the above-cited cases, of a Treaty provision.
137 Craig and de Búrca (n 124) 277.

5.7.4 THE DIRECT EFFECT OF ARTICLE 4(3) TEU

We have seen that for a Treaty provision to have direct effect, it needs to be sufficiently precise, unconditional, and intended to confer rights on individuals. In the following, I assess whether Article 4(3) meets these criteria. Before doing so, I briefly examine the scholarly debate on the direct effect of Article 4(3).

5.7.4.1 *The scholarly debate*

Article 4(3) is generally said not to have direct effect in itself.[138] According to Temple Lang, the provision itself never creates duties,[139] but only together with some other rule of EU law, or some principle or objective of EU policy which is to be facilitated or, at least, not jeopardised.[140] As the ECJ indicated in the *Deutsche Grammophon* case,[141] the legal consequences of Article 4(3) depend on the rule of EU law with which it is combined.[142] Temple Lang has argued that Article 4(3) has direct effects only when the other rule of EU law has direct effect.[143] If the other rule does not have direct effect, Article 4(3) may create some direct effects but does not necessarily do so.[144]

Temple Lang has contended that Article 4(3) applies only in combination with some other rule or policy determining the objectives to be achieved, which is incomplete and insufficient in itself.[145] If the provision in question is comprehensive enough, there is no need to rely on Article 4(3).[146] In support, Temple Lang has referred to four cases.[147] In three of those four cases, the ECJ consistently ruled that there were no grounds for holding that there had been a failure to fulfil the general obligations in Article 4(3) TEU separate from a previously noted failure to fulfil more specific EU obligations under secondary

---

[138] See e.g. Lenaerts and Van Nuffel (n 68) 148 and Temple Lang, 'The Development' (n 73) 1517–18.

[139] Temple Lang argues that the terms of the article make this clear. See Temple Lang, 'The Duties' (n 78) 91. See also Laurence W Gormley, 'The Development of General Principles of Law Within Article 10 (ex Article 5) EC' in Ulf Bernitz and Joakim Nergelius (eds), *General Principles of European Community Law* (Kluwer Law International, 2000) 113, 114 and 115.

[140] Temple Lang, 'The Development' (n 73) 1517.

[141] Case 78/70 *Deutsche Grammaphon* [1971] ECR 00487, para 5.

[142] Temple Lang, 'The Development' (n 73) 1518.

[143] Ibid. Temple Lang used the term direct *applicability* here, but he probably meant to refer to the concept of direct *effect*.

[144] Temple Lang, 'The Development' (n 73) 1518.

[145] Temple Lang, 'The Duties' (n 78) 91.

[146] Temple Lang, 'The Development' (n 73) 1517.

[147] These are: Case C-392/02 *Commission v. Denmark* [2005] ECR I-09811, Case C-464/02 *Commission v. Denmark* [2005] ECR I-07929, Case C-84/04 *Commission v. Portugal* [2006] ECR I-09843, and Case T-60/05 *UFEX* [2007] ECR II-03397.

EU legislation.[148] Therefore, these cases seem to warrant the conclusion that there is no need to rely on a breach of Article 4(3) if there is an apparent breach of a more specific obligation under secondary EU legislation.

However, it could be questioned whether it is right to conclude from these cases that the provision applies in combination with some other rule or principle *only*. In the three cases mentioned earlier, the ECJ did not explicitly say so. What is more, I doubt whether the ECJ's statement in these cases is to be interpreted in that way. Rather than declaring a general rule of application of Article 4(3), the ECJ seems to have merely stated the obvious: if a breach of a more specific provision of secondary law has been found, it is not necessary to investigate a possible separate breach of Article 4(3).

This reading appears to be supported by AG Trstenjak's interpretation of the ECJ's approach in one of these cases, as provided in her Opinion in Case C-19/05 (*Commission v. Denmark*).[149] According to the AG, the idea behind the ECJ's statement was that Article 4(3) is a general provision relative to Decision 94/728/EC:[150]

> [W]here an infringement of the special provision is established, it is not necessary also to establish the infringement of a general provision. In accordance with the principle *specialia generalibus derogant*, an infringement of a special rule also entails an infringement of a general provision.[151]

Likewise, in the case of *Commission v. Ireland*, the ECJ held that it is not necessary to find that there has been a failure to comply with the general obligations contained in Article 4(3) TEU if a failure to comply with a Member State's more specific obligation, which must be understood as a specific expression of the duty of loyalty under Article 4(3), has already been established.[152]

As for the fourth case cited by Temple Lang (*UFEX*), in this case the CFI did state that Article 4(3) (by itself) does not confer on a complainant seeking the annulment of a Commission decision under Article 263 TFEU a right of action

---

[148] See Case C-392/02 *Commission v. Denmark* (n 146) para 69, Case C-464/02 *Commission v. Denmark* (n 146) para 84, and Case C-84/04 *Commission v. Portugal* (n 146) para 40.

[149] Case C-19/05 *Commission v. Denmark* [2007] ECR I-08597, Opinion of AG Trstenjak, paras 85–86.

[150] The ECJ had found Denmark to fall short of its obligations under Articles 2 and 8 of Council Decision 94/728/EC of 31 October 1994 on the system of the European Communities' own resources [1994] OJ L293. See Case C-392/02 *Commission v. Denmark* (n 147) para 68.

[151] Case C-19/05 *Commission v. Denmark*, Opinion of AG Trstenjak (n 149) para 86.

[152] Case C-459/03 *Commission v. Ireland* [2006] ECR I-04635, paras 169–71. In this respect, see also Case C-31/00 *Dreessen* [2002] ECR I-00663, para 30.

against decisions falling within the scope of ex Article 86 TEC (now Article 106 TFEU).[153] However, in my view, it would (again) be a bit far-fetched to extract from this the general rule that Article 4(3) can *only* be applied in combination with another rule or policy, especially since the ECJ's statement was in reaction to an attempt by the appellant to stretch procedural boundaries.

De Witte has referred to three older cases[154] in which the ECJ (implicitly) ruled on the direct effect of Article 4(3) TEU.[155] In the case of *Hurd*, the ECJ held that the more specific negative obligation for Member States – deriving from Article 4(3) TEU – not to impose domestic taxation on the European supplement of members of the teaching staff of a European School did not produce direct effects, since it was not sufficiently precise.[156] Accordingly, Mr Hurd could not rely on that negative obligation before UK courts and tribunals.[157] In *Vreugdenhil*, the ECJ, confronted with the argument that the Commission had acted in breach of the division of powers between the institutions, ruled that:

> [T]he aim of the system of the division of powers between the various Community institutions is to ensure that the balance between the institutions provided for in the Treaty is maintained, and not to protect individuals. Consequently, a failure to observe the balance between the institutions cannot be sufficient on its own to engage the Community's liability towards the traders concerned.[158]

Yet, de Witte has argued that it is not clear whether this restrictive approach is limited to liability cases, where the ECJ requires the violation of a rule of law for the protection of individuals, or whether it also extends to actions for annulment, where the violation of any higher norm of EU law can be invoked.[159] In a third case, *Bettati v. Safety Hi-Tech*, the ECJ had to rule on the argument, raised in the course of national proceedings, that the relevant secondary EU legislation was incompatible with Article 4(3). The ECJ held that Article 4(3) imposed on Member States and the EU institutions mutual duties of sincere cooperation and that:

> [C]onsequently, that provision cannot relate to a measure adopted by the Community legislature in the environmental field which might possibly entail advantages or disadvantages for certain undertakings. Accordingly, it is unnecessary to rule on the

---

153   Case T-60/05 *UFEX* (n 147) para 192.
154   Cases 44/84 *Hurd* [1986] ECR 00029, paras 46–49; C-282/90 *Vreugdenhil* [1992] ECR I-01937, paras 20–21, and C-341/95 *Bettati v. Safety High Tech* [1996] ECR I-04631, para. 77.
155   De Witte (n 30) 157.
156   Case 44/84 *Hurd* (n 154) paras 48 and 49.
157   Ibid.
158   Case C-282/90 *Vreugdenhil* (n 154) paras 20–21.
159   De Witte (n 30) 157.

compatibility of Article 5 of the Regulation with Article 5 of the Treaty (Article 4(3) TEU).[160]

Of the three cases cited by de Witte, *Bettati v. Safety Hi-Tech* seems to most explicitly rule out the possibility of Article 4(3) having direct effect in itself and being used by private litigants to challenge the validity of EU secondary legislation. The case appears to set the general rule that Article 4(3) cannot be relied on to challenge the validity of secondary EU legislation. Nevertheless, as de Witte has argued, all 'institutional' general principles by definition impose obligations on the Member States and EU institutions only.[161] Drawing on the logic of *Bettati v. Safety Hi-Tech*, the conclusion would have to be that these principles can therefore never be invoked by individuals.[162]

De Witte states that such a radical conclusion should not be drawn from this one case as there have been cases in which the ECJ recognised the possibility for individuals to invoke what he calls 'written institutional norms' in support of actions for annulment of EU measures.[163] According to de Witte, institutional principles, to the extent that they are laid down in the Treaty,[164] would seem to be available to individuals as a ground for requesting the annulment or invalidation of EU acts, either in an Article 263 TFEU-action (for annulment) or in an Article 267 TFEU-validity reference (preliminary ruling).[165]

Following de Witte's line of reasoning, private parties should be able to rely on the Treaty-based principle of loyalty (Article 4(3) TEU) to, directly (263 TFEU) or indirectly (267 TFEU), challenge the validity of secondary EU legislation on the ground of its incompatibility with Article 4(3). Even if this line of reasoning is not followed, it is questionable whether the ECJ's ruling in *Bettati v. Safety Hi-Tech* provides a basis for concluding that Article 4(3) does not have direct effect (in itself).[166] Since the latter case, the Courts were in several cases presented with the opportunity to confirm the ECJ's approach in *Bettati v. Safety*

---

[160] Case 341/95 *Bettati v. Safety Hi-Tech* (n 154) para 77.

[161] De Witte (n 30) 157.

[162] Ibid.

[163] Ibid 157–158.

[164] De Witte mentions the EC Treaty (now TFEU), but his argument is assumed to likewise apply to the TEU.

[165] De Witte (n 30) 158.

[166] On the legal effects of Article 4(3), see also Case C-9/99 *Echirolles* [2000] ECR I-08207, Opinion of AG Alber, para 21, and Case C-206/04 P *Mulhens* [2006] ECR I-02717, Opinion of AG Colomer, para 71. An example of a case in which a Member State's obligations stemmed solely from Article 4(3) and in which the Member State concerned failed to meet the negative obligation to refrain from any measure which could jeopardise the attainment of the Union's objectives is Case C-246/07 *Commission v. Sweden* [2010] ECR I-03317, paras 103–05.

Hi-Tech, but did not do so.[167] Even though the Courts seem to have had practical reasons for not ruling on the direct effect of Article 4(3) in these cases,[168] they could have seized the opportunity had they wanted to make a statement about the provision's legal effects. The European Courts, after all, are no strangers to judicial activism.

What is more, there are other reasons to question the alleged general rule that Article 4(3) does not have direct effect in itself and can be applied only in combination with some other rule or policy. First, as Lenaerts and Gutiérrez-Fons have contended, it is settled case law that a general principle of EU law may be invoked vertically against the state, i.e. has *vertical* direct effect.[169]

Second, a number of cases, both older and more recent, have been said to illustrate the point that general principles of EU law can also have *horizontal* direct effect.[170] Lenaerts and Gutiérrez-Fons have cited the *Defrenne*[171] and *Angonese*[172] cases, in which the ECJ ruled that the general principles of equal pay for equal work and of free movement of workers may produce horizontal direct effect.[173] Amtenbrink and Vedder have argued that more recent cases like *Mangold*[174] and *Kücükdeveci*[175] – both on the principle of non-discrimination on grounds of age – indicate that the general principles of EU law *do* have (horizontal) direct effect.[176]

Nevertheless, the problem with cases like *Defrenne*, *Angonese*, *Mangold* and *Kücükdeveci* is that they concerned general principles of EU law that could be classified as fundamental rights principles. It is sensible for the ECJ to declare

---

[167] Case T-339/04 *France Télécom* [2007] II-00521, para 39, and Case C-470/03 *AGM COSMET* [2007] ECR I-02749, para 40. In Joined Cases T-172/98, 175/98 to 177/98 *Salamander* [2000] II-02487, the CFI explicitly excluded the possibility of Article 4(3) having *inverse* vertical direct effect (paras 56–57). Article 4(3) does not create obligations for private entities and cannot be invoked as such.

[168] In *France Télécom*, the CFI argued that, contrary to the Commission's intention, it was clear that the applicant was not relying on an autonomous infringement of Article 4(3) TEU (para 46). In *AGM COSMET*, the ECJ did not answer the question posed as it considered the subject matter at hand to have been exhaustively regulated and therefore assessed the national measures concerned in the light of the applicable Directive and not of the provisions of the Treaty.

[169] Lenaerts and Gutiérrez-Fons (n 40) 1639.

[170] Horizontal direct effect refers to the situation in which an EU rule (or in this case principle) has legal effects as between private parties.

[171] Case 43/75 *Defrenne* (n 126).

[172] Case C-281/98 *Angonese* [2000] ECR I-04139.

[173] Lenaerts and Gutiérrez-Fons (n 40) 1647–48.

[174] Case C-144/04 *Mangold* [2005] ECR I-09981.

[175] Case C-555/07 *Kücükdeveci* [2010] ECR I-00365.

[176] Amtenbrink and Vedder (n 14) 177. See also Lenaerts and Gutiérrez-Fons (n 41) 1641–42 and 1649.

principles like these to have horizontal direct effect in situations where individuals would otherwise be deprived of fundamental rights, such as the right not to be discriminated against on grounds of nationality or age. The principle of loyalty, however, is a general principle of EU law of a wholly different nature. It does not concern individual rights, but rather the relationships between the Member States themselves and between the Member States and the EU.[177] As such, it is often classified as an 'institutional' or 'administrative' principle.[178]

The conclusions drawn from cases like *Defrenne, Angonese, Mangold* and *Kücükdeveci*, cannot therefore automatically be extrapolated to institutional principles of EU law like the principle of loyalty. It is uncertain whether the ECJ would be willing to go as far with these principles as it went with fundamental rights principles in the aforementioned cases. Still, these cases do indicate that general principles of EU law can, in principle, even have horizontal direct effect. This at least questions the common conception that Article 4(3) does not have direct effect in itself and can be applied in combination with some other rule or policy only.

5.7.4.2 Article 4(3) TEU and the Courts' criteria

As indicated earlier, the first criterion to be met for a Treaty provision to have direct effect is that it sufficiently precise. Article 4(3) TEU provides that:

> [P]ursuant to the principle of sincere cooperation, the Union and the Member States shall, in full mutual respect, assist each other in carrying out the tasks which flow from the Treaties. The Member States shall take any appropriate measure, general or particular, to ensure fulfilment of the obligations arising out of the Treaties or resulting from the acts of the institutions of the Union. The Member States shall facilitate the achievement of the Union's tasks and refrain from any measure which could jeopardise the attainment of the Union's objectives.

On the face of it, the wording of Article 4(3) does not appear to be very precise and clear. The duties on the EU and the Member States are vague and general. The EU and the Member States shall assist each other in carrying out the tasks which flow from the Treaties. Furthermore, the Member States shall take any appropriate measures to make sure that they meet their EU obligations and they shall facilitate the achievement of the EU's tasks and not take any measures which could jeopardise the attainment of the EU's objectives. As indicated in section 5.4, Member States have a whole range of specific obligations

---

[177]   In this respect, see also de Witte (n 30) 155.
[178]   See e.g. Tridimas (n 29) 4.

under Article 4(3), but those obligations cannot be directly inferred from the wording of the provision.

Yet, in *Defrenne*, the ECJ was not bothered by the lack of precision and clarity of the relevant provision. Instead, it identified and isolated the principle behind the relevant provision and – on that basis – declared it to have direct effect. The principle behind Article 4(3) is obviously the general EU law principle of loyalty. However, unlike the principle underlying Article 141 TEC in *Defrenne* – the principle of equality of men and women – the principle of loyalty does not seem to be intended to confer right on individuals. Being an institutional principle, the principle of loyalty concerns the relationships between the Member States themselves and between the Member States and the EU.

Still, as argued by de Witte, the institutional character of certain 'written institutional norms' did not prevent the ECJ from recognising the possibility for individuals to invoke these norms in support of actions for annulment of EU measures. According to de Witte, institutional principles – to the extent that they are laid down in the Treaty – would seem to be available to individuals as a ground for requesting the annulment or invalidation of EU acts.

On the basis of de Witte's reasoning, I would, however, not maintain that Article 4(3) has direct effect. In the cases examined above, the underlying criterion for a provision to have direct effect appeared to be the intention to confer rights on individuals. The ECJ seems to allow individuals to invoke EU provisions in the national legal order so as to make sure that no lacunas in their protection under EU law arise. Again, Article 4(3) TEU does not appear to be intended to confer rights on individuals. Even though the provision is unconditional and its lack of precision and legal clarity do not, based on *Defrenne*, appear to constitute insurmountable obstacles, I would therefore argue that Article 4(3) TEU does not in itself have direct effect. This makes the question of whether Article 194(1)(c) does have direct effect all the more relevant.[179]

---

[179] To this extent see also Temple Lang, 'The Duties' (n 78) 84; Temple Lang, 'The Development' (n 73) 1517; See also Gormley, 'Some Further Reflections on the Development of General Principles of Law Within Article 10 EC', in Ulf Bernitz, Joakim Nergelius and Cecilia Cardner (eds), *General Principles of EC Law in a Process of Development* (Kluwer Law International, 2008) 303, 312, and Gormley, 'The Development' (n 138) 115. In relation to the direct effect of Article 4(3) and the alleged necessity to be linked to some other EU rule of law, policy or objective, Temple Lang has argued that such rule, policy or objective needs to be precise enough to be justiciable. Likewise, Gormley has contended that a concrete duty is needed and that Article 4(3) must be linked to the infringement of or guaranteeing of a concrete and specific right or obligation founded in the Treaty.

5.7.5 THE DIRECT EFFECT OF ARTICLE 194(1)(C) TFEU

Article 194(1)(c) provides that:

> [I]n the context of the establishment and functioning of the internal market and with regard for the need to preserve and improve the environment, Union policy on energy shall aim, in a spirit of solidarity between Member States, to promote energy efficiency and energy saving and the development of new and renewable forms of energy.

The provision seems to require EU energy policy to have a hotchpotch of different objectives and points of attention: the establishment and the functioning of the internal market, the need to preserve and improve the environment, a spirit of solidarity between Member States, and the advancement of energy efficiency and energy saving and of new and renewable forms of energy. Article 194(1)(c) appears to be a typical example of a compromise text resulting from the wishes of various Treaty negotiating partners.

The wording of the various objectives and points of attention listed in Article 194(1) is vague and general. The meaning of the phrase 'in the context of the establishment and functioning of the internal market' is unclear. Slightly more precise, but still vague, is the phrase 'with regard for the need to preserve and improve the environment'. The objective 'to promote energy efficiency and energy saving and the development of new and renewable forms of energy' at first glance appears to be quite clear but, on second thought, raises a number of questions. What exactly, for instance, does the promoting of energy efficiency and energy saving and the development of new and renewable forms of energy mean?[180]

Furthermore, it is not clear what the exact difference is between energy efficiency and energy saving. However, most indefinite of all, is arguably the article's solidarity clause. As argued in section 5.7.2, 'in a spirit of solidarity between Member States' can mean many things, depending on one's definition of the term solidarity. Even though the concept of solidarity is frequently referred to in the Treaties, it is nowhere defined. Solidarity, as stated before, is one of the most elusive concepts of EU law.

---

[180] Energy efficiency and the development of renewable energy can be promoted in many different ways. Energy efficiency and the development of renewable energy can be encouraged by, for instance, introducing subsidy schemes and tax incentives. Alternatively, both can also be encouraged by taxing the use of energy and related CO_2 emissions, as is the case with a greenhouse gas emissions trading scheme like the EU ETS.

Considering the lack of precision in the wording of Article 194(1)(c), I would argue that the provision does not have direct effect. Solidarity arguably being the basic principle behind the provision, this finding is confirmed when looking at the principle underlying Article 194(1(c). Solidarity is too indistinct and vague a concept for Article 194(1)(c) to have direct effect.

What is more, Article 194(1)(c) can also not be said to be intended to confer rights on individuals. In this regard, it is useful to briefly turn to the case of *Zaera*, mentioned earlier. I argued that the underlying reason for Article 2 TEEC not to have direct effect might have been the provision's apparent lack of intention to confer rights on individuals. As a provision containing broad political objectives, Article 2 TEEC was not intended to create rights for individuals, but rather to give political direction.

Even though Article 194(1)(c) TFEU is not located at the beginning of the Treaty,[181] it does, like Article 2 TEEC, contain broad political objectives. Similarly, the article can be said to contain 'expressions of intent, purpose and motive, rather than rules that are of direct operative effect'. Article 194(1)(c) outlines part of the policy objectives which are to be strived for within the framework of EU energy policy. As the level of generality and vagueness of the article's wording indicates, such provisions are not intended to confer rights on individuals and therefore probably do not have direct effect.

Based on the above, it is unlikely that individuals would be able to rely on Article 4(3) in conjunction with Article 194(1)(c) to challenge the validity of national measures implementing Article 21(2)(b) before a national court.

5.8 LESSONS FOR THE DESIGN OF MEMBER STATES' REGIMES FOR THIRD-PARTY ACCESS TO CO_2 TRANSPORT AND STORAGE INFRASTRUCTURE

The case law of the ECJ – the cases of *Socridis* and *Wachauf* – indicates that (a provision of) a Directive does not infringe the general principles of EU law if it leaves Member States enough discretion to implement it into national law in a way that does not infringe the general principles of EU law. Since Article 21(2)(b) of the CCS Directive gives Member States the *possibility* to require the operators of CO_2 transport and storage infrastructure to refuse to grant access to the

---

[181] Nevertheless, the underlying general principle of solidarity can, as mentioned above, be found in Article 2 TEU, which lists the values on which the EU is founded.

relevant infrastructure when the capacity concerned is needed to meet part of a Member State's international and EU greenhouse gas emission reduction obligations, the provision arguably leaves Member States sufficient discretion to implement it in a way that would not infringe the principle of loyalty enshrined in Article 4(3) TEU. Accordingly, the Courts would likely not find Article 21(2)(b) to infringe Article 4(3) TEU in conjunction with Article 194(1)(c) TFEU.

Yet this does not relieve the Member States of the obligation to ensure that the national measures implementing Article 21(2)(b) respect the boundaries set by the general principles of EU law and the Treaty provisions. National implementing measures must be in conformity with Articles 4(3) TEU and 194(1)(c) TFEU. An analysis of both provisions and the relevant case law reveals that national implementing measures are unlikely to infringe both provisions of primary EU law. The obligation under Article 4(3) TEU for Member States to mutually assist each other arguably only applies when the Member State in need of assistance is (trying to) meeting its obligations under EU law. The latter does not seem to be the case when third parties are seeking access to CO_2 transport and storage infrastructure in other Member States as the installations most likely to deploy CCS participate in the EU ETS.

Nevertheless, should a Member State nonetheless seek access to another Member State's CO_2 transport and/or storage infrastructure for meeting its EU obligations, the mutual assistance obligation under Article 4(3) TEU might arguably oblige the host Member State to require the operator refusing access to expand its capacity when it is economically sensible to do so or when the potential customer is willing to pay for that expansion. In relation to CO_2 transport infrastructure, capacity expansion would appear to be relatively straightforward; existing pipeline and compression infrastructure can be modified or new infrastructure can be constructed.

However, with regard to CO_2 storage infrastructure things might be more difficult. As the capacity of any given CO_2 storage reservoir is fixed, capacity enlargement can normally only take place through the development of a new storage reservoir. Considering the high costs involved with storage site characterisation, it might not automatically be economically sensible to develop a new storage reservoir and it can be questioned whether a potential customer would be willing to pay for such costs. Finally, Article 194(1)(c) TFEU does not seem to impose any obligations on Member States, but rather obliges the EU legislature to draft EU energy policies and legislation 'in a spirit of solidarity between Member States'.

Apart from both provisions apparently not narrowing the scope of Member States for implementing Article 21(2)(b) of the CCS Directive, there also seem to be procedural hurdles for challenging the validity of national measures implementing Article 21(2)(b) on the ground of an infringement of Articles 4(3) and 194(1)(c). Both provisions arguably cannot be invoked by individuals to challenge the validity of national implementing measures before a national court as they appear to be insufficiently precise and clear and do not seem to be intended to confer rights on individuals.

Based on the above, Member States do not have to worry about the validity of Article 21(2)(b) of the CCS Directive when implementing the provision into national law. In view of the relatively broad margin of discretion given to the Member States by Article 21(2)(b), the Courts would likely not find Article 21(2)(b) to infringe 4(3) TEU in conjunction with Article 194(1)(c) TFEU. What is more, individuals would not seem able to invoke both provisions to challenge the validity of Article 21(2)(b) before a national court. Equally, Member States do not have to worry about Articles 4(3) and 194(1)(c) narrowing their scope for implementing Article 21(2)(b) in their national legal order.

Nevertheless, just to be sure, Member States would be wise to require CO_2 transport or storage operators refusing access on the ground mentioned in Article 21(2)(b) to expand their capacity when it is economically sensible to do so or when the potential customer is willing to pay for that expansion. As we will see in Chapter VII, this will also prevent CO_2 transport and storage operators from running into problems with Article 102 TFEU.

CHAPTER VI

REFUSING ACCESS TO CCS
INFRASTRUCTURE AND
ARTICLE 102 TFEU

- CO_2 transport and storage operators should be careful when refusing access to their infrastructure on technical grounds. CO_2 storage facilities, in particular, are likely to be considered indispensable under Article 102 TFEU

- In order to avoid liability under Article 102, CO_2 transport and storage operators are advised to think of a precautionary defence for refusing access to their infrastructure

- Such a defence would have to identify potentially negative consequences for public health/the environment, assess those risks on the basis of the most reliable and recent international research and make sure that the refusal is non-discriminatory and objective

- As at least CO_2 storage operators would likely still have the freedom to adapt the CO_2 stream requirements in such a way as to prevent the infringement of Art. 102 from occurring, CO_2 transport and storage operators are advised against relying on the state action defence

6.1   INTRODUCTION

Like the previous chapter, this chapter deals with Article 21 of the CCS Directive on the access of third parties to CO_2 transport and storage infrastructure (third-party access). Article 21(2)(c) provides that Member States shall take into account the need to refuse access where there is 'an incompatibility of technical specifications which cannot be reasonably overcome'. This provision would seem to provide the operators of CO_2 transport and storage infrastructure the possibility to refuse to grant access to the relevant infrastructure when the technical specifications of the specific CO_2 stream are incompatible with the required technical standard and the incompatibility 'cannot reasonably be overcome'. As differences in national technical standards for CCS infrastructure are likely to appear, this provision has the potential to form an obstacle to the cross-border trade of captured CO_2.

The question, however, is whether access refusal on technical grounds would be compatible with EU antitrust provisions. Article 102 of the Treaty on the Functioning of the EU (TFEU) provides that any abuse by one or more undertakings of a dominant position within the internal market or a substantial part of it shall be prohibited as incompatible with the internal market in so far as it may affect trade between Member States. One of the possible abuses under Article 102 TFEU is the refusal by a dominant undertaking to grant access to infrastructure which is considered indispensable for entering a certain market. Depending on the specific market characteristics, operators of CO_2 transport and storage infrastructure could be in a position of dominance in these respective markets, possibly restricting the scope to refuse to grant third parties access to such infrastructure.

As primary EU (antitrust) law takes precedence over both secondary EU law and national legislation, the ground for refusal of third-party access in Article 21(2)(c) of the CCS Directive would not seem to allow a CO_2 transport or storage operator to refuse access to the relevant infrastructure if such refusal would constitute an abuse of a dominant position under Article 102 TFEU. This raises questions as to the room under Article 102 TFEU for CO_2 transport and storage operators to refuse to grant third-party access to their infrastructure.

In the following sections, I answer the question of to what extent there is scope under Article 102 TFEU for CO_2 transport and storage operators to refuse to grant access to their infrastructure on technical grounds. Section 6.2 briefly examines Article 21(2)(c) of the CCS Directive and the concept of third-party access to CO_2 transport and storage infrastructure. Section 6.3 deals with Article 102 TFEU in relation to the refusal to grant access to CO_2 transport and storage infrastructure. In this section, I link the possible structure of future EU CO_2 transport and storage markets to the constitutive elements of Article 102 TFEU. Section 6.4 outlines a number of possible justifications and exceptions; suggestions are made for avoiding liability under Article 102 TFEU when refusing third-party access to CO_2 transport and storage infrastructure. Finally, section 6.5 draws several lessons for the operators of CO_2 transport and storage infrastructure.

6.2   ARTICLE 21(2)(C) OF THE CCS DIRECTIVE

Article 21 addresses third-party access to CO_2 transport networks and storage sites. By virtue of Article 21(2), it is up to each Member State to determine the manner in which to design their regime for third-party access to CO_2 transport and storage infrastructure,[1] as long as such access is provided in a fair, open and non-discriminatory manner and it takes into account:

a) the storage/transport capacity which is or can reasonably be made available;
b) the proportion of its CO_2 reduction obligations pursuant to international legal instruments and to Community legislation that it intends to meet through CCS;[2]
c) the need to refuse access where there is an incompatibility of technical specifications which cannot reasonably be overcome;
d) the need to respect the duly substantiated reasonable needs of the owner or operator of the storage site or of the transport network and the interests of all other users of the storage or the network or relevant processing or handling facilities who may be affected.

As argued above, Article 21(2)(c) would seem to provide the operators of CO_2 transport and storage infrastructure the possibility to refuse to grant access to the relevant infrastructure when the technical specifications of the specific CO_2 stream are incompatible with the required technical standard and the incompatibility 'cannot reasonably be overcome'. Article 21(2) clearly states that it is up to the Member States to determine the overall regime for third-party access to transport networks and storage sites. In the network-bound EU electricity and gas sectors, several types of access regimes exist, varying from a rudimentary regime of negotiated third-party access to a regime of strictly regulated third-party access.[3] Under a negotiated access regime, the majority of binding access terms and conditions are set by the parties themselves through negotiation.[4] Under a regime of regulated third-party access, access terms and conditions are set externally by, for instance, independent regulatory bodies, state ministries or competition authorities.[5]

---

[1]   See also recital 38 of the preamble to the CCS Directive.
[2]   In this regard, see the previous chapter on refusal of access to CCS infrastructure and the general EU law principle of loyalty.
[3]   Martha M Roggenkamp, 'The Concept of Third Party Access Applied to CCS' in Martha M Roggenkamp and Edwin Woerdman (eds), *Legal Design of Carbon Capture and Storage* (Intersentia 2009) 273, 281.
[4]   Thomas W Waelde and Andreas J Gunst, 'International Energy Trade and Access to Energy Networks' (2002) 36 Journal of World Trade 191, 197.
[5]   Ibid 198.

Brockett and Roggenkamp have both argued that the CCS Directive's provisions on third-party access to transport networks and storage sites intend to apply the principles of negotiated rather than regulated access.[6] According to Roggenkamp, this follows, among other things, from the resemblance between the wording in Article 21 of the CCS Directive (in particular Article 21(2)) and Article 34 (in particular Article 34(2)) of Directive 2009/73/EC (Gas Directive).[7, 8] The rather 'loose' wording and conditions in Article 21 indeed do not seem to hint at a regime of regulated third-party access. However, nor do they appear to preclude Member States from introducing a regime of regulated third-party access to CO_2 transport and storage infrastructure.[9]

Issues of third-party access can, for instance, occur if spare capacity exists in a CO_2 pipeline, network of pipelines or storage site and the relevant authority has an interest in optimising the use of that capacity.[10] The impact assessment on implementing the third-party access requirements of the CCS Directive of the UK Department of Energy and Climate Change (DECC) mentions several likely scenarios for third-party access to CO_2 transport and storage infrastructure, such as a new capture plant linking in with a nearby pipeline passing the facility, an extension of an existing pipeline to reach a new capture plant and an increase of the capacity of a pipeline prior to construction to accommodate another project in the same area.[11]

---

[6]   Scott Brockett, 'The EU Enabling Legal Framework for Carbon Capture and Storage and Geological Storage' (2009) 1 Energy Procedia 4433, 4439 and Roggenkamp (n 3) 293. See also DECC, 'Towards Carbon Capture and Storage: Government Response to Consultation' (2009) 81 <http://webarchive.nationalarchives.gov.uk/+/www.berr.gov.uk/files/file51115.pdf> accessed 21 March 2011. According to the UK Department of Energy and Climate Change (DECC), the UK government should intervene to ensure third-party access to CO_2 transport networks and storage sites only in the event it proves impossible for parties to reach commercial agreement.

[7]   European Parliament and Council Directive 2009/73/EC of 13 July 2009 concerning common rules for the internal market in natural gas and repealing Directive 2003/55/EC [2009] OJ L211/94 (Gas Directive).

[8]   Roggenkamp (n 3) 293. Article 34 of the Gas Directive addresses third-party access to upstream gas pipeline networks and prescribes a regime of negotiated third-party access, like its predecessor, Article 20 in European Parliament and Council Directive 2003/55/EC of 26 June 2003 concerning common rules for the internal market in natural gas and repealing Directive 98/30/EC [2003] OJ L176/57.

[9]   See also Roggenkamp (n 3) 293.

[10] International Energy Agency (IEA), 'Carbon Capture and Storage – Model Regulatory Framework' (2010) 46 <www.iea.org/ccs/legal/model_framework.pdf> accessed 18 March 2011.

[11] DECC, 'Implementing the Third Party Access Requirements of the CCS Directive – Impact Assessment' (2010) 6 <www.decc.gov.uk/en/content/cms/consultations/ccs_3rd_party/ccs_3rd_party.aspx> accessed 21 March 2011.

Within the framework of Directive 2009/72/EC (Electricity Directive)[12] and the Gas Directive, the right of third parties to access infrastructure controlled by others essentially entails the right to receive the service provided by the controlling operator.[13] In the case of *Sabatauskas*, the European Court of Justice (ECJ) indicated that this access right, at least in relation to EU provisions on the internal electricity market,[14] entailed the right to use the system, rather than the right to be physically connected to it in a certain way.[15]

Depending on the characteristics of the applicable third-party access regime, the relevant operator will be under a duty to either supply the desired service or capacity (duty to deal/contract – regulated third-party access) or to at least enter into negotiations over the service/capacity (duty to negotiate – negotiated third-party access).[16] As the CCS Directive leaves it to Member States to determine the manner in which third-party access to CO_2 transport networks and storage sites is provided, national law will determine whether the relevant operator will have the duty to supply the transport or storage service or at least enter into negotiations over the service.

The provisions on third-party access in the CCS Directive constitute so-called 'sector-specific regulation', that is, regulation tailored and applicable to the needs of a specific sector. The relationship between sector-specific regulation and another instrument available for ensuring third-party access to CO_2 transport and storage infrastructure, EU competition law, has been extensively discussed in antitrust literature. To a certain extent, the two regulatory instruments are complementary.[17] In sectors where market players cannot be active without

---

12   European Parliament and Council Directive 2009/72/EC of 13 July 2009 concerning common rules for the internal market in electricity [2009] OJ L211.
13   Ketil Bøe Moen, 'The Gas Directive: Third Party Transportation Rights – But to what Pipeline Volumes?' (2002) 13 The Centre for Energy, Petroleum and Mineral Law and Policy Internet Journal 3, 8, and Alexander Kotlowksi, 'Access Rights to European Energy Networks – A Construction Site Revisited' in Bram Delvaux and others (eds), *EU Energy Law and Policy Issues: Energy Law Research Forum Collection* (Euroconfidential 2009) 8. In the case of CCS, these will be CO_2 transport and storage services respectively. See e.g. Laetitia Birkeland (Bellona), 'Burying CO_2: The New EU Directive on Geological Storage of CO2 from a Norwegian Perspective' (2009) <www.bellona.org/filearchive/fil_Bellonas_paper_-_Burying_CO2-_The_New_EU_Directive_on_Geological_Storage_of_CO2_from_a_Norwegian_Perspective.pdf> accessed 23 March 2011. On third-party access in the EU energy sector, see also Alexander Kotlowksi, 'Third-party Access Rights in the Energy Sector: A Competition Law Perspective' (2007) 16 Utilities Law Review 101.
14   European Parliament and Council Directive 2003/54/EC of 26 June 2003 concerning common rules for the internal market in electricity and repealing Directive 96/92/EC [2003] OJ L176, now repealed by Directive 2009/72/EC.
15   Case C-239/07 *Sabatauskas* [2008] ECR I-07523. On this case, see also Kotlowksi (n 13).
16   Moen (n 13) 3.
17   See, for instance, Alberto Heimler, 'Is a Margin Squeeze an Antitrust or a Regulatory Violation?' (2010) 6 Journal of Competition Law and Economics 879, 885, and Damien

access to highly capital-intensive physical infrastructure, so-called 'network-bound sectors'[18] such as the telecommunications, railway or electricity and gas sectors, both competition law and sector-specific regulation discipline the behaviour of undertakings.[19]

However, there are also various differences between the instruments of sector-specific regulation and competition law.[20] The two regulatory instruments are, for instance, said to have different objectives.[21] As a consequence, conflicts can arise between the two instruments when they are applied concurrently.[22] This raises the question which of the two instruments should in that case prevail.

An obvious and straightforward answer is that EU competition rules, as provisions of primary EU law, always prevail over 'secondary' EU sector-specific rules. Likewise, EU competition law provisions take precedence over national sector-specific regulation, even if the national law implements EU sector-specific rules.[23] As argued by Monti, the precedence of EU competition rules over national sector-specific rules was confirmed by the *O2* decision (Article 101 TFEU), in which the Commission provided that:

> [S]ubject to the principle of the primacy of Community law, the national regulatory framework and the EU competition rules are of parallel and cumulative application. National rules may neither conflict with the EU competition rules nor can compatibility with national rules and regulations prejudice the outcome of an assessment under the EU competition rules.[24]

Geradin and Robert O'Donoghue, 'The Concurrent Application of Competition Law and Regulation: The Case of Margin Squeeze Abuses in the Telecommunications Sector' (2005) 1 Journal of Competition Law and Economics 355, 409 and 425.

[18] Slot and Skudder describe network-bound sectors as sectors that crucially depend on a fixed network that cannot, or cannot easily, be duplicated because of the enormous investments involved and the constraints imposed by environmental, safety and zoning restrictions. See Piet Jan Slot and Andrew Skudder, 'Common Features of Community Law Regulation in the Network-bound Sectors' (2001) 38 Common Market Law Review 87, 87.

[19] Heimler (n 17) 885.

[20] For a list of examples of differences between the two, see Pierre-André Buigues and Robert Klotz, 'Margin Squeeze in Regulated Industries: The CFI Judgment in the *Deutsche Telekom* Case' (2008) 7 CPI Antitrust Chronicle 16.

[21] Geradin and O'Donoghue (n 17) 424 and Damien Geradin, 'Refusal to Supply and Margin Squeeze: A Discussion of why the "Telefonica Exceptions" Are Wrong' (2010) TILEC Discussion Paper 8 <www.ssrn.com/abstract=1762687> accessed 29 March 2011.

[22] Geradin and O'Donoghue (n 17) 424.

[23] See also See G Monti, 'Managing the Intersection of Utilities Regulation and EC Competition Law' (2008) 4 The Competition Law Review 123, 124.

[24] *T-Mobile Deutschland/O2 Germany* (Case COMP/38.369) Commission Decision 2004/207/EC [2004] OJ L75, para 22. This principle was not challenged on appeal to the CFI. See Case T-328/03 *O2 Germany* [2006] ECR II-01231.

The principle of precedence of primary EU (competition) law over both secondary (sector-specific) EU law and national (sector-specific) rules would seem to apply no less to situations where conflicts arise as a consequence of the concurrent application of Article 21 of the CCS Directive and Article 102 TFEU. Accordingly, the grounds for refusal of third-party access to transport networks or storage sites mentioned in Article 21(2) would not appear to allow an operator to refuse access to the relevant infrastructure if such refusal would constitute an abuse of a dominant position under Article 102. A refusal of access in the case of an incompatibility of technical specifications which cannot be reasonably be overcome (Article 21(2)(c)), would therefore be unlawful if it breaches Article 102. This possible type of abuse of a dominant position is further discussed in the following section.

6.3   ARTICLE 102 TFEU: ABUSE OF A DOMINANT POSITION

Contrary to Article 101 TFEU and Regulation 139/2004 (Merger Regulation),[25] Article 102 TFEU applies to *unilateral* anti-competitive conduct of undertakings. The essence of Article 102 is the control of market power.[26] It prohibits any abuse by one or more undertakings dominant within the internal market or a substantial part of it in so far as the abuse may affect trade between Member States. Article 102 mentions several examples of 'abusive behaviour':

a) directly or indirectly imposing unfair purchase or selling prices or other unfair trading conditions;
b) limiting production, markets or technical development to the prejudice of consumers;
c) applying dissimilar conditions to equivalent transactions with other trading parties, thereby placing them at a competitive disadvantage; or
d) making the conclusion of contracts subject to acceptance by the other parties of supplementary obligations which, by their nature or according to commercial usage, have no connection with the subject of such contracts.

---

[25]   Council Regulation (EC) 139/2004 of 20 January 2004 on the control of concentrations between undertakings (the EC Merger Regulation) [2004] OJ L24. Article 101 TFEU and Regulation 139/2004 both deal with agreements between undertakings.
[26]   Craig and de Búrca, *EU Law: Text, Cases and Materials* (3rd edn, Oxford University Press 2003) 992.

The wording of Article 102 requires a number of cumulative conditions to be satisfied before a violation can be established.[27] These conditions follow from the article's central notions:

1) *Undertaking.* Article 102 applies to undertakings only;[28]
2) *Dominance.* The undertaking(s) concerned must be in a position of dominance on the relevant market;
3) *Substantial part of the internal market.* This market should comprise at least a substantial part of the internal market;
4) *Abuse.* The conduct of the undertaking(s) in question has to amount to an 'abuse' of the dominant position;
5) *Affect inter-state trade.* The abuse should affect trade between Member States.[29]

In the enforcement of Article 102 TFEU, conditions two and four usually lead to significant problems of definition, while the other three are generally more easily met. Therefore, in the following, the central notions of dominance and abuse are further discussed in relation to the refusal to grant access to CO_2 transport and storage infrastructure. The question of whether a refusal of access on the ground of technical incompatibility of the captured CO_2 stream with the relevant infrastructure infringes Article 102 can (obviously) only be answered by assessing the facts and characteristics of each specific case.

Yet, on the basis of the relevant decision practice and case law of the Commission and the Courts, it is possible to sketch the general scope for CO_2 transport and storage operators to refuse to grant access to their infrastructure based on technical incompatibility. The following sections do so by exploring the concepts of dominance and abuse in relation to this specific scenario.

---

[27] Robert O'Donoghue and A Jorge Padilla, *The Law and Economics of Article 82 EC* (Hart Publishing 2006) 2.

[28] In the case of *Höfner and Elser*, the ECJ defined an undertaking as an 'entity engaged in an economic activity, regardless of the legal status of the entity and the way in which it is financed'. See Case C-41/90 *Höfner and Elser* [1991] ECR I-1979, para 21. It is not certain that transport and storage operators will in practice constitute undertakings within the meaning of Article 102. However, in the following I would, (partly) for the sake of conciseness, like to focus on the often much more controversial concepts of dominance and abuse.

[29] As Article 102 has direct effect, enforcement actions can be initiated by both public and private parties. See Case 155/73 *Sacchi* [1974] ECR 00409, para 18. The EU concept of direct effect allows EU citizens, when the applicable provision of EU law meets a number of conditions, to invoke that provision in the national legal order. The principle was first mentioned by the ECJ in Case 26/62 *Van Gend en Loos* [1963] ECR 00001.

6.3.1 DOMINANCE AND CO_2 TRANSPORT AND STORAGE OPERATORS

6.3.1.1 A position of 'dominance'

The requirement that an undertaking is in a position of dominance on the relevant market is an absolute pre-requisite under Article 102. If the undertaking concerned does not hold a dominant position on the relevant market, Article 102 is not applicable. It needs to be stressed that the holding of a dominant position as such is not prohibited under Article 102.[30] There is no guidance in the Treaties as to when an undertaking is considered to be in a position of dominance. In the *United Brands* case (*UBC*), the ECJ provided that the term dominance 'relates to a position of economic strength enjoyed by an undertaking which enables it to prevent effective competition being maintained on the relevant market by giving it the power to behave to an appreciable extent independently of its competitors, customers and ultimately of its consumers'.[31]

In the 2009 guidance document on its enforcement priorities in applying Article 102 (guidance notice),[32] the Commission indicates that this notion of independence is related to the degree of competitive constraint exerted on the undertaking in question.[33] Dominance entails that these competitive constraints are not sufficiently effective and that the undertaking in question enjoys substantial market power over a period of time.[34] Assessing dominance requires an analysis of whether the firm under investigation faces significant competitive constraints, to be conducted in two steps.

First, the relevant market needs to be defined, comprising all those products (and their geographic locations) that impose an effective competitive constraint on the product(s) of the firm whose unilateral practices are under scrutiny. Second, an assessment has to be made of the competitive position of the allegedly

---

[30] Case 322/81 *Michelin* [1983] ECR 3461, para 10.
[31] Case 27/76 *United Brands* [1978] ECR 00207 para. 65.
[32] It is important to underline that the Commission cannot through the guidance notice depart from the Courts' case law on Article 102. As the title suggests, the guidance notice lists the Commission's enforcement priorities and intentions with regard to certain abuses under Article 102. The status of the guidance notice will be discussed further in section 6.3.2.2.
[33] Commission, 'Communication from the Commission – Guidance on the Commission's Enforcement Priorities in Applying Article 82 of the EC Treaty to Abusive Exclusionary Conduct by Dominant Undertakings' 2009/C 45/02, 8. On the guidance paper, see, for instance: Frances Dethmers and Heleen Engelen, 'Fines under Article 102 of the Treaty on the Functioning of the European Union' (2011) 32 European Competition Law Review 86, and Ariel Ezrachi, 'The Commission's Guidance on Article 82 EC and the Effects Based Approach: Legal and Practical Challenges' in Ariel Ezrachi (ed), *Article 82 EC: Reflections on its Recent Evolution* (Hart Publishing 2009).
[34] Commission, 'Enforcement Priorities' (n 33) para 10.

dominant undertaking on the relevant market, i.e. its ability to raise prices or reduce output in relation to their competitive levels for a sustained period of time.[35]

This two-stage procedure can be problematic as the process of market definition may be hard to separate from what is supposed to be the second step, assessing the undertaking's power on that market.[36] It can be difficult to determine which factors should be taken into account when defining markets and which factors should be taken into account when considering the undertaking's position on the market.[37] In practice, the Commission and the Courts do not always maintain a strict division between the two.[38]

The relevant market is typically defined in a product and a geographic dimension. In its Notice on the definition of the relevant market for the purposes of Community competition law (market definition notice), the Commission defines the relevant *product* market as comprising 'all those products and/or services which are regarded as interchangeable or substitutable by the consumer, by reason of the product's characteristics, their prices and their intended use'.[39] The interchangeability of products is assessed from both the demand and the supply sides of the market. Demand-side substitution analyses whether consumers can switch their consumption to interchangeable products. Supply-side substitution assesses whether competitors can rapidly supply products that consumers consider to be interchangeable with the product concerned.

The market definition notice describes the relevant *geographic* market as:

> the area in which the undertakings concerned are involved in the supply and demand of products or services, in which the conditions of competition are sufficiently homogenous and which can be distinguished from neighbouring areas because the conditions of competition are appreciably different in those areas[40]

The relevant geographic market may be anything from a regional/local market to a global market, depending on market conditions. As a rule of thumb, the easier it is to bring goods or services in from a distance, the wider the likely

---

[35]   O'Donoghue and Padilla (n 27) 63.

[36]   A Jones and B Suffrin, *EU Competition Law: Text, Cases and Materials* (4th edn, Oxford University Press 2011) 292.

[37]   Ibid 292.

[38]   To that extent, see e.g. Stothers's comments on Case C-7/97 *Bronner* [1998] ECR I-07791 in Christoph Stothers, 'Refusal to Supply as Abuse of a Dominant Position: Essential Facilities in the European Union' (2001) 22 European Competition Law Review 256.

[39]   Commission, 'Commission Notice on the Definition of the Relevant Market for the Purposes of Community Competition Law' 97/C 372/03, para 7.

[40]   Ibid para. 8.

market will be.[41] There are several factors for determining the scope of the geographic market, including, for instance, past evidence of changes in prices between different areas and consequent reactions by customers (diversion of orders to other areas), the current geographic pattern of consumer purchases, trade flows/patterns of shipments, and barriers and switching costs associated to diverting orders to other areas.[42]

High transport costs are an important factor narrowing the scope of the relevant geographic market.[43] In the case of *Suikerunie v. Commission*, the ECJ indicated that transport costs constitute one of the most significant factors in defining the concept of the relevant market.[44] According to the Commission, the impact of transport costs and transport restrictions arising from legislation or the nature of the relevant products is perhaps the clearest obstacle for a customer to divert his orders to other areas.[45]

The assessment of dominance will take into account the competitive structure of the relevant product and geographic market, and in particular the following factors:

1) the market position of the relevant undertaking and its competitors, initially indicated by market shares;
2) barriers to entry or expansion of potential or actual competitors respectively (e.g. transports costs, capacity constraints or regulatory barriers such as technical standards);
3) the bargaining strength of the undertaking's customers (countervailing buyer power).[46]

The concept of dominance is highly subjective and there have been quite a few cases where a finding of dominance by the Commission and/or the Courts received strong criticism from antitrust economists and lawyers.[47] The criticism of the application of the dominance test by the Commission and the Courts is

---

[41] DG Goyder, *EC Competition Law* (4th edn, Oxford University Press 2003) 274.
[42] Commission, 'Market Definition Notice' (n 39) paras 44–52.
[43] Jones and Suffrin (n 36) 316 and Craig and de Búrca (n 26) 999.
[44] Joined Cases 40/73 to 48/73, 50/73, 54/73 to 56/73, 111/73, 113/73 and 114/73 *Suiker Unie v. Commission* [1975] ECR 1663, para 372.
[45] Commission, 'Market Definition Notice' (n 39) para. 50. Padilla and O'Donoghue even consider transport costs the most important factor in the definition of the relevant geographic market. See O'Donoghue and Padilla (n 27) 94.
[46] Commission, 'Enforcement Priorities' (n 33) para 12.
[47] See, for instance, Frances Dethmers and Ninette Dodoo, 'The Abuse of Hoffman-La Roche: The Meaning of Dominance under EC Competition Law' (2006) 27 European Competition Law Review 537, 537, Brian Sher, 'The Last of the Steam-powered Trains: Modernising Article 82' (2004) 25 European Competition Law Review 243, 246, and Adrian Majumdar, 'Whither Dominance?' (2006) 27 European Competition Law Review 161.

part of a wider critique on what has been described as a 'form-based analysis' that pays insufficient attention to the actual (economic) effects in the relevant market. In its guidance paper, the Commission has tried to counter some of this criticism by the introduction of a more (economic) effects-based analytical approach.[48] In relation to the relevance of market shares, the guidance notice, for instance, states that it will interpret market shares in the light of the relevant market conditions, including the dynamics of the market and the trend or development of market shares.[49]

However, it is questionable whether the Commission has thereby fully relinquished its traditional approach in which market shares play a central role. First, it has been questioned whether the Commission indeed no longer applies the traditional dominance threshold of 40%[50] above which undertakings (combined with some indications of entry barriers) are automatically considered to be in a dominant position.[51] Second, even if the Commission no longer applies this threshold, it is still bound by the case law of the Courts that have traditionally relied heavily on market shares.[52] Guidelines are rules of practice, not rules of law.[53] Whether the guidelines will indeed change the enforcement practice under Article 102 very much depends on the willingness of the Courts to follow the Commission's new approach and depart from their own case law. The role of market shares is therefore not played out yet, even more so since the Commission still considers market shares to provide a useful first indication of the market structure and of the relative importance of the various undertakings active on that market.[54]

6.3.1.2 The relevant CCS product markets

In Chapter I, I outlined the three main parts of the CCS value chain: the capture (including compression), transport, and storage and monitoring of CO_2. The question is whether these three activities can, in general, be seen as constituting

---

[48] Ezrachi (n 33) 53. In the remainder of this chapter, the term effects-based approach refers to an approach that pays (more) attention to the *actual* economic effects in the relevant market in terms of, for instance, a loss of consumer welfare.

[49] Commission, 'Enforcement Priorities' (n 33) para 13.

[50] In its guidance notice, the European Commission states that its experience suggests that dominance is not likely (but not impossible) if the undertaking's market share is below 40% in the relevant market. Commission, 'Enforcement Priorities' (n 33) para 14.

[51] James Kavanagh, Neil Marshall and Gunnar Niels, 'Reform of Article 82 EC – Can the Law and Economics be Reconciled?' in Ariel Ezrachi (ed), *Article 82 EC: Reflections on its Recent Evolution* (Hart Publishing 2009) 1, 4.

[52] See Jones and Suffrin (n 36) 326 and Liza Lovdahl Gormsen, 'Why the European Commission's Enforcement Priorities on Article 82 EC Should be Withdrawn' (2010) 31 European Competition Law Review 45, 47.

[53] Gormsen (n 52) 49.

[54] Commission, 'Enforcement Priorities' (n 33) para 13.

different product markets in the EU CCS sector. Considering the functional similarity between the gas sector and the CCS sector and the likelihood of some players being simultaneously active in the CCS sector and in either the electricity or the gas sector, an analogy with the latter two sectors might be helpful in answering this question.[55]

In the *EDF/London Electricity* case, the Commission distinguished four separate electricity product markets, which required different assets and were characterised by a different market structure and conditions of competition:

1) the production of electricity (generation);
2) the transport of electricity over high-tension cables (transmission);
3) the transport of electricity over low-tension cables (distribution);
4) the delivery of electricity to the final consumer (supply).[56]

In the later *Electrabel/E.ON* case, the Commission further specified this rough division into five broad electricity product markets:

1) generation and wholesale supply;
2) transmission;
3) distribution;
4) ancillary services; and
5) retail supply of electricity for end-users.[57]

In the same year, the Commission identified the following gas product markets in the *Rreef Fund/Endesa/UFG/Saggas* case:

1) the production and exploration for natural gas;
2) gas wholesale supply;
3) gas transmission (via high-pressure systems);
4) gas distribution (via low-pressure systems);
5) gas storage;
6) gas trading;
7) gas supply to end-customers; and
8) the market for infrastructure operations for gas imports.[58]

---

[55]   In this regard, see also Chapter VII, which deals with the development and management of CO_2 transport infrastructure and EU antitrust law.
[56]   *EDF/London Electricity* (Case IV.M/1346) Commission Decision of 27 January 1999, para 12.
[57]   *Electrabel/E.ON* (Case COMP/M.5512) Commission Decision of 16 October 2009, para 14.
[58]   *Rreef Fund/Endesa/UFG/Saggas* (Case COMP/M.5649) Commission Decision of 21 October 2009, para 11.

Obviously, the exact relevant CCS product market will have to be determined in each individual case. Nevertheless, by analogy, the three main activities in the CCS value chain could be said to constitute different product markets. Based on an analogy with the natural gas market, Andrews, in an Australian context, also reaches the conclusion that 'each stage of the CCS process likely would qualify as a separate market'.[59] As no well-developed CO_2 capture, transport and storage markets exist in the EU at the time of writing, it is difficult to say anything as to their (possibly differing) respective market structures and conditions of competition (ref. *EDF/London Electricity*). Yet, it is fairly easy to see that the three main CCS activities mostly require very different assets and resources, due to the different nature of the activities. The only assets and resources that the three activities have in common are probably the installations for compression of the captured CO_2 stream.

What is more, when looking at the, from an economic point of view, most immediate and effective disciplinary force on the suppliers of a given product/service – demand substitutability[60] – the three types of activity do not appear to be interchangeable. The services provided are inherently different and stand in a vertical relation to each other; they complement each other. Nevertheless, as I argue in section 6.3.1.4 below, the same undertaking could be active in more than one of these product markets by, for instance, capturing CO_2 and simultaneously providing transport services.

Cases like *Electrabel/E.ON* and *Rreef Fund/Endesa/UFG/Saggas* illustrate that electricity and gas product markets can, in practice, be narrower than the archetypal product markets of production, transmission, distribution and supply.[61] Likewise, the relevant CCS product markets might be more specific than the relatively broad markets for the capture, transport, and storage and monitoring of CO_2. CO_2 capturers using different technologies for the sequestration of CO_2 could,[62] for instance, be in different markets for the capture of CO_2. Due to the specifics of the production process or facility, a capturer might not be able to switch to another technology or be able to do so only against prohibitively high investments in the adjustment of existing or the development of new infrastructure. Likewise, in certain geographic areas or situations, the only (economically or physically) viable option of transporting

---

59   Adam M Andrews, 'Picking up on What's Going Underground: Australia Should Exempt Carbon Capture and Geo-Sequestration from Part IIIA of the Trade Practices Act' (2008) 17 Pacific Rim Law and Policy Journal 407, 418.
60   Commission, 'Market Definition Notice' (n 39) para 13.
61   Next to these cases, see also e.g. *E.ON/MOL* (Case COMP/M.3696) Commission Decision 2006/622/EC [2006] OJ L253, para 12 and *Centrica/Venture Production* (Case COMP/M.5585) Commission Decision of 21 August 2009, paras 8–20.
62   See section 1.1 of the first chapter for the different capture technologies.

captured CO_2 might be transport by pipeline.[63] In that case, the relevant product market will be the market for the transport of CO_2 *by pipeline*, excluding, for instance, the transport of CO_2 by ship.

Finally, as the long-term safety of the geological storage of CO_2 will to a great extent be determined by a combination of the specific composition of the captured CO_2 stream (CO_2 content and (level of) contaminants) and the geological characteristics of the relevant storage formation,[64] it could well be that the relevant product market for CO_2 storage in practice consists of a number of smaller markets. At the early stage of CCS development, it is likely that a distinct CO_2 stream from a particular plant will be linked to an appropriate CO_2 storage site.[65] As there will likely be a range/number of storage formations with suitable geological characteristics for the different types of CO_2 streams resulting from the various production processes and capturing technologies, different product markets for CO_2 storage (arranged by geophysical characteristics) could develop.[66]

An interesting question is whether the relevant product market for the storage and monitoring of CO_2 includes enhanced hydrocarbon recovery[67] or other CO_2 reuse activities. From a functional point of view the thought seems appealing. In a 2011 report on the industrial use of captured CO_2, the Global CCS Institute considered several enhanced hydrocarbon recovery and other reuse technologies to be an alternative form of CCS since they result in permanent CO_2 storage.[68] According to the International Energy Agency (IEA),

---

63   Calculations by the Intergovernmental Panel on Climate Change (IPCC) indicate that the transport of CO_2 by ship becomes cost-competitive with transport by pipeline only if CO_2 needs to be transported over larger distances. See IPCC, 'IPCC Special Report on Carbon Dioxide Capture and Storage. Prepared by Working Group III of the Intergovernmental Panel on Climate Change' (Bert Metz and others (eds) Cambridge University Press 2005) 345. See also Joris Morbee, Joana Serpa and Evangelos Tzimas, 'The Evolution of the Extent and the Investment Requirements of a Trans-European CO_2 Transport Network' (2010) 1 <http://publications.jrc.ec.europa.eu/repository/bitstream/111111111/15100/1/ldna24565enn.pdf> accessed 7 December 2011.
64   Commission, 'Implementation of Directive 2009/31/EC on the Geological Storage of Carbon Dioxide – Guidance Document 2: Characterisation of the Storage Complex, CO_2 Stream Composition, Monitoring and Corrective Measures' (2011) 74–77 <http://ec.europa.eu/clima/policies/lowcarbon/docs/gd2_en.pdf> accessed 8 April 2011.
65   Ibid 67.
66   In its guidelines on CO_2 stream composition, the Commission states that geomechanical reactions are highly site specific, depending on the precise mineralogy, fluid chemistry, pressure and temperature of the host formation. See Commission, 'Guidance Document 2' (n 64) 76.
67   See n 56 Chapter I.
68   The report mentions technologies such as enhanced oil recovery, enhanced coal bed methane recovery, concrete curing (CO_2 stored in concrete) and potentially algae cultivation. See Global CCS Institute, 'Accelerating the Uptake of CCS: Industrial Use of Captured Carbon

joint CO_2 storage/enhanced hydrocarbon recovery and 'pure' enhanced hydrocarbon recovery projects are both likely to result in significant quantities of CO_2 being geologically stored.[69]

Based on the expected volumes of CO_2 that can be stored through reuse activities and the current low level of technological development of most of these technologies,[70] it can, however, be doubted whether capturers will actually consider reuse to constitute a real alternative to geological storage. In its impact assessment of the CCS Directive, the Commission estimates some 160 $MtCO_2$ (equal to 5% of total CO_2 emissions from energy) to be captured in 2030.[71] It is unlikely that such quantities can be stored through the reuse of CO_2. In 2005, the Intergovernmental Panel on Climate Change (IPCC) concluded that 'although the precise figures are difficult to estimate and even their sign is questionable, the contribution of these technologies to CO_2 storage is negligible'.[72] This is confirmed by the findings of the Global CCS Institute.[73] Nevertheless, this does not exclude CO_2 reuse as an alternative to geological storage in certain specific individual situations, especially in the demonstration phase.[74]

6.3.1.3   The relevant CCS geographic markets

For each of the above-mentioned markets for the capture, transport, and storage and monitoring of CO_2, there will be a relevant geographic market. As in the case of the relevant product market, the relevant geographic market will have to be determined on a case-by-case basis. Above, I indicated that transport costs are one of the most important factors in defining the relevant geographic market. High transport costs narrow the scope of the relevant geographic market.

The crucial question is when transport costs are considered high enough to constitute a barrier for customers to switch to providers in other areas. Obviously, the answer to this question will depend on the relevant facts of each

---

Dioxide' (2011) 38 <www.globalccsinstitute.com/resources/publications/accelerating-uptake-ccs-industrial-use-captured-carbon-dioxide> accessed 22 April 2011.

[69]   From a functional point of view, it could, however, also be argued that enhanced hydrocarbon recovery and the geological storage of CO_2 are inherently different. Enhanced hydrocarbon recovery projects are optimised to maximise hydrocarbon production and minimise the CO_2 stored. CO_2 is injected for the sole purpose of enhancing hydrocarbon recovery and any CO_2 stored is incidental to the aim of the project. See IEA (n 10) 114.

[70]   Global CCS Institute (n 68) 102.

[71]   This is in a scenario where CCS is enabled under the EU ETS by means of CO_2 prices in the range of €40–45/tCO_2. See Commission, 'Accompanying Document to the Proposal for a Directive of the European Parliament and of the Council on the Geological Storage of Carbon Dioxide: Impact Assessment' COM (2008) 18 final 52.

[72]   IPCC (n 63) 335.

[73]   Global CCS Institute (n 68) 102.

[74]   Ibid.

individual case. Transport costs differ from case to case, depending on variables such as the particular transport mode used, the distance to be covered and the quantities transported. Nevertheless, the Commission's decision practice on mergers provides some guidance as to the way in which to assess whether transport costs are high and might constitute a barrier to switching to providers in other areas.

The first point of relevance is the benchmark for determining the level of transport costs. The latter cannot be determined in the abstract; costs can only be high or low in relation to something else. The decision practice of the Commission is not very consistent in this regard. Often, the value or, what is de facto the same, the price of the transported product forms the benchmark for determining the level of transport costs.[75] However, there are also cases in which the Commission determines the level of transport costs in relation to total product costs[76] or sales revenue.[77] The differences are perhaps not shocking, but transport costs as a percentage of total product costs are not the same as transport costs as part of product price/value.

Where transport costs are high in relation to the value of the product (or any of the other benchmarks), the geographic market will normally be limited in scope, and vice versa, as this will deter suppliers in one area from selling in another.[78] The Commission does not seem to have developed any threshold or minimum level above which transport costs are considered high or capable of constituting a barrier to switching to providers in other areas. However, in general it seems to regard transport costs of up to at least 10% as low or marginal.[79] On the basis of percentages ranging from 16–20%, the Commission concluded in the *Kali-Salz* case that transport costs were not 'of a level capable of preventing significant

---

[75] *KNP/Bührmann-Tetterode and VRG* (Case IV/M.291) Commission Decision of 4 May 1993, para 44; *Amcor/Danisco/Ahlstrom* (Case COMP/M.2441) Commission Decision of 6 November 2001, para 14; *Sovion/HMG* (Case IV-M.3605) Commission Decision of 21 December 2004, para 20; *Greek Lignite and Electricity Markets* (Case 38700) Commission decision of 5 March 2008, para 170; *Sun Capital/DSM Special Products* (Case COMP/M.5785) Commission Decision of 12 February 2010, paras 81 and 84; and *Triton III Holding 6/Wittur Group* (Case COMP/M.5991) Commission Decision of 12 February 2010, para 25.

[76] *Teka/Finatlantis/Holdivat* (Case COMP/M.2313) Commission Decision of 6 April 2001, para 17, and *Sun Capital/DSM Special Products* (n 75) para 89.

[77] *Kali-Salz/MdK/Treuhand* (Case IV/M.308) Commission Decision 94/449/EC [1994] L186/38, para 42, and *MEI/Phillips* (Case COMP/M.2386) Commission Decision of 29 May 2001, para 12.

[78] Ivo van Bael and Jean-François Bellis, *Competition Law of the European Community* (Kluwer Law International 2005) 791.

[79] See e.g. *Teka/Finatlantis/Holdivat* (n 76) para 17 and *Sun Capital/DSM Special Products* (n 75) paras 81 and 84.

trade flows' between the Member State where the relevant producers where based (Germany) and other Member States.[80]

Several cases illustrate the obvious point that the relevance of the level of transport costs increases as the value of the transported product decreases. In the *Avesta/British Steel* case, the Commission stated that 'for stainless steel transport costs are less important than for carbon steel because of the higher value of the product'.[81] In the case of *VIAG/Continental Can*, it concluded that 'because empty cans and glass bottles have only a small value and take up a great deal of space, they cannot economically support transport costs over a long distance'.[82] Similarly, the market definition notice provides that 'the impact of transport costs will usually limit the scope of the geographic market for bulky, low-value products'.[83]

In the case of the relevant CCS product markets, high CO_2 transport costs will likely lead to the distance from the capturing plant to potential storage locations having a stronger influence on the decision of a capturer to have the captured CO_2 stored in a particular (nearby) formation.[84] This will probably exclude certain storage facilities and the providers of these storage services from the geographic range of a particular capturing facility and thus narrow the relevant geographic market for the storage of CO_2. On the relevant product market for CO_2 transport, high transport costs will raise the price of the service itself, thus likely decreasing demand for the particular mode(s) of transport. Instead of constituting a barrier for customers to switch to providers in other areas, the high transport costs (of e.g. transport by ship) might lead to customers seeking alternative modes of transport and thereby broaden the relevant product market.

As for the geographic scope of the market for the capture of CO_2, CO_2 transport costs are not likely to play a role. Instead, the costs of transporting the particular technological equipment to (soon to be) capturing installations are more likely

---

[80]   *Kali-Salz/MdK/Treuhand* (n 77) para 42.
[81]   *Avesta/British Steel/NCC/AGA/Axel Johnson* (Case IV/M.239) Commission Decision of 4 September 1992, para 25.
[82]   *VIAG/Continental Can* (Case IV/M.081) Commission Decision of 6 June 1991, para 16.
[83]   Commission, 'Market Definition Notice' (n 39) para 50.
[84]   However, general conclusions as to the effect of CO_2 transport costs on the choice of capturers for a particular storage formation should not be drawn too easily. Economic research conducted in the US, for instance, indicates that the assumption that electricity generation companies will construct *new* power plants geographically near storage sites in order to minimise CO_2 transport costs does not hold as electricity transmission costs generally outweigh CO_2 pipeline costs in new construction. See Adam Newcomer and Jay Apt, 'Implications of Generator Siting for CO_2 Pipeline Infrastructure' (2008) 36 Energy Policy 1776, and Jeff Bielicki, 'Carbon Capture and Storage and the Location of Industrial Facilities' (presentation held at the Research Experience in Carbon Sequestration 2007, Montana State University, 2 August 2007) <belfercenter.ksg.harvard.edu/> accessed 13 April 2011.

to be of relevance. However, considering the high cost of the capturing technology itself, the relative importance of such transport costs can be questioned.[85]

In most CCS systems capture represents by far the largest cost component. Generally, transport costs are considered to be small compared to the overall capture costs.[86] Yet this does not automatically mean that CO_2 transport costs are also low in relation to the above-mentioned benchmarks nor that they will not, in practice, constitute a factor narrowing the scope of the relevant geographic market. The level of CO_2 transport costs is likely to be primarily important for determining the geographic scope of the product market for the storage and monitoring of CO_2. It therefore makes sense to measure transport costs in terms of storage costs, even though these costs concern two different relevant product markets. Strictly speaking, this approach differs from the above outlined traditional approach where costs of transporting a certain product to other areas are used to determine the geographic range of a single relevant product market. However, if transport costs are high in relation to storage costs, the former are likely to narrow the geographic range of the relevant market for CO_2 storage and vice versa.

Both CO_2 transport and CO_2 storage costs consist of different components, basically divided between capital, operating and other costs.[87] CO_2 transporters and storage operators will likely want to incorporate all three kinds of cost as much as possible in their transport, and storage and monitoring tariffs, as they will probably seek to not only cover variable costs, but also to recoup sunk costs.[88] Whether they will actually be able to do so depends on the tariff structure of the applicable third-party access regime. The case of *Gaz de France/ Suez* suggests that differing transport tariffs might point at separate geographic markets.[89]

---

[85] McKinsey and Company, 'Carbon Capture and Storage: assessing the economics' (2008) 18 <http://assets.wwf.ch/downloads/mckinsey2008.pdf> accessed 23 October 2012. According to McKinsey, additional capital expenditure (capex) for capture-specific equipment in general contributes more than half of the CO_2 capture cost. In most CCS systems, the cost of capture (including compression) is the largest cost component. See Howard Herzog, 'CO2 Capture and Storage (CCS): Costs and Economic Potential' (presentation at the Joint SBSTA/IPCC side-event, COP11, Montreal, 30 November 2005).

[86] IEA, 'Prospects for CO_2 Capture and Storage' (2004) 79.

[87] IPCC (n 63) 259.

[88] Sunk costs are costs that have been incurred and cannot be reversed. See The Economist, 'Economics A-Z' <www.economist.com/research/economics/> accessed 18 April 2011. In the case of CO_2 transport, these will be, among other things, the costs made in laying pipelines and other physical infrastructure such as compressor stations.

[89] *Gaz de France/Suez* (Case COMP-M.4180) Commission Decision of 14 November 2006, para 385.

In its 2005 special report on CCS, the IPCC gives several general estimates of both CO_2 transport and storage costs. Notwithstanding the fact that the actual level of both costs depends on a number of case-specific variables,[90] these estimates give a rough idea of the order of magnitude of CO_2 transport and storage costs. According to IPCC estimates, *transport* costs for onshore CO_2 pipelines could range from approximately \$1–6/tCO_2/250km, depending on the yearly transport volume.[91] At a transport volume of 1 Mt/year, transport costs would amount to at least \$6/tCO_2/250km.[92] In comparison, IPCC estimates for costs of CO_2 *storage* in depleted oil and gas fields and saline aquifers range from \$0.5–8/tCO_2 stored (excluding monitoring costs).[93] It should be stressed that the full range of cost estimates for individual options is very large,[94] and that storage is the component with the highest relative cost variability due to the range of possible characteristics of storage locations and the potential for enhanced hydrocarbon recovery.[95] Monitoring is estimated to add \$0.1–0.3/tCO_2 stored.

Based on these figures it is difficult to make any general statements as to the level of transport costs in comparison with storage and monitoring costs. The estimates have a fairly broad range and the exact numbers will depend on the specifics of each case. Nevertheless, both cost ranges seem to be pretty comparable. The estimates in the range of transport costs are not much lower than the estimated range of CO_2 storage and monitoring costs. This picture seems to be confirmed by a more recent study by consulting firm McKinsey.[96] McKinsey estimates total cost of CO_2 transport by onshore pipeline to be around €4/tCO_2, while total onshore storage cost is estimated at €4/tCO_2 for depleted oil and gas fields and €5/tCO_2 for saline aquifers.[97] Should transport costs in practice indeed be comparable to storage costs, this could be considered as a factor significantly narrowing the relevant geographic market for the storage of CO_2.

The question of transport costs and the extent to which these may hinder trade between different areas will generally be addressed by an analysis of trade flows/patterns of shipments.[98] Substantial cross-border trade flows are an indication

---

[90]   Storage costs, for instance, differ according to the geological characteristics of the particular formation (depleted oil or gas field or saline aquifer) and the depth at which the captured CO_2 is to be stored. See IPCC (n 64) 345.

[91]   Ibid.

[92]   Ibid.

[93]   Ibid.

[94]   Ibid.

[95]   McKinsey and Company (n 85) 28.

[96]   McKinsey and Company (n 85).

[97]   Ibid 19–20.

[98]   Commission, 'Market Definition Notice' (n 39) para 31.

of a wider than national geographic market. The relevance of cross-border trade is that it will normally make prices in the different Member States converge (by competitive pressure) and thus homogenise conditions of competition.[99] In its 2005 report on progress in creating the internal gas and electricity market, the Commission noted that the absence of price convergence across the EU and the low level of cross-border trade were key indicators for the lack of integration between national markets.[100] In 2004, cross-border trade flows of electricity, for instance, stood at around 10.7% of total consumption.[101] For an idea of the possible future geographic scope of the relevant product markets for CO_2 transport and storage and monitoring, it could therefore be useful to briefly address expectations of cross-border CO_2 transport and storage.

In 2008, McKinsey developed several scenarios for the development of EU CO_2 transport infrastructure in the demonstration phase (until 2020), early commercial phase (2020–2030) and commercial phase (post-2030).[102] In the demonstration phase, it expects CO_2 to be transported locally and cost-effectively through point-to-point or hub and spoke connections.[103] Likewise, it foresees clusters of local cost effective CO_2 transport through hub and spoke connections for the early commercial phase. In the commercial phase, CCS could develop in two different ways, depending on the regional availability of storage. If widely distributed local storage is proven, McKinsey expects the CCS roll-out to remain largely local, with regional capture-storage clusters. If Europe will not have enough widespread, accessible local storage or public discussions were to lead to a mainly offshore solution, the necessary transport network would have to increase significantly in size. In that case, a pan-European transport network could be developed to connect regional clusters with large international storage locations such as offshore deep saline aquifers in the North Sea area.

In a 2010 report commissioned by the North Sea Basin Task Force,[104] Element Energy consultancy drafted several scenarios for the deployment of CCS in the

---

[99]   In the *United Brands* case, the ECJ first introduced the criterion of 'a clearly defined geographic area (...) where the conditions of competition are sufficiently homogenous for the effect of the economic power of the undertaking concerned to be evaluated'. See Case 27/76 *United Brands* (n 31) para 11.

[100] Commission, 'Communication from the Commission to the Council and the European Parliament: Report on Progress in Creating the Internal Gas and Electricity Market' COM (2005) 568 final 2.

[101] Ibid 3.

[102] McKinsey and Company (n 85) 37–39.

[103] The concept of hub and spoke connections refers to captured CO_2 from several installations being transported to a central hub from which it is consequently transported to a storage formation.

[104] On the North Sea Basin task force, see its website at <http://nsbtf.squarespace.com/> accessed 18 April 2011.

North Sea area by four European countries with CCS ambitions (the UK, Norway, the Netherlands and Germany).[105] The report concludes that cross-border transport of CO_2 between the members of the North Sea Basin Task Force may occur before 2020 (demonstration phase), but will likely not be necessary or important.[106] This finding is in line with the earlier mentioned expectations of McKinsey and the Commission for the same period.[107] The report further finds that cross-border transport of CO_2 will not be a central feature of transport networks in 2030.[108] In a 'very high' scenario (270 Mt CO_2/yr captured and stored in 2030), cross-border transport of CO_2 could comprise 10–15% of overall CO_2 storage in the Member States by 2030.[109] In the longer term (up to 2050), the report considers cross-border transport to be essential under the same scenario, with Germany and the Netherlands storing CO_2 in UK and Norwegian storage formations.[110]

The expectations of both McKinsey and Element Energy indicate that EU cross-border trade flows of captured CO_2 could be small until at least around 2030.[111] By 2030, McKinsey expects CO_2 to be transported through local clusters, while Element Energy estimates cross-border transport of CO_2 to constitute 10–15% of overall CO_2 storage in the most optimistic scenario for the four countries concerned. These are levels comparable to those that the Commission found to be low in determining the scope of the geographic electricity market in 2005 (national at most). After 2030, chances of substantial cross-border trade flows of captured CO_2 may, under several uncertain circumstances, increase. Accordingly, the expected trade flows of captured CO_2 in both reports do not seem to hint in the direction of a geographic range of the relevant product markets for the transport and storage and monitoring of CO_2 wider than, at most, a single Member State before at least 2030.

---

[105] Element Energy, 'One North Sea: A Study into North Sea Cross-border CO_2 Transport and Storage' (2010) <www.npd.no/Global/Engelsk/3%20-%20Publications/Reports/OneNorthSea/OneNortSea_Final.pdf> accessed 16 November 2011.

[106] Element Energy (n 105) 57.

[107] See Commission, 'Guidance Document 2' (n 64) 67.

[108] Element Energy (n 105) 63.

[109] Ibid 62.

[110] Ibid 67.

[111] See also Commission, 'Impact Assessment – Accompanying Document to the Communication from the Commission to the European Parliament, the Council, the European Economic and Social Committee and the Committee of the Regions: Energy Infrastructure Priorities for 2020 and Beyond – A Blueprint for an Integrated European Energy Network' SEC (2010) 1395 final 21. In the impact assessment of its 2010 Communication on EU energy infrastructure priorities for 2020 and beyond the Commission states that without intervention, CO_2 pipelines will remain geographically remote (unconnected) from one another. According to the Commission, without intervention, CO_2 pipelines installed during 2014–2020 will be relatively short in length, associated with specific projects and tailored to the (particular) needs of these projects.

However, as I have argued in the Introduction to this thesis, there are indications that captured CO_2 could, at least in North-West Europe, mainly be stored offshore in the coming years. For some Member States, this might make cross-border CO_2 transport and storage necessary. France and Poland, for instance, do not have offshore storage capacity and would, in such a scenario, need to transport their CO_2 abroad.[112] In a report on the development of a large-scale CO_2 transport infrastructure in Europe, Neele and others estimate that about 7% of total captured CO_2 will transported cross-border in 2020 in an offshore storage-only scenario an about 19% in and enhanced oil recovery scenario.[113] For 2030, the estimates for both scenarios are 70% and 71% respectively.[114] Should the predominantly offshore storage of CO_2 in the EU materialise in the coming years, then there is a fair chance of the geographic scope of CO_2 transport and storage markets being wider than a single Member State, somewhere between 2020 and 2030.

6.3.1.4   The competitive position of CO_2 transport and storage operators

As we have seen in section 6.3.1.1, market shares still play a considerable role in the assessment of dominance. The higher the market share, the more likely it is to constitute an important preliminary indication of the existence of a dominant position.[115] An undertaking with a statutory monopoly[116] will obviously be in a dominant position.[117] In the absence of statutory monopoly, the Commission and the Courts will begin the assessment of market power by looking at market shares.[118] The *E.ON Gas* and the *RWE/Essent* cases seem to suggest that an undertaking with a natural monopoly[119] in the relevant market of a network-bound sector[120] like the electricity and gas sectors is likely to be in a dominant position.[121] The Commission has held in several cases that operators controlling

---

[112]   Filip Neele and others (CO2Europipe), 'Development of a Large-scale CO_2 Transport Infrastructure in Europe: Matching Captured Volumes and Storage Availability' (2010) 32 <www.co2europipe.eu/Publications/D2.2.1%20-%20CO2Europipe%20Report%20CCS%20 infrastructure.pdf> accessed 5 October 2012.

[113]   Neele (n 112) 33.

[114]   Ibid.

[115]   European Commission, 'Enforcement Priorities' (n 33) para 15.

[116]   A statutory monopoly, also known as a legal monopoly, is a monopoly granted by the state.

[117]   Jones and Suffrin (n 36) 326.

[118]   Ibid.

[119]   A natural monopoly is a monopoly in a market that can be served at lower cost by having only one producer rather than many producers. See William W Sharkey, *The Theory of Natural Monopoly* (Cambridge University Press 1982) 2.

[120]   See n 18.

[121]   See *E.ON Gas* (Case COMP/39.317) Commission Decision of 4 May 2010, paras 23–24, and *RWE/Essent* (Case COMP/M.5467) Commission Decision of 23 June 2009, paras 181 and 186.

gas and electricity transport infrastructure (both distribution and transmission) hold a natural monopoly in the relevant product market.[122]

According to Roggenkamp, the CCS Directive appears to be based on the concept that 'both underground storage facilities and CO_2 transport infrastructure are natural monopolies'.[123] Roggenkamp argues that this is not surprising in view of 'the relative scarcity of storage capacity and the costs involved in developing suitable storage and transportation facilities'.[124] Similarly, Andrews finds that 'some markets for CCS services will exhibit characteristics of natural monopoly'.[125] However logical the general qualification of CO_2 transport and storage infrastructure as natural-monopoly infrastructure may seem, its practical validity can be doubted.

First, a general qualification of infrastructure as natural-monopoly infrastructure disregards the fact that natural monopolies do not exist in the abstract, but in concrete real-life markets, the characteristics of which can greatly vary and change over time.[126] Even in the gas sector, where the use of the term natural monopoly arguably is much more appropriate than in the CCS sector,[127] it can be questioned whether, for instance, transportation is a

---

[122] See e.g. *Verbund/EnergieAllianz* (Case COMP/M.2947) Commission Decision of 11 June 2003, para 27; *E.ON/MOL* (Case COMP/M.3696) Commission Decision of 21 December 2005, paras 97, 98, 212 and 215; *ENEL/EMS* (Case COMP/M.4841) Commission Decision of 20 December 2007, para 12; *CASC/JV* (Case COMP/M.5154) Commission Decision of 14 August 2008, para 19; *RWE/Essent* (n 119) para 181; *TenneT/E.ON* (Case COMP/M.5707) Commission Decision of 4 February 2010, para 6; *E.ON Gas* (n 119) para 24; and *TenneT/Elia/ Gasunie/APX-Endex* (Case COMP/M.5911) Commission Decision of 15 September 2010, paras 39 and 45.

[123] Roggenkamp (n 3) 276.

[124] Ibid.

[125] Andrews (n 59) 437.

[126] In this regard, see also Paul L Joskow and Roger G Noll, 'The Bell Doctrine: Applications in Telecommunications, Electricity, and Other Network Industries'(1999) 51 Stanford Law Review 1249, 1251–52. Joskow and Noll argue that the fact that components of the network have natural monopoly characteristics does not necessarily imply that any part of the industry ought to be a monopoly. See also Adam Candeub, 'Trinko and Re-grounding the Refusal to Deal Doctrine' (2005) 66 University of Pittsburgh Law Review 821, 832. Candeub argues that 'the concept of a natural monopoly is only clear in economic theory and whether one, in fact, exists is difficult to show'.

[127] This is due to the fact that extensive (state-controlled and financed) infrastructure already was in place when markets where liberalised. As a Commission official in the Directorate General for Competition once pointed out in relation to the incumbent telecommunications networks, 'in the fixed network field the new entrants are faced with a situation where the incumbents hold fixed network assets built over one hundred years of monopoly. None of the new entrants can, in the short term, build parallel networks in the local loop which could rival these assets worth 200–300 billions of euros of investment'. See Nikos T Nikolinakos, 'Access Agreements in the Telecommunications Sector: Refusal to Supply and the Essential Facilities Doctrine under E.C. Competition Law' (1999) 20 European Competition Law Review 399, 404. See also Neelie Kroes, 'The Interface Between Regulation and Competition

monopoly per se.[128] There might occasionally be a realistic possibility of a competing pipeline being established,[129] as has indeed happened in the past.[130] One should therefore be careful in generally characterising CO_2 transport and storage infrastructure as natural-monopoly infrastructure. Whether an operator really holds a natural monopoly in the market for CO_2 transport or storage, depends on the specifics of the relevant market.

Second, in the case of an alleged natural monopoly the question is whether it is, from a societal point of view, economically efficient to create an alternative to the alleged natural-monopoly infrastructure/facility. Only if the answer is no does a natural monopoly exists. In relation to CO_2 transport by pipeline there might in certain cases be an economically efficient alternative. IPCC calculations indicate that CO_2 transport by ship becomes cost-competitive with CO_2 transport by pipeline when CO_2 is transported over longer distances.[131] It would be difficult to maintain that the operator of a long-distance CO_2 transport pipeline holds a natural monopoly in the market for CO_2 transport if transport by ship would indeed be able to compete with transport by pipeline over long distances.

As indicated earlier, along with market share, the Commission will also assess other factors, notably the credible threat of future expansion by actual competitors or entry by potential competitors (expansion and entry). In its guidance paper, the Commission notes that for it to consider expansion or entry likely, it must be sufficiently profitable for the competitor or entrant, taking into account factors such as the barriers to expansion or entry.[132] The Commission's decision practice and the guidance notice suggest that there are a number of relevant types of barriers to entry, among which are economies of scale and sunk costs, opportunity costs, vertical integration and regulatory barriers (statutory monopolies/legal regulation).[133] In the following, I briefly examine these four

---

Law' 2 (speech at the Bundeskartellamt Conference on Dominant Companies, Hamburg, 28 April 2009) <http://ec.europa.eu/competition/speeches/index_theme_13.html> accessed 10 May 2011, and Jones and Suffrin (n 37) 494–5.

[128] Moen (n 13) 4. See also Commission, 'DG Competition Report on Energy Sector Inquiry' SEC (2006) 1724, 26 <http://ec.europa.eu/competition/sectors/energy/inquiry/full_report_part1.pdf> accessed 7 November 2011. In its sector inquiry report, the Commission states that '*in most cases*, the construction of competing parallel gas networks is not economically viable: the network operator on a given transport market *can*, therefore, *often be considered* to be in control of a natural monopoly' (emphasis added). On the issue of competition in gas transmission markets, see also Annelieke Beukenkamp, 'Pipeline-to-pipeline Competition: An EU Assessment' (2009) 27 Journal of Energy and Natural Resources Law 5, 32 and 41.

[129] Moen (n 13) 4. See also Joskow and Noll (n 126) 1293–94.

[130] Waelde and Gunst (n 4) 196. See also Kim Talus, *Vertical Natural Gas Transportation Capacity, Upstream Commodity Contracts and EU Competition Law* (Kluwer Law International 2011) 211–14.

[131] See n 63.

[132] Commission, 'Enforcement Priorities' (n 33) para 16.

[133] Ibid para 17, and Jones and Suffrin (n 36) 342–49.

types of potential barriers to entry in relation to the CO_2 transport, and storage and monitoring markets.

Sunk costs and economies of scale constitute perhaps the most obvious potential barrier to entry to both markets. Economies of scale refer to the situation in which an increase of output can be achieved at a lower average cost due to fixed costs[134] being spread out over more units of production.[135] As economies of scale are most likely to be found in industries with high fixed costs[136] and the fixed costs of CO_2 pipelines and storage facilities are considered to be high,[137] economies of scale can be expected to be of relevance for CO_2 transport, and storage and monitoring markets.

Both the 2005 IPCC special report on CCS as well as more recent research confirm that there generally seems to be considerable potential for economies of scale in both markets, particularly in the CO_2 transport market.[138] However, according to Bielicki and NERA Economic Consulting, returns to scale[139] will not only arise, as indicative of economies of scale; at a certain scale of capturing, transporting and storing CO_2, diseconomies of scale will arise.[140] Returns to scale vary with the CO_2 capacity of the capture-transport-storage system or the scale of deployment.[141] At small scales, returns to scale are increasing, suggesting

---

[134] Fixed costs are those costs that must be incurred even if production were to drop to zero; they do not vary with production as do variable costs. See Steven M Suranovic, *International Trade Theory and Policy* (2010) 80–81.

[135] Ibid and The Economist (n 88).

[136] Suranovic (n 134) 80–81.

[137] See e.g. Richard Haigh, 'A Framework for the Future Development of CCS' (2010) Carbon Capture Journal <www.carboncapturejournal.com/displaynews.php?NewsID=583> accessed 27 April 2011; Michael J Kuby, Richard S Middleton and Jeffrey M Bielicki, 'Analysis of Cost Savings From Networking Pipelines in CCS Infrastructure Systems' (2011) 4 Energy Procedia 2808, 2811; and Jason Mann, Jens Perner and Cristoph Riechmann, 'Carbon Capture & Storage (CCS) – Supporting by Carrots or Sticks?' (presentation held at the eight Conference on Applied Infrastructure Research, Berlin, 10 October 2009) <http://wip.tu-berlin.de/typo3/fileadmin/documents/infraday/2009/presentation/09a_mann_1pdf.pdf> accessed 27 April 2011.

[138] IPCC (n 63) 225, 259, 263 and 344, and Kuby, Middleton and Bielicki (n 137) 2811. See also DECC (n 11) 6.

[139] The return to scale is the increase in output relative to the increase in resource usage. Economies of scale in production is equivalent to increasing returns to scale. See Suranovic (n 136) 80–81.

[140] Jeffrey M Bielicki, 'Returns to Scale in Carbon Capture and Storage Infrastructure and Deployment' (2008) Harvard Kennedy School Energy Technology Innovation Policy Discussion Paper 2008–04 26 <http://belfercenter.ksg.harvard.edu/files/Bielicki_CCSReturnsToScale.pdf> accessed 27 April 2011, and NERA Economic Consulting, 'Developing a Regulatory Framework for CCS Transportation Infrastructure (Vol. 1 of 2)' (2009) 48 <www.decc.gov.uk/assets/decc/what%20we%20do/uk%20energy%20supply/energy%20mix/carbon%20capture%20and%20storage/1_20090617131338_e_@@_ccsreg1.pdf> accessed 27 April 2011.

[141] Bielicki (n 140) 26.

that it is efficient to expand the system to capture and store more CO_2.[142] At large scales, however, the returns to scale are decreasing (diseconomies of scale), suggesting that it would be more efficient to scale the system back.[143]

In line with the economies of scale, the sunk costs that have to be incurred in order to enter the markets for the transport, and storage and monitoring of CO_2 are generally considered to be high, particularly in relation to CO_2 transport.[144] As indicated above, these sunk costs will, among other things, consist of costs of constructing pipelines and other physical infrastructure such as compressor stations. Sunk costs that have to be incurred in order to enter the market for CO_2 storage and monitoring are, for instance, the costs of initial exploration, site assessment and site preparation (e.g. drilling). According to McKinsey, CAPEX (capital expenditure), which includes the sunk costs of storage equipment, pumps and the drilling of wells, represents around 75–80% of total storage costs per tonne of CO_2 abated for onshore storage and about 80–90% for offshore storage (platforms).[145]

In the *Intel* case, the Commission found several barriers to entry resulting from sunk investment[146] associated with fixed costs for a number of activities.[147] It stated that 'in general, a high share of fixed costs is indicative of significant barriers to entry and expansion'.[148] In view of this case and the above, fixed costs, sunk costs and economies of scale could constitute (significant) barriers to entry to the markets for CO_2 transport, and storage and monitoring.

---

[142] Bielicki (n 140) 26.

[143] Ibid.

[144] See e.g. Diederik Frans Apotheker, 'The Design of A Regulatory Framework for a Carbon Dioxide Pipeline Network' (MSc thesis, Technical University of Delft 2007) 2; Zhou and others, 'Uncertainty Modelling of CCS Investment Strategy in China's Power Sector' (2010) 97 Applied Energy 2392, 2393; Roman Mendelevitch and others, 'CO_2 Highways for Europe: Modelling a Carbon Capture, Transport and Storage Infrastructure for Europe' (2010) Centre for European Policy Studies Working Document No. 340 1 <http://aei.pitt.edu/15200/1/WD_340_CO2_Highways.pdf> accessed 27 April 2011, and Hans Schokkenbroek and others, 'Elements for a National Masterplan for CCS: Lessons Learnt' (2011) 4 Energy Procedia 5810, 5810.

[145] McKinsey and Company (n 85) 20.

[146] Even though the concepts of fixed costs and sunk costs appear to be similar, they are not the same. The difference between the two is that fixed costs can be reduced or eliminated if the undertaking is willing to exit the market entirely; sunk costs cannot be avoided or changed even by discontinuing production entirely. See Utilityregulation.com, 'Costing Definitions and Concepts' 3 <www.utilityregulation.com/content/essays/t1.pdf> accessed 28 April 2011.

[147] *Intel* (Case COMP/C-3/37.990) Commission Decision of 13 May 2009, para 877.

[148] Ibid.

Opportunity costs[149] might constitute a barrier to entry to the CO_2 storage and monitoring market.[150] In their 2010 CO_2 Transport and Storage Strategy, Energie Beheer Nederland (EBN)[151] and Gasunie[152] indicate that the exploration and production companies controlling nearly depleted gas fields in the Netherlands are, in principle, willing to cooperate in an early transfer of those fields, provided that proper compensation is paid for the remaining recoverable gas.[153] The market value of the remaining recoverable gas is indicative of the opportunity costs of freeing up nearly depleted gas fields. These costs are likely to constitute a barrier to entry to the CO_2 storage market as exploration and production companies will probably not consider storing captured CO_2 in these fields if their respective loss of income is not compensated. The same would likely hold for nearly depleted oil fields suitable for CO_2 storage.

Likewise, opportunity costs related to natural gas storage in (nearly) depleted oil or gas fields or even saline aquifers[154] may constitute a barrier to entry to the CO_2 storage and monitoring market. To operators of nearly depleted oil or gas fields or even saline aquifers suitable for gas storage, CO_2 storage and natural gas storage represent competing uses. They will likely only consider CO_2 storage an interesting alternative to natural gas storage if the business case is at least as strong as it is for natural gas storage. In both of the above examples (remaining recoverable natural gas or oil and natural gas storage) opportunity costs are increased by the uncertainties surrounding the commercial viability of CCS and corresponding future demand for CO_2 storage. It is questionable whether this uncertainty will diminish in the (short-term) future as it is to a large extent caused by the inherent volatility of CO_2 prices under the EU greenhouse gas emissions trading scheme (EU ETS).

---

[149]   Opportunity costs are the value of something which must be given up in order to achieve or acquire something else. See Jones and Suffrin (n 36) 348.

[150]   For an example of a case in which opportunity costs were considered a barrier to entry, see *British Midland* (Case IV/33.544) Commission Decision 92/213/EEC [1992] OJ L96/34. They could perhaps also form a barrier to entry to the CO_2 transport market as pipelines used for the transport of natural gas can also be used for the transport of CO_2 streams.

[151]   On behalf of the Dutch government EBN participates in the exploration and production of oil, natural gas and condensate in the Netherlands.

[152]   Gasunie N.V. is the Dutch gas transmission system operator.

[153]   EBN and Gasunie, 'CO2 Transport and Storage Strategy' (2010) 21.

[154]   In its special report on CCS, the IPCC notes that 'the majority of [natural] gas storage projects are in depleted oil and gas reservoirs and *saline formations*' (emphasis added). See IPCC, 'IPCC Special Report on Carbon Dioxide Capture and Storage. Prepared by Working Group III of the Intergovernmental Panel on Climate Change' (Bert Metz and others (eds) Cambridge University Press 2005) 211.

Future EU CCS markets could well be characterised by another barrier to entry, namely vertical integration.[155] In a 2010 paper, researchers from the Dresden University of Technology examined ownership structures of major enhanced oil recovery projects in the US, finding most market players to be active on all three levels of the CCS value chain.[156] Even though they believe EU CCS market players to be faced with risks different from those experienced by parties active in the US enhanced oil recovery market, Herold et al. also expect a tendency towards vertical integration in future EU CO_2 transport markets.[157] This would be caused by the high risk and uncertainty about e.g. available storage capacity and the future regulatory regime.[158]

Likewise, Vedder has argued that it is likely that some form of vertical integration in the form of ownership or long-term contracts will exist in future EU CCS markets.[159] According to Vedder, the considerable investments required in CO_2 capture, transport and storage will lead to parties seeking mutual assurances that the installations will be used at optimal capacity.[160] Vedder has argued that given that the entities most likely to be interested in CCS (electricity generators and the chemical industry) do not possess geological storage capacity, such vertical integration will probably involve entities currently active in the production of oil and gas.[161]

Finally, the 2010 report on a strategy for CO_2 transport and storage in the Netherlands by Gasunie and EBN expects Dutch CO_2 transport operators to seek

---

[155]   See also Jennifer Skougard, 'Getting from Here to There: Devising an Optimal Regulatory Model for CO2 Transport in a New Carbon Capture and Sequestration Industry' (2010) 30 Journal of Land, Resources, and Environmental Law 357, 357, and Element Energy, 'CO_2 Pipeline Infrastructure: An Analysis of Global Challenges and Opportunities' (2010) 86 <www.ccsassociation.org.uk/docs/2010/IEA%20Pipeline%20final%20report%20270410.pdf> accessed 2 January 2012. Vertical integration has in the past been found to indicate dominance. See e.g. Case 27/76 *United Brands* (n 31), *Soda Ash* (Case COMP/33.133-C) Commission Decision 2003/6/EC [2003] OJ L10/10 and *Distrigaz* (Case COMP/B-1/37966) Commission Decision of 11 October 2007.

[156]   Johannes Herold and others, 'Vertical Integration and Market Structure along the Extended Value Added Chain including Carbon Capture, Transport and Sequestration (CCTS)' (2010) <www.feem-project.net/secure/plastore/Deliverables/SECURE_deliverable_5%203%202.pdf> accessed 12 August 2010. Herold and others also found long-term take-or-pay contracts to be common in the enhanced oil recovery business. A take-or-pay contract is an agreement between a purchaser and a seller that requires the purchaser to either purchase a minimum volume of a product or service at a set price ('take'), or pay the minimum volume without taking delivery ('pay'). See Carl R Schultz, 'Modelling Take-or-Pay Contract Decisions' (1997) 28 Decision Sciences 213, 213.

[157]   Herold and others (n 156) 1.

[158]   Ibid.

[159]   Hans Vedder, 'An Assessment of Carbon Capture and Storage under EC Competition Law' (2009) 29 European Competition Law Review 586, 593–96.

[160]   Ibid 596.

[161]   Ibid.

long-term contracts with substantial financial guarantees in the demonstration (before 2020) and pre-commercial (2020–2030) phases.[162] This appears to lend support to the claims by Herold et al. and Vedder that future EU CO_2 (transport) markets could tend towards vertical integration.[163]

The long-term contracts mentioned above could be beneficial for both the carbon capturer and the storage operator, as the capturer would have the certainty that captured CO_2 is stored at predictable prices, whereas the storage operator will have a predictable flow of carbon dioxide.[164] However such long-term contracts could form a barrier to entry to the market for the storage and monitoring of CO_2, as they would make it more difficult for potential new entrants to develop a customer base. Cases like *Delimitis*,[165] *Distrigaz* (gas)[166] and *EDF* (electricity)[167] indicate that long-term supply contracts can be a means to foreclose competitors from their customer base.[168] For the same reason, one could imagine long-term CO_2 transport contracts forming a potential barrier to entry to the market for CO_2 transport. Likewise, vertical integration in the form of ownership either across the full chain or parts of the CCS chain (e.g. capture and transport or capture and storage) might make it more difficult for potential competitors in the CO_2 transport and storage and monitoring markets to develop a customer base.

Finally, there might be legal and regulatory barriers to entry, such as licensing requirements or a statutory monopoly.[169] In order to enter the CO_2 transport and storage and monitoring markets, undertakings will have to obtain a number of national permits such as environmental permits, permits related to spatial planning like the Dutch spatial planning and development permit (*omgevingsvergunning*) and possibly other permits. Depending on the length and costliness of permitting procedures, such permit requirements could constitute substantial barriers to entry to both markets.

---

[162]   EBN and Gasunie (n 153).

[163]   On the likelihood of future EU CO_2 transport markets tending towards vertical integration, see also NERA (n 141) 46. According to NERA, it is a likely scenario that UK CCS demonstration projects will see early CO_2 pipelines being developed by vertically integrated developers of CO_2 capturing plants, storage sites and pipelines.

[164]   Vedder (n 159) 593.

[165]   Case C-234/89 *Delimitis v. Henninger Bräu* [1991] ECR I-00935.

[166]   *Distrigaz* (n 155) paras 21–4.

[167]   *EDF* (Case COMP/39.386) paras 29–30.

[168]   Ricardo Cardoso and others, 'The Commission's GDF and E.ON Gas Decisions Concerning Long-term Capacity Bookings: Use of Own Infrastructure as Possible Abuse under Article 102 TFEU' (2010) European Commission Competition Policy Newsletter 2010–3, 4 <http://ec.europa.eu/competition/publications/cpn/cpn_2010_3_2.pdf> accessed 29 April 2011.

[169]   For Article 102 cases concerning statutory monopolies, see e.g. Case C-241–2/91 *Magill* [1995] ECR I-00734 and Case 226/84 *British Leyland* [1986] ECR 03263.

In addition, the CCS Directive provides that no storage site may be operated without a storage permit.[170] Article 8 of the Directive lists the conditions that have to be met before a storage permit may be issued by the competent authority. In practice, these are probably not easily met, as the would-be operator must be financially sound, technically competent and reliable to operate and control the site, and professional and technical development and training of the operator and all staff have to be provided.[171] Furthermore, a permit application shall contain, among other things, a (costly)[172] assessment of the storage site and storage complex as well as proof of a valid and effective financial security.[173]

By virtue of Article 6(2), the storage permitting procedures have to be open to all entities 'possessing the necessary capacities'. This means that, for instance, capturers can in principle also apply for a storage permit. Based on the requirement that the applicant possesses the necessary capacities, the primary candidates for CO_2 storage seem to be the oil and gas exploration and production companies, since they already possess the required technical competence and experience and often even hold valuable knowledge of the geological characteristics of a potential storage formation (e.g. a [nearly] depleted oil or gas field). Capturers could, however, easily make up for their shortcomings in this regard by establishing a joint venture with a production company. Nevertheless, the generally onerous requirements[174] that have to be met before a storage permit is granted could prove to be a barrier to entry to the CO_2 storage and monitoring market, particularly for parties other than exploration and production companies.

As for the CO_2 transport market, the CCS Directive contains only a bare minimum of regulation in this regard (Article 31). Further regulation, such as rules on who is to operate CO_2 pipelines, is left to the Member States.[175] Member States could decide to draft new legislation to this end, but they could also choose

---

[170] Article 6.

[171] Article 8(1)(b).

[172] See IEA (n 10) 47.

[173] Article 7(10). Article 19 of the Directive provides that the potential operator needs to present proof of a financial security or an equivalent as part of the application for a storage permit. In its guidance document on Articles 19 and 20 of the CCS Directive, the Commission states that a bottom-line contingency of at least 25% should be required (except for surrender of allowances in case of leakage). This means that the full cost of all possible obligations mentioned in Article 19 of the Directive (surrender of allowances in case of leakage excepted), would have to be supplemented by a margin of 25%, in case the obligations were underestimated. This makes the financial security obligation a potential barrier to entry to the CO_2 storage and monitoring market.

[174] For examples in the Dutch context see L Zima and G Vriezen, 'Wijziging van de Mijnbouwwet: Mooie Nazomer of Nieuwe Lente?' (2011) 5/6 Nederlands Tijdschrift voor Energierecht, 242.

[175] Commission, 'Impact Assessment' (n 71) 2.

to have the operation of CO_2 pipelines fall under existing regulation of, for instance, the gas sector. In this case, the applicable regulatory regime will likely depend on the qualification of CO_2 pipelines under national law.[176] If CO_2 pipelines are considered to be similar to high-pressure natural gas pipelines (transmission), the operation of those pipelines would normally fall under the (in the EU not uncommon) statutory monopoly of the gas transmission system operator.[177] Such statutory monopoly would obviously be a major barrier to entry to the market for the transport of CO_2.

The above shows that there is a real likelihood of undertakings active in the future markets for CO_2 transport and storage and monitoring being in a dominant position. Whether an undertaking really holds a dominant position in the relevant market will obviously have to be determined through the facts of each specific case. Expectations of the geographic scope of the relevant product markets vary (and are thus difficult to base expected market shares on), but, as we have seen above, both markets could well be characterised by several barriers to entry.

6.3.2   ABUSE AND DOMINANT CO_2 TRANSPORT AND STORAGE OPERATORS

One of the possible abuses under Article 102 is the refusal to deal/supply a product or service, including the refusal to grant access to infrastructure/facilities.[178] In the following, this type of abuse under Article 102 is further discussed in the light of the relevant decision practice and case law of the Commission and the Courts. Thereafter, the changes following the Commission's guidance notice are addressed. Finally, the consequences of all of this for the refusal to grant access to CO_2 transport and storage infrastructure based on a technical incompatibility of the captured CO_2 stream with the relevant infrastructure are discussed.

6.3.2.1   'Abuse' of a dominant position

In the early case of *Hoffmann-La Roche*, the ECJ gave a very wide definition of what constitutes an abuse under Article 102.[179] By virtue of this definition, conduct which does not qualify as 'normal competition' and negatively affects

---

[176]   In this regard, see also Roggenkamp (n 3) 293. Roggenkamp argues that the geological storage of CO_2 could be characterised as a reversed downstream activity, to which the upstream laws and regulations apply.

[177]   The transmission system operator is the entity operating the gas transmission network.

[178]   See also Jones and Suffrin (n 36) 479.

[179]   Case 85/76 *Hoffmann-La Roche* [1979] ECR 00461, para 91.

competition is considered an abuse under Article 102. In the *Continental Can* case, the ECJ provided that the list of abuses in Article 102 is not limitative as it 'merely gives examples, not an exhaustive enumeration of the sort of abuses of a dominant position prohibited by the Treaty'.[180]

The types of abuse under Article 102 are generally classified into two categories: exploitative and exclusionary abuses.[181] The difference between the two types of abuse, in essence, is that the first is directly harmful to *consumers*, while the second type of abuse directly affects the undertaking's *competitors*. In the end, however, both lead to a loss of consumer welfare. The division should therefore not be treated too rigidly, especially not since the same conduct may constitute both an exploitative and an exclusionary abuse.[182] Nevertheless, O'Donoghue and Padilla argue that, for analytical reasons, it is important to be clear about which clause in Article 102 applies, as in each case the legal and economic principles are somewhat different.[183]

Article 102 mentions, in order of paragraphs, (a) *exploitative* abuses (excessive prices and unfair terms), (b) *exclusionary* abuses (limiting of production, markets or technical development), (c) *discriminatory* abuses (subjecting equivalent transactions to dissimilar conditions) and (d) *tying* abuses (making the conclusion of contracts subject to other obligations). In practice, limiting production, markets and technical development is the most frequent and important category of abuse, since it broadly covers any type of exclusionary conduct that limits rivals' possibilities and causes harm to consumers, such as refusal to deal, discrimination against downstream rivals and abuses in connection with the adoption of standards or other specifications.[184] The vast majority of infringement decisions on abuse of a dominant position have concerned exclusionary abuses under Article 102(b). The wording of Article 102(b) indicates that only conduct that results in output limitation and causes prejudice to consumers is considered exclusionary under the article.[185]

---

[180]   Case 6/72 *Continental Can* [1973] ECR 00215, para 26. See, for instance, also Case C-333/94 P *Tetra Pak II* [1996] ECR I-5951, para 37, and Case C-95/04 P *British Airways* [2007] ECR I-2331, para 57.

[181]   Some also include a third category, 'reprisal abuses'. This type of abuse refers to dominant undertakings disciplining competitors. See, for instance, John Temple Lang, 'Abuse of a Dominant Position in European Community Law, Present and Future: Some Aspects' in Barry E Hawk (ed), *Fifth Fordham Corporate Law Institute* (Law and Business 1979) 25.

[182]   Craig and de Búrca (n 26) 1007. An example is discriminatory behaviour. According to Goyder, the European Commission regards the distinction as not of great importance. See Goyder (n 41) 283.

[183]   O'Donoghue and Padilla (n 27) 194–95.

[184]   Ibid 175 and 197.

[185]   Ibid 197–98.

The following section focuses on one of these types of exclusionary abuse under Article 102.

6.3.2.2 Refusing to deal

a)   The concept of refusal to deal

The point of departure for all undertakings, including dominant ones, is that they are generally free to deal with whomever they choose. However, it is well established under Article 102 that undertakings in a dominant position may, in limited circumstances, be required to deal with third parties with whom they do not wish to enter into or continue contractual relations.[186] The phrase 'in limited circumstances' is of great importance. Article 102 does not impose upon dominant undertakings an absolute duty to deal with all those requesting them to do so.[187]

The basic rationale for a duty to deal is that if an input is essential for the provision of a product or service in a downstream market – in the sense that it is either impossible or prohibitively expensive to duplicate – it would, if denied to an undertaking operating in the downstream market, effectively remove that undertaking as a competitor.[188] The duty to deal is highly controversial, since it interferes with the freedom of contract and basic property rights and is therefore only applied in extraordinary circumstances.[189]

For conduct of a dominant undertaking to constitute a refusal to deal, it does not have to be an outright refusal to supply the relevant product, service or grant access to the facility concerned. 'Constructive' refusals, i.e. dealing on unreasonable (or discriminatory) terms only, might also constitute an abuse of a dominant position under Article 102.[190]

In its guidance notice, the Commission states that the concept of refusal to supply covers a broad range of practices, such as refusal to supply products to existing or new customers, refusal to licence intellectual property rights or refusal to grant access to an essential facility or network.[191] Another common

---

186   O'Donoghue and Padilla (n 27) 407.
187   Jones and Suffrin (n 36) 480.
188   O'Donoghue and Padilla (n 27) 409.
189   Ibid 407.
190   *Deutsche Post* (Case COMP/C-1/36.915) Commission Decision 2001/892/EC [2001] OJ L331/40, para 141. See also *Sea Containers* (Case IV/34.689) Commission Decision 94/19/EEC [1994] OJ L15/8, discussed below. See also Jones and Suffrin (n 36) 479–80.
191   Commission, 'Enforcement Priorities' (n 33) para 78.

distinction is that between refusal to deal with competitors and refusal to deal with customers.[192]

The abuse of a dominant position through a refusal to deal came to the fore in a number of early Commission and Court cases on Article 102.[193] All of these early cases, except the *United Brands* case, concerned the situation in which upstream facilities/infrastructures were used by an undertaking dominant in the relevant upstream market to prevent (new) competition on a related downstream market.[194] According to Capobianco and Anderman, the early Commission decisions led to fears over opportunistic conduct of competitors of incumbent undertakings, causing the ECJ to try and minimise risks of such behaviour in its judgment in the *Bronner*[195] case.[196]

In this seminal Article 102 case, the ECJ specified a limited number of circumstances in which a duty to grant access to a facility exists. In *Bronner*, the ECJ was asked whether it constituted an abuse for a newspaper group to refuse to allow access to its delivery network to the publisher of a competing newspaper.[197] It provided that a refusal could constitute an abuse under Article 102 only if (1) it would be likely to eliminate all competition in the downstream market from the person requesting the service, (2) the service in itself is indispensable to carrying on that person's business as there is no actual or potential substitute for it and (3) it cannot be objectively justified.[198] The ECJ further indicated that, in assessing 'indispensability', the question was whether

---

[192]   See, for instance, O'Donoghue and Padilla (n 27) 423 and 463.

[193]   See, for instance, Joined Cases 6 and 7/73 *Commercial Solvents* [1974] ECR 00223; Case 27/76 *United Brands Company* (n 31); *London European* (Case IV/32.318) Commission Decision 88/589/EEC [1988] OJ L317/47; *British Midland* (n 148); *Sea Containers* (n 188); and *Port of Rødby* Commission Decision 94/119/EC [1994] OJ L55/52.

[194]   I use the terms upstream and downstream in the 'antitrust way', which differs from the way in which both terms are used in EU energy law. There, upstream refers to the gas/electricity production sector, whereas downstream sectors are the gas/electricity transport and sales sectors. In EU antitrust law, the upstream market is the market for the infrastructure access to which is sought by a party not controlling that infrastructure, for instance the gas transport market, and the downstream market is the market in which that input is essential, for instance the gas sales market.

[195]   Case C-7/97 *Bronner* [1997] ECR I-07791.

[196]   Antonio Capobianco, 'The Essential Facilities Doctrine: Similarities and Differences Between the American and the European Approach' (2001) 26 European Competition Law Review 548, 558, and Steven Anderman, 'The Epithet That Dares Not Speak its Name: The Essential Facilities Concept in Article 82 EC and IPRs after the Microsoft Case' in Ariel Ezrachi (ed), *Article 82 EC: Reflections on its Recent Evolution* (Hart Publishing 2009) 87, 90.

[197]   Frances Murphy, 'Abuse of Regulatory Procedures: The AstraZeneca Case: Part 3' (2009) 30 European Competition Law Review 314, 322.

[198]   Para 41. Jones and Suffrin mention *four* factors, splitting up the above second factor in the requirement that the access must be indispensable to carrying on the other person's business and the separate condition that there must be no actual or potential substitute for it. However, the ECJ's use of the connecting word 'inasmuch' suggests that the two in fact constitute one

---

there were 'technical, legal or even economic obstacles' to developing an alternative facility.[199]

Stothers has, in my view, convincingly argued that the ECJ in the *Bronner* case confused the issues of dominance and abuse and that conditions (1) and (2) in reality are facets of the determination of dominance as they essentially ask whether there are substitutes available to the person requesting the service.[200] They are in fact part of the assessment of demand substitutability. According to Stothers, the underlying test applied by the ECJ in the *Bronner* case is identical to that applied in the *Commercial Solvents* case[201] and therefore, in a roundabout way, says that a refusal to supply without an objective justification will constitute abuse of a dominant position.[202] However, the ECJ's judgment in *Bronner* is generally seen as a more restrictive, less interventionist approach to refusals to deal under Article 102,[203] making clear that an obligation to grant access to a facility will only arise in exceptional circumstances.[204]

b)   The requirement of two separate but vertically related markets

One of the most interesting questions related to Article 102 is whether the dominant undertaking controlling the facility/infrastructure to which access is requested must be active in the associated downstream market for a refusal to grant access to constitute an abuse.[205] Under the heading of refusal to supply, the Commission's guidance notice only mentions the classic situation of the dominant undertaking competing on the downstream market with the customer whom it refuses to supply.[206] The guidance notice seems to suggest that the Commission does not regard refusals to deal as an enforcement priority where the dominant undertaking does not compete on the downstream market.[207]

---

condition. In its judgment, the ECJ followed the conservative approach that AG Jacobs took in his opinion. See Case C-7/97 *Bronner* [1998] ECR I-07791, Opinion of AG Jacobs.

[199]   Para 44. See O'Donoghue and Padilla (n 27) 426.
[200]   Stothers (n 38) 258–59.
[201]   See n 192.
[202]   Stothers (n 38) 259.
[203]   See e.g. Capobianco (n 196) 559; Mats A Bergman, 'The Bronner Case: A Turning Point for the Essential Facilities Doctrine?' (2000) 21 European Competition Law Review 59; Slot and Skudder (n 18) 95; O'Donoghue and Padilla (n 27) 427; and Jones and Suffrin (n 36) 494.
[204]   Jones and Suffrin (n 36) 494.
[205]   See also Vassilis G Hatzopoulos, 'The Essential Facilities Doctrine (EFD) in EU Antitrust Law' (2006) <www.concorrencia.pt/Download/Essential_Facilities-Hatzopoulos.pdf> accessed 4 May 2011.
[206]   Commission, 'Enforcement Priorities' (n 33) para 76.
[207]   See also Gravengaard and Kjaersgaard who argue that this could indicate that the Commission would normally not consider a refusal to supply from non-vertically integrated undertakings as being contrary to Article 102 TFEU, or at least that the likelihood of anticompetitive foreclosure is so small that these cases generally have no priority. See Martin Andreas Gravengaard and Niels Kjaersgaard, 'The EU Commission Guidance on

A question preceding the above question is whether it is at all necessary to have two separated but vertically related markets in order for a refusal to deal to constitute an abuse under Article 102. According to O'Donoghue and Padilla, it has always been understood that the duty to deal under Article 102 only applied in vertical situations, that is an upstream market for the input in question and a downstream market in which that input is essential.[208] Fine, however, has argued that the ECJ's judgment in the *Bronner* case and the Commission's decision in the *IMS Health* case[209] indicate that the existence of two markets is not necessary for the application of Article 102.[210]

However, the Commission's statement in the *IMS Health* case that the fact that the Courts' case law on refusal to deal involved two markets 'does not preclude the possibility that a refusal to licence an intellectual property right can be contrary to Article 82 [now Article 102 TFEU]'[211] was corrected by the ECJ in its preliminary ruling[212] in the same case. In this ruling, the ECJ held that it was relevant, in order to assess whether the refusal to grant access to a product or service indispensable for carrying on a particular business activity was an abuse, to distinguish an upstream market, constituted by the product or service, and a (secondary) downstream market, on which the product or service in question is used for the production/supply of another product/service.[213] The ECJ added that it sufficed, to this end, to identify a 'potential' or 'hypothetical' upstream market.[214]

The ECJ thereby also invalidated Fine's reading of the *Bronner* judgment, who had argued that the fact that there was no upstream 'access market' in the Bronner case – as the dominant undertaking, Mediaprint, had never licensed access to it – demonstrated that the absence of two markets was not an impediment to the application of Article 102.[215] In its guidance notice, the Commission still adheres to the ECJ's ruling in *IMS Health*. It states that 'the Commission does not regard it as necessary for the refused product to have been already traded: it is sufficient that there is demand from potential purchasers and

---

[208] Exclusionary Abuse of Dominance and its Consequences in Practice' (2010) 31 European Competition Law Review 285, 303.
[208] O'Donoghue and Padilla (n 27) 435.
[209] *IMS Health* (Case COMP D3/38.044) Commission Decision of 3 July 2001.
[210] Frank Fine, 'NDS/IMS: A Logical Application of the Essential Facilities Doctrine' (2002) 23 European Competition Law Review 457, 459–60.
[211] Para 184.
[212] Article 267 TFEU contains the preliminary reference procedure by which the ECJ can give guidance to national courts on the interpretation of the Treaties and the validity and interpretation of acts.
[213] Case C418/01 *IMS Health* [2004] ECR I-05039, para 42.
[214] Ibid para 44.
[215] Fine (n 210) 459.

that a potential market for the input at stake can be identified'.[216] It therefore seems necessary to have at least two vertically related markets, even though one of these is merely potential (upstream market),[217] for a refusal to grant access to an indispensable facility to be abuse of a dominant position under Article 102.[218]

The second question then is, as stated above, whether the dominant undertaking controlling the facility to which access is requested has to be active in the related downstream market for a finding of abuse under Article 102 to be possible. Based on the *Ladbroke* case,[219] Temple Lang and Doherty have contended that this question must, at least in intellectual property cases, be answered in the positive.[220]

However, this can be doubted on the basis of the later (post-*Bronner*) *Aéroport de Paris*[221] case.[222] In this case, the dominant undertaking, Aéroport de Paris, had argued that its behaviour did not constitute an abuse of a dominant position as it was not present in the downstream market for ground handling services, for which, since it had a legal monopoly over the management of Paris airports, it granted concessions.[223] The Court of First Instance (CFI) rejected the applicant's argument and stated that:

> [W]here the undertaking in receipt of the service is on a separate market from that on which the person supplying the service is present, the conditions for the applicability of Article 86 [now Article 102 TFEU] are satisfied provided that, owing to the dominant position occupied by the supplier, the recipient is in a situation of economic dependence vis-à-vis the supplier, *without their necessarily having to be present on the same market*. It is sufficient if the service offered by the supplier is necessary to the exercise by the recipient of its own activity.[224]

The CFI's statement was not challenged in the subsequent appeal before the ECJ.[225] As Evrard argues, the *Aéroport de Paris* case is an example of the duty to

---

[216]   Commission, 'Enforcement Priorities' (n 33) para 79.
[217]   For a criticism of this standard see e.g. Damien Geradin, 'Limiting the Scope of Article 82 EC: What Can the EU Learn From the US Supreme Court's Judgment in *Trinko*, in the Wake of *Microsoft*, *IMS*, and *Deutsche Telekom*?' (2004) 41 Common Market Law Review 1519, 1530.
[218]   See Anderman (n 196) 88.
[219]   Case T-504/93 *Ladbroke* [1997] ECR II-00923.
[220]   See John Temple Lang, 'The Principle of Essential Facilities in European Community Competition Law: The Position since Bronner' (2000) 14 <http://lawcourses.haifa.ac.il/antitrust_s/index/main/syllabus/lang_article.pdf> accessed 5 May 2011, and Barry Doherty, 'Just What are Essential Facilities?' (2001) 38 Common Market Law Review 397, 426.
[221]   Cases T-128/98 *Aéroport de Paris* [2000] ECR II-03929.
[222]   See Sébastien J Evrard, 'Essential Facilities in the European Union: Bronner and Beyond' (2004) 10 Columbia Journal of European Law 491, 515.
[223]   Case T-128/98 *Aéroport de Paris* (n 221) para 163.
[224]   Para 165 (emphasis added).
[225]   C-82/01 P *Aéroport de Paris* [2002] ECR I-09297.

deal of an owner vis-à-vis his customers, with whom he does not compete in the related downstream market.[226] Another example is the Commission's decision in the 2010 *Swedish Interconnectors* case.[227] In this case, the Commission did not consider the fact that the dominant Swedish electricity transmission system operator, Svenska Kraftnät, was not active in the downstream markets of retail and wholesale of electricity in Sweden, an impediment to finding that Svenska Kraftnät had abused its dominant position in the market for electricity transmission services in Sweden.

As the *Aéroport de Paris* and the *Swedish Interconnectors* cases appear to contradict the *Ladbroke* case, it seems difficult to give a definitive answer to the question of whether the dominant undertaking controlling the facility/ infrastructure to which access is requested must be active in the related downstream market for a refusal to grant access to constitute an abuse. Yet, even if the above question must be answered positively, there are several cases to suggest that the upstream dominant undertaking is not required to be *dominant* in the related downstream market for its conduct to possibly be abusive under Article 102.[228]

c)   Changes following the Commission's guidance notice

For the assessment of all types of exclusionary conduct, the most important principle in the guidance notice allegedly is that[229] 'the Commission will normally intervene under Article 82 where, on the basis of cogent and convincing evidence, the allegedly abusive conduct is likely to lead to anti-competitive foreclosure'.[230] 'Anti-competitive foreclosure' is described as 'a situation where effective access of actual or potential competitors to supplies or markets is hampered or eliminated (...) to the detriment of consumers'.[231] As Gravengaard and Kjaersgaard and Kellerbauer indicate, the Commission will therefore probably not intervene over conduct which might result in the elimination of a competitor if such conduct is not likely to result in harm to consumers.[232] This view is supported by one of the paragraphs on the purpose of the guidance notice, in which the Commission states that:

---

[226]   Evrard (n 222) 515.

[227]   *Swedish Interconnectors* (Case COMP/39351) Commission Decision of 14 April 2010.

[228]   See e.g. Case T-219/99 *British Airways* [2003] II-05917, para 127, and Case C-52/09 *TeliaSonera* [2011] ECR I-00527, paras 83–89.

[229]   Commission, 'Enforcement Priorities' (n 33) para 20.

[230]   Gravengaard and Kjaersgaard (n 207) 288.

[231]   Commission, 'Enforcement Priorities' (n 33) para 19.

[232]   Gravengaard and Kjaersgaard (n 207) 288 and Manuel Kellerbauer, 'The Commission's New Enforcement Priorities in Applying Article 82 EC to Dominant Companies' Exclusionary Conduct: A Shift Towards a More Economic Approach?' (2010) 31 European Competition Law Review 175, 176.

[T]he Commission is mindful that what really matters is protecting an effective competitive process and not simply protecting competitors. This may well mean that competitors who deliver less to consumers in terms of price choice, quality and innovation will leave the market.[233]

According to Gravengaard and Kjaersgaard, this has raised the threshold which competitors of a dominant undertaking must exceed before the Commission will intervene against allegedly exclusionary conduct, both in terms of the proof required to establish an abuse and in terms of the competitor's own efficiency.[234] They argue that it is difficult to interpret the guidance paper otherwise than that the Commission's shift in approach will provide dominant undertakings with increased commercial freedom without risking intervention by the Commission under Article 102.[235] The least that can be said is that the Commission has the proclaimed intention to raise the practical burden of proof for a finding of an exclusionary abuse under Article 102.

The guidance notice mentions a number of factors, common to Article 102, which the Commission generally considers to be relevant for its assessment: (1) the position of the dominant undertaking, (2) the conditions on the relevant market (including barriers to entry and expansion), (3) the position of the dominant undertaking's competitors, (4) the position of that undertaking's customers and suppliers, (5) the extent, duration and frequency of the dominant undertaking's conduct, (6) possible evidence of actual foreclosure and (7) direct evidence of any exclusionary strategy.[236]

In relation to refusal to supply practices, the notice states that the Commission will consider refusal to supply practices as an enforcement priority (and probably also as an abuse under Article 102)[237] if the refusal relates to a product or service that is objectively necessary to be able to compete effectively on a downstream market, is likely to lead to the elimination of effective competition on the downstream market and likely to lead to consumer harm.[238]

The Commission indicates that the words 'objectively necessary' must be understood to mean that an input is indispensable where there is no actual or potential substitute on which competitors in the downstream market could rely so as to counter – at least in the long term – the negative consequences of the

---

[233] Commission, 'Enforcement Priorities' (n 33) para 6. This is a clear reaction to the often uttered criticism that EU competition law protects competitors instead of competition.
[234] Gravengaard and Kjaersgaard (n 207) 289. See also Kellerbauer (n 232) 185.
[235] Ibid 305.
[236] Commission, 'Enforcement Priorities' (n 33) para 20.
[237] Cardoso and others (n 168) 3. See also Gravengaard and Kjaersgaard (n 207) 304.
[238] Commission, 'Enforcement Priorities' (n 33) para 81.

refusal.[239] In this regard, the Commission will normally make an assessment of whether competitors could in the foreseeable future effectively duplicate the input produced by the dominant undertaking.[240] This condition is clearly reminiscent of the above-mentioned second criterion from the *Bronner* ruling.

As for the second condition mentioned in the notice, the Commission states that if the first criterion is met, it considers a dominant undertaking's refusal to supply to be generally liable to eliminate, immediately or over time, effective competition in the downstream market.[241] It considers the likelihood of effective competition being eliminated generally to be greater the higher the market share of the dominant undertaking in the downstream market.[242]

In relation to possible consumer harm, the Commission will examine whether, for consumers, the likely negative consequences of the refusal to deal in the relevant market over time outweigh the negative consequences of imposing an obligation to deal.[243] The new requirement of the likelihood of consumer harm is said to represent a breach with standing case law on refusals to deal tangible property, in which such considerations were not taken into account.[244] This is illustrative of the guidance notice's perceived move from a formalistic approach to an economic-based approach,[245] whereby the Commission, as it states in the notice, 'will focus on those types of conduct that are most harmful to consumers'.[246]

In certain specific cases, the Commission will not consider it necessary to conduct a detailed analysis involving the three above-mentioned conditions, but will analyse the undertaking's conduct solely in the light of the seven factors used in assessing potential anti-competitive foreclosure.[247] In these cases, the

---

[239]   Commission, 'Enforcement Priorities' (n 33) para 83.
[240]   Ibid.
[241]   Ibid para 85.
[242]   Ibid.
[243]   Ibid para 86.
[244]   Eirik Østerud, *Identifying Exclusionary Abuses by Dominant Undertakings under EU Competition Law* (Wolters Kluwer Law and Business 2010) 180.
[245]   Ezrachi (n 33) 53. See also Ioannis Lianos, 'Categorical Thinking in Competition Law and the "Effects-based" Approach in Article 82 EC' in Ariel Ezrachi (ed), *Article 82 EC: Reflections on its Recent Evolution* (Hart Publishing 2009) 19, 20 and 44; Gravengaard and Kjaersgaard (n 205) 285 and 304, and Kavanagh and others (n 51) 8.
[246]   Commission, 'Enforcement Priorities' (n 33) para 5.
[247]   Commission, 'Enforcement Priorities' (n 33) para 82, and Cardoso and others (n 168) 3. These are cases where regulation already imposes a duty to deal on the dominant undertaking and it is clear from the considerations underlying such regulation that the necessary balancing of incentives has already been made by the public authority when imposing such a duty to deal and cases in which the upstream market position of the dominant undertaking has been developed under the protection of special or exclusive rights or has been financed by state resources.

Commission argues, it is clear that imposing a duty to deal is 'manifestly not capable of having negative effects on the input owner's and/or other operators' incentives to invest and innovate upstream, whether ex ante or ex post'.[248]

Important as the guidance notice may be, the Commission's effects-based agenda and the prioritising exercise at the heart of the guidance, can only exist within the boundaries of Article 102 as set by the Courts.[249] The guidance notice explicitly states that the document is not intended to constitute a statement of the law and is without prejudice to the interpretation of Article 102 by the Courts.[250] Even though the Commission can impose a limit at its own discretion from which it may not be able to depart without breaching general principles of EU law,[251] guidelines are legally non-binding on Member States[252] and must be in accordance with the Courts' standing case law. The latter is still to a large extent form-based and also seeks to protect the structure of the market and competition as such.[253]

As stated before, whether the guidance notice will indeed change the enforcement practice under Article 102 very much depends on the willingness of the Courts to follow the Commission's new approach and depart from standing case law. In a 2009 paper, Ezrachi reviewed a number of recent judgments to see whether there was yet any support from the Courts for the Commission's new effects-based analysis.[254] Ezrachi concluded that the judgments[255] revealed little support for the shift from a form-based to an effects-based approach.[256] These judgments, however, were all rendered before both the informal and the formal (OJ) release of the guidance notice. It is therefore not surprising that the judgments revealed little support for an effects-based analysis. Nevertheless, two judgments from 2010 and 2011 also appear to suggest that the Courts have not

---

248 Commission, 'Enforcement Priorities' (n 33) para 82.
249 Ezrachi (n 33) 56.
250 Commission, 'Enforcement Priorities' (n 33) para 3.
251 Kellerbauer (n 232) 185.
252 Article 288 TFEU lists the legal acts of the Union. Accordingly, Regulations, Directives and Decisions are binding, while recommendations and opinions are not. Guidelines are, however, administratively binding on the Commission itself. See to that extent e.g. Case C-351/98, *Spain v. Commission* [2002] ECR I-08031, para 53. They form rules of practice from which the administration may not depart in an individual case without giving reasons that are compatible with the principle of equal treatment. See Joined Cases C-189/02 P, C-202/02 P, C/205–02 P to C/208–02 P and C/213–02 P *Dansk Rørindustri* [2005] ECR I-05425, para 209.
253 See e.g. Gormsen (n 52) 47; Kavanagh and others (n 51) 17 and Kellerbauer (n 232) 176.
254 Ezrachi (n 33) 57–60.
255 Case C-95/04 *British Airways* [2007] ECR I-02331, Case T-201/04 *Microsoft* [2007] ECR II-03601, Case T-340/03 *France Télécom* [2007] ECR II-00107, and Case C-468/06–478/06 *GlaxoSmithKline* [2008] ECR I-07139, Opinion of AG Colomer.
256 Ezrachi (n 33) 57.

yet adopted a supportive stance towards the Commission's new effects-based approach.

In the *Deutsche Telekom* case,[257] Deutsche Telekom appealed a Commission decision which found it to have breached Article 102 by charging its competitors prices for access to its local network that were higher than the prices it charged final consumers for access to the same network ('margin squeeze').[258, 259] One of the pleas brought forward by Deutsche Telekom was that the margin squeeze identified had no effect on the market. This plea was, however, rejected by the CFI which stated that 'the anti-competitive effect which the Commission is required to demonstrate relates to the *possible* barriers which the applicant's pricing practices *could have created* for the growth of competition in that market'.[260] Furthermore, it contended that 'a margin squeeze between the applicant's wholesale and retail charges will *in principle* hinder the growth of competition in the downstream markets'.[261]

Jones and Suffrin have argued that the CFI's judgment in *Deutsche Telekom* was notable for its formalistic approach and the absence of an approach based on an analysis of the effects of the impugned practices on consumers.[262] As the CFI's judgment was delivered six months before the release of the Commission's guidance notice, it does not come as a surprise that the CFI did not adopt a more effects-based approach. However, this form-based approach was upheld in the ECJ's ruling in the appeal brought by Deutsche Telekom against the CFI's judgment, which was delivered some two years after the release of the guidance notice.[263]

In its ruling, the ECJ, in response to a reiteration of Deutsche Telekom's earlier claim of the lack of actual effect on the market, noted that:

> [T]he CFI therefore held in paragraph 235 of the judgment under appeal, *without any error of law*, that the anti-competitive effect which the Commission is required to demonstrate (...) relates to the *possible barriers* which the appellant's pricing practices *could have created* for the growth of products on the retail market in end-user access services.[264]

---

[257] See further section 6.4.2.1.

[258] A margin squeeze occurs where a vertically integrated undertaking dominant in the upstream market for an input sells the input to its downstream competitors at a price which does not allow them to operate profitably. See Jones and Suffrin (n 36) 413.

[259] Case T-271/03 *Deutsche Telekom* [2008] ECR II-00477.

[260] Ibid para 235 (emphasis added).

[261] Ibid para 237 (emphasis added).

[262] Jones and Suffrin (n 36) 418.

[263] Case C-280/08 P *Deutsche Telekom* [2010] ECR I-09555.

[264] Ibid para 252 (emphasis added).

What is more, it held that:

> [T]he CFI was entitled to hold in paragraph 237 of the judgment under appeal (…)
> that a margin squeeze resulting from the spread between wholesale prices for local
> loop access services and retail prices for end-user access services, in principle,
> hinders the growth of competition in the retail markets in services to end-users.[265]

The ECJ therefore rejected this part of one of Deutsche Telekom's grounds of appeal.[266]

According to Vedder, the ECJ's ruling in the *Deutsche Telekom* case breathes the language of an effects-based approach.[267] It is true that the ECJ in parts of the judgment uses language which points at an effects-based approach. For instance, it holds that the CFI correctly rejected the Commission's argument that the very existence of a pricing practice leading to a margin squeeze constituted an abuse, without it being necessary for an anti-competitive effect to be demonstrated.[268] Nevertheless, this effects-based language has not prevented the ECJ from approving (and thereby adopting) the formalistic approach taken by the CFI and thus seems to pay mere lip-service. In the end, the ECJ did not accept Deutsche Telekom's argument that for it to have abused its dominant position in the upstream (network) market there had to be an actual effect on the downstream market.

In another telecommunications case, *TeliaSonera* (2011), a Swedish court had asked the ECJ whether the abusive nature of the pricing practice in question depended on there actually being an anti-competitive effect.[269] The ECJ, referring to its ruling in the *Deutsche Telekom* case, started out with effects-based language similar to that used in the latter case by stating that:

> [T]he Court has ruled out the possibility that the very existence of a pricing practice
> of a dominant undertaking which leads to the margin squeeze of its equally efficient
> competitors can constitute an abuse within the meaning of Article 102 TFEU without
> it being necessary to demonstrate an anti-competitive effect.[270]

---

265   Case C-52/09 *Teliasonera* (n 269) para 255 (emphasis added).
266   Ibid para 261.
267   Hans Vedder, 'Competition in the EU Energy Sector – An Overview of Developments in 2009 and 2010' (2011) 9 <http://papers.ssrn.com/sol3/papers.cfm?abstract_id=1734639> accessed 12 May 2011.
268   Case C-280/08 P. *Deutsche Telekom* (n 263) para 250.
269   See Case C-52/09 *Teliasonera* [2011] ECR I-00527, para 60.
270   Ibid para 61.

However, it went on to hold that:

> In order to establish whether such a practice is abusive, that practice must have an anti-competitive effect on the market, *but the effect does not necessarily have to be concrete*, and it is sufficient to demonstrate that there is an anti-competitive effect which *may potentially exclude* competitors who are at least as efficient as the dominant undertaking.[271]

Like in the *Deutsche Telekom* case, the ECJ in *TeliaSonera* adopted a form-based approach. Even though it left the appraisal of the question whether of the effect of TeliaSonera's pricing practice was '*likely* to hinder'[272] the ability of its competitors to trade on the retail market for broadband connection services to the Swedish court, it did not require an *actual* effect on competition in the relevant downstream market for TeliaSonera's conduct to constitute an abuse. It was sufficient for there to be a likelihood of such an effect in practice. The *TeliaSonera* case, in line with the *Deutsche Telekom* case, therefore appears to suggest that the Courts have not (yet) abandoned the approach whereby certain forms of conduct seem to be an abuse without the existence of any anti-competitive effect being demonstrated.

6.3.2.3 Refusing to grant access to CO₂ transport and storage infrastructure

As I have indicated above, Article 102 does not require an outright refusal to deal for there to be an abuse of a dominant position. 'Constructive' refusals, i.e. dealing on unreasonable or discriminatory terms only, can also constitute an abuse of a dominant position under Article 102. A refusal to grant access to CO_2 transport or storage infrastructure on the basis of a technical incompatibility of the captured CO_2 stream with the relevant infrastructure would not constitute an outright refusal to deal. Rather, it could be a constructive refusal to deal if the CO_2 stream-purity requirements are unreasonable or discriminatory. The question of whether such requirements will be unreasonable or discriminatory in essence asks whether they can be objectively justified.[273] The (objective) need to guarantee the safety and the integrity of CO_2 transport or storage infrastructure can be expected to play a crucial role in this regard.

Case law demonstrates that for a refusal to grant access to CO_2 transport or storage infrastructure to constitute abuse under Article 102 there must at least be a (potential) upstream market and a related downstream market. In the event of a refusal to grant access to CO_2 transport or storage infrastructure, there likely are several relevant upstream and downstream markets. First, there are the

---

[271]   Case C-52/09 *Teliasonera* [2011] ECR I-00527, para 64 (emphasis added).
[272]   Emphasis added.
[273]   The concept of objective justification is discussed in section 6.4.1.

markets for CO_2 transport and storage and the internal electricity market respectively. As recital 38 of the preamble to the CCS Directive states, 'access to CO_2 transport networks and storage sites (...) could become a condition for entry into or competitive operation within the internal electricity and heat market, depending on the relative prices of carbon and CCS'.

However, as greenhouse gas emission reduction targets become more stringent, CCS technologies will have to be deployed more extensively outside of the electricity sector.[274] For industries like the cement industry, CCS might well be the only viable means of achieving deep emissions cuts.[275] Depending on the costs of CCS and EU ETS emission allowances,[276] access to CO_2 transport and storage infrastructure could therefore similarly become a pre-condition for entry/competitiveness in such industries. These downstream markets will likewise be vertically related with the upstream markets for the transport and storage of CO_2. Furthermore, the markets for the transport and storage of CO_2 also stand in a vertical relation to one another. In order to enter the market for CO_2 storage, a capturer will need to access to CO_2 transport infrastructure. Similarly, a party looking to deploy enhanced hydrocarbon recovery activities will need to have his captured CO_2 transported to the desired location.

The Courts and the Commission have not consistently answered the question of whether the upstream dominant undertaking must be active in the related downstream market, but case law seems to suggest that downstream *dominance* at least is not required. If the presence of the upstream dominant undertaking on the related downstream market is indeed required, there will have to be some form of vertical integration (ownership structures) between the CO_2 transport or storage operator (or both) on the one hand and the capturer on the other. When it comes to the upstream and downstream markets for the transport and storage of CO_2, the same goes for CO_2 transport and storage operators. Should the presence of the upstream dominant undertaking on the downstream market not be required, a refusal to grant access by a dominant CO_2 transport or storage operator who is active in the market for CO_2 transport or storage only could constitute an abuse under Article 102.

---

[274]   See Kamiel Bennaceur and Dolf Gielen, 'Energy Technology Modelling of Major Carbon Abatement Options' (2010) 4 International Journal of Greenhouse Gas Control 309, 312–13.

[275]   United Nations Industrial Development Organization (UNIDO), 'Carbon Capture and Storage in Industrial Applications: Technology Synthesis Report' (2010) 60 <www.unido.org/fileadmin/user_media/Services/Energy_and_Climate_Change/Energy_Efficiency/CCS/synthesis_final.pdf> accessed 18 May 2011.

[276] CO_2 prices under the EU ETS will be determined by the scarcity of the allowances, a factor that in turn depends, inter alia, on the stringency of the underlying target for the reduction of greenhouse gases.

There might be an incentive for a vertically integrated undertaking (e.g. capture and storage) not to grant access to its upstream storage facility in order to minimise competition in the downstream market. Such anti-competitive intent will most likely be absent in the case of an upstream dominant undertaking that is not active in a related downstream market. By virtue of the *Clearstream* case,[277] anti-competitive intent is not a prerequisite for finding an abuse of a dominant position, but its presence may reinforce the conclusion that the relevant conduct constitutes an abuse under Article 102.[278] In the case of *AstraZeneca*, the CFI stated that 'intention none the less also constitutes a relevant factor which may, should the case arise, be taken into consideration by the Commission'.[279] The Commission's resolve to do so is illustrated by the guidance notice, in which it mentions direct evidence of any exclusionary strategy as one of the seven relevant factors for its assessment under Article 102.

As the law stands, the relevant legal test for determining whether a refusal by a dominant undertaking to grant access to CO_2 transport or storage infrastructure constitutes an abuse under Article 102 is that of the *Bronner* case. The refusal must be likely to eliminate all competition in the downstream market from the entity requesting access, there must be no actual or potential substitute for the relevant CO_2 transport or storage infrastructure (it must be indispensable) and the refusal cannot be objectively justified. The indispensability of the facility concerned is indicated by the presence of technical, legal or economic obstacles to developing an alternative facility.

Roggenkamp has argued that CO_2 transport (pipeline) and storage facilities can be characterised as essential facilities[280] per se.[281] This would imply that these facilities are indispensable and that the *Bronner* case's second criterion will automatically be met when access to CO_2 transport or storage infrastructure is refused. However, in line with the reasoning on the natural monopoly character of CO_2 transport and storage infrastructure in section 6.3.1.4, such facilities are better characterised as *potential* essential facilities.[282] The analysis of the

---

[277] See also Case 85/76 *Hoffmann-La Roche* (n 177).

[278] Case T-301/04 *Clearstream* [2009] ECR II-03155 para 142.

[279] Case T-321/05 *AstraZeneca* (2010) ECR II-02805 para 359.

[280] An essential facility can be described as a network or other type of infrastructure to which access is necessary to compete on a given market. See Cardoso and others (n 168) 3. It has been claimed that the Commission and the Courts apply an essential facilities doctrine under Article 102, according to which an undertaking which has a dominant position in the provision of facilities which are essential for the supply of goods or services on another market abuses its dominant position where, without objective justification, it refuses access to those facilities. In this writing, I will not further discuss the essential facilities doctrine, since the perceived doctrine does not change the fact that the legal requirements of Article 102 still need to be met.

[281] Roggenkamp (n 3) 279.

[282] See also Vedder (n 159) 586.

competitive position of CO_2 transport and storage operators in section 6.3.1.4 indicates that there is a real possibility of, primarily, CO_2 storage facilities being indispensable. There are several potential technical, legal and economic barriers to entry to the market for the storage and monitoring of CO_2. As these barriers to entry will make it difficult for potential new market entrants to develop new facilities, access to existing storage facilities might well be indispensable for capturers to competitively operate on related downstream markets.

The Commission's guidance notice seems to have raised the overall burden of proof for a finding of an abusive refusal to deal. This is reflected in two of the three conditions for finding an abusive refusal to deal, which differ somewhat from those in the *Bronner* case. The Commission explicitly requires the likelihood of consumer harm and will weigh the likely consequences of a refusal to deal for static efficiency with those of a duty to deal for dynamic efficiency.[283] Furthermore, the refusal to deal must be likely to eliminate effective competition on the downstream market instead of all downstream competition from the entity requesting access. If the input to which is requested is indispensable, the Commission will consider a dominant undertaking's refusal to deal to be generally liable to eliminate, immediately or over time, effective competition in the downstream market.[284]

However, as I have indicated earlier, there are circumstances in which the Commission will not apply this more detailed test, but will rather analyse the undertaking's conduct in the light of the seven factors used in assessing potential anti-competitive foreclosure: existing regulation, based on the necessary balancing of incentives, already imposes a duty to deal or the dominant undertaking's position has been developed by means of exclusive rights or state resources. In these cases, a duty to deal cannot, according to the Commission, negatively affect the dominant undertaking's incentives to invest in the relevant upstream infrastructure. The Commission does not explain what the necessary balancing of incentives entails, but it is assumed here that this constitutes an assessment of the likely effects of a duty to deal on both static and dynamic efficiency.

---

[283] Static efficiency refers to *current* efficiency in terms of the sum of consumer and producer surplus, and production costs ('allocative' and 'productive' efficiency respectively). Dynamic efficiency refers to *future* efficiency measured in terms of the present value of the future stream of static total welfare, which can be influenced by, for instance, product or process innovations. See Machiel van Dijk and Machiel Mulder, 'Regulation of Telecommunication and Deployment of Broadband' (2005) CPB Netherlands Bureau for Economic Policy Analysis Memorandum 6–7 <www.cpb.nl/sites/default/files/publicaties/download/memo131.pdf> accessed 30 March 2011.

[284] See Commission 'Enforcement Priorities' (n 33) para 85.

The circumstances cited in the guidance notice seem to typically hint at network-bound sectors like the electricity and gas sectors and possibly also the CCS sector. It is not unlikely that CO_2 transport and storage infrastructure will (partly) be constructed with public means. The investments required as well as the risks present are considerable. The financing of EU CCS demonstration projects through the EU ETS new entrants reserve[285] indicates that the construction of CO_2 transport and storage infrastructure with public means is currently already happening indirectly. The projects receiving subsidies from the new entrants reserve are required to implement the full CCS chain, including the transport and storage of CO_2.[286]

As for existing sector-specific regulation, Article 21 of the CCS Directive requires Member States to ensure third-party access to CO_2 transport and storage infrastructure for the purpose of the geological storage of the captured CO_2. Yet unlike the Gas Directive,[287] the CCS Directive does not contain an exemption provision for new infrastructure. Whereas the Gas Directive contains an exemption provision that is clearly based on considerations of dynamic efficiency,[288] neither the CCS Directive nor the directive's impact assessment[289] address the potential consequences of requiring third-party access to CO_2 transport and storage infrastructure for incentives to invest in such infrastructure.

It could be argued that the difference is not surprising considering that the Gas Directive prescribes regulated third-party access to transmission infrastructure while the CCS Directive does not require Member States to introduce this most stringent form of third-party access. However, the Directive does also not prevent Member States from adopting such a regime of regulated third-party access.

---

[285]   See Article 10a(8) of the Directive revising the EU ETS, European Parliament and Council Directive 2009/29/EC of 23 April 2009 amending Directive 2003/87/EC so as to improve and extend the greenhouse gas emission allowance trading scheme of the Community [2009] OJ L140/63.

[286]   See Commission, 'Commission Decision of 3.11.2010 Laying down Criteria and Measures for the Financing of Commercial Demonstration Projects that Aim at the Environmentally Safe Capture and Geological Storage of CO2 as well as Demonstration Projects of Innovative Renewable Energy Technologies under the Scheme for Greenhouse Gas Emission Allowance Trading within the Community established by Directive 2003/87/EC of the European Parliament and of the Council' C (2010) 7499 final 17.

[287]   Directive 2009/73/EC. Article 36 of the Gas Directive provides that major new natural gas (transport and storage) infrastructure, upon request, may temporarily be exempted from a number of provisions in the Directive, including those on third-party access (Articles 32–34).

[288]   When I say that Article 36 of the Gas Directive is based on considerations of dynamic efficiency, I mean that the provision accommodates the potential negative effects that favourable conditions for access to network infrastructures can have on investment in the development of new infrastructure or the upgrading of existing infrastructure.

[289]   See n 71.

Based on the above, it looks like there is a real possibility of the Commission performing a less detailed analysis in the case of a refusal to grant access to CO_2 transport or storage infrastructure on grounds of technical incompatibility of the captured CO_2 stream with the relevant infrastructure. However, as I have argued in section 6.3.2.2, two judgments from 2010 and 2011 provide little support for the assumption that the Courts are willing to support the Commission's new approach. It might therefore very well be that the *Bronner* criteria will constitute the prescribed legal test for refusals to grant access to CO_2 transport or storage infrastructure for some time to come.

6.4 JUSTIFICATIONS AND EXCEPTIONS

6.4.1 OBJECTIVE JUSTIFICATION

Article 102 TFEU does not contain a provision exempting certain abusive behaviour. In order to provide some flexibility in what would otherwise be too draconian an application of Article 102, the Courts have developed the concept of objective justification.[290] As we have seen above, the ECJ determined in the *Bronner* case that a refusal to deal constitutes an abuse under Article 102 only if it cannot be objectively justified. The concept of objective justification was developed by the ECJ in several cases long before the *Bronner* case.[291] Several defences are possible under the concept of objective justification: (1) reasonable steps to protect one's commercial interest ('meeting competition defence'), (2) defences of objective necessity (e.g. concerns about quality, security, or safety in refusal to deal cases) and (3) efficiency defences (efficiency gains outweigh the anti-competitive effects).[292]

According to O'Donoghue and Padilla, and Jones and Suffrin, there have been very few cases in which objective justification was accepted by courts and competition authorities, making the chances of a dominant undertaking successfully relying on one of these defences very slim indeed.[293] While the legal burden of proving an infringement of Article 102 is borne by the Commission, it is for the dominant undertaking to put forward evidence that its conduct is objectively justified.[294]

---

[290] Craig and de Búrca (n 26) 1030 and Jones and Suffrin (n 36) 376.
[291] O'Donoghue and Padilla (n 27) 227. O'Donoghue and Padilla mention cases like Case 24/67 *Park, Davis and Co* [1968] ECR 00055, Case 40/70 *Sirena* [1971] ECR 00069, and Case 78/70 *Deutsche Grammophon* [1971] ECR 00487.
[292] O'Donoghue and Padilla (n 27) 227–34 and Jones and Suffrin (n 36) 376–80.
[293] O'Donoghue and Padilla (n 27) 232 and Jones and Suffrin (n 36) 376.
[294] (n 239) 27. See also Østerud (n 244) 248.

As argued before, Article 21(2)(c) of the CCS Directive would seem to provide the operators of CO_2 transport and storage infrastructure the possibility to refuse to grant to access to the relevant infrastructure when the technical specifications of the specific CO_2 stream are incompatible with the required technical standard and the incompatibility 'cannot reasonably be overcome'. The provision, therefore, already seems to require an operator refusing access to either CO_2 transport or storage infrastructure on technical grounds to demonstrate that the refusal is reasonable, i.e. that there is an objective necessity related to, for instance, considerations of security or safety (the second category of defences of objective necessity as mentioned above).

A dominant operator refusing to grant access to CO_2 transport or storage infrastructure on the basis of a technical incompatibility of the captured CO_2 stream with the facility concerned will therefore likely already have given an objective justification for this refusal in the context of Article 21 of the CCS Directive. Article 21 does not specify the procedure for assessing the reasonableness of a refusal to grant access on the basis of an incompatibility of technical specifications. Depending on the kind of third-party access regime through which each Member State will implement the obligation to ensure open and non-discriminatory access to CO_2 transport and storage infrastructure, such assessment will likely be done by the relevant competent authorities on either an ex-ante or an ex-post basis.[295]

A dominant undertaking's conduct will not automatically be considered justified simply because it furthers one of the abovementioned legitimate objectives.[296] Cases like *United Brands*[297] and *Atlantic Container Line*[298] indicate that in order to steer clear of the prohibition in Article 102, the conduct must also be proportionate, i.e. comply with the general EU law principle of proportionality.[299] This principle requires a measure, first, to be suitable for achieving the objective pursued ('suitability test') and, second, not to go beyond what is necessary to achieve this objective ('necessity test'). When assessing the proportionality of a refusal to grant access to CO_2 transport or storage infrastructure on the basis of a technical incompatibility of the captured CO_2

---

[295] In the former case, the competent authorities could create an exhaustive list of technical grounds on which access to CO_2 transport and storage infrastructure may be refused.
[296] Østerud (n 244) 267.
[297] Case 27/76 *United Brands Company* (n 31) para 158.
[298] Joined Cases T-191/98, T-212/98 to 214/98 *Atlantic Container Line* [2003] ECR II-03275, para 1120.
[299] Østerud (n 244) 267. See also Stothers (n 38) 260.

stream with the facility concerned, yet another general principle of EU law, the precautionary principle,[300] might be of relevance.[301]

The precautionary principle can be found in Article 191(2) TFEU. Neither the TFEU nor the Treaty on European Union (TEU) define the principle. The most common (international environmental law) version of the precautionary principle provides that where there is a threat to human health or the environment, a lack of full scientific certainty should not be used as a reason to postpone measures that would prevent or minimise such a threat.[302] The traditional role of the principle is to justify (trade-) restrictive regulation – predominantly EU but also national – in areas that are characterised by scientific uncertainty. In the *National Farmers' Union* case, the ECJ stated that 'where there is uncertainty as to the existence or extent of risks to human health, the institutions may take protective measures without having to wait until the reality and the seriousness of those risks become fully apparent'.[303]

In a 2006 article, Heyvaert surveyed the EU case law on the precautionary principle and concluded that the majority of cases reviewed the legality of binding EU instruments, issued either by the Council and Parliament, the Council or the Commission, with reference to precautionary prescriptions.[304] In a second category of cases discerned by Heyvaert, the precautionary principle functioned as a yardstick to review the legality of Member States' rules implementing or derogating from EU obligations.[305]

I argue here that as the Commission and the Courts review the proportionality of any justification given by a dominant undertaking for a refusal to deal, there might be scope for a dominant CO_2 transport or storage operator to bring forward a precautionary defence. To date, there is hardly any experience with the large-scale application of CCS technologies[306] and there seems to be limited

---

[300] See e.g. Joined Cases T-74/00, T-76/00, T-83/00, T-84/00, T-85/00, T-132/00, T-137/00 and T-141/00 *Artegodan* [2002] II-04945, para 184.
[301] This principle has been said to apply the principle of proportionality to conditions of scientific uncertainty. See Joseph Corkin, 'Science, Legitimacy and the Law: Regulating Risk Regulation Judiciously in the European Union' (2008) 33 European Law Review 359, 374.
[302] See Elizabeth Fisher, 'Opening Pandora's Box: Contextualising The Precautionary Principle in the European Union' (2007) University of Oxford Faculty of Law Legal Studies Research Paper Series Working Paper No 2/2007 4 <http://papers.ssrn.com/Abstract=956952> accessed 23 May 2011.
[303] Case C-157/96 *National Farmers' Union* [1998] ECR I-02211, para 63.
[304] Veerle Heyvaert, 'Facing the Consequences of the Precautionary Principle in European Community Law' (2006) 31 European Law Review 185, 189.
[305] Ibid.
[306] For the relevance of the state of technology for the precautionary principle, see e.g. Elizabeth Fisher, 'Is the Precautionary Principle Justiciable?' (2001)13 Journal of Environmental Law 315, 320.

scientific understanding on the impacts of impurities in the CO_2 stream on the safety and integrity of, primarily, storage reservoirs.[307] One of the potential risks related to geological CO_2 storage, for instance, consists of heavy metals in the captured CO_2 stream leaking from the storage complex and contaminating underground sources of drinking water.[308]

It could be argued that the precautionary principle is not intended to justify anti-competitive conduct of undertakings since it was developed as a principle justifying the adoption of precautionary regulatory standards by the EU legislature. The latter is illustrated by Article 191(2) TFEU, which states that 'Union policy on the environment shall (…) be based on the precautionary principle'. Yet, in my opinion, this line of reasoning would be too narrow and formalistic. In recent years, private entities have, as a consequence of both public pressure and government regulation, increasingly played a role in tackling complex environmental problems such as climate change. As a result, undertakings are more likely to be confronted with situations in which they, due to their often (in comparison with government) superior technical knowledge and practical expertise, will need to take environmental decisions in the presence of (major) scientific uncertainty. In such circumstances, it makes sense for the precautionary principle to be applicable, provided, of course, that the specific area is regulated by EU environmental legislation.[309]

However, in a situation where national regulation on CO_2 stream purity leaves CO_2 transport and storage operators no margin of discretion, a dominant operator trying to justify a refusal to deal by pointing at scientific uncertainty[310] in fact defends the precautionary approach taken by the relevant Member State in adopting the rule on the basis of which access is refused.[311] In that case, the question is whether the Member State's precautionary approach in the implementation of or derogation from Article 12 of the CCS Directive is well-founded. In this regard, it is important that there is said to be a well-established

---

[307]   See e.g. Tom Mikunda and Heleen de Coninck (CATO2), 'Possible Impacts of Captured CO_2 Stream Impurities on Transport Infrastructure and Geological Storage Formations: Current Understanding and Implications for EU Legislation' (2011) 5.

[308]   Commission, 'Guidance Document' (n 64) 80.

[309]   On the concept of exhaustion, see section 3.4.1 in Chapter III. In a 2011 article on the regulation of chemicals in the EU, Fleurke and Somsen give an example of the applicability of the precautionary principle to private entities through EU environmental legislation. See Floor Fleurke and Han Somsen, 'Precautionary Regulation of Chemical Risk: How REACH Confronts the Regulatory Challenges of Scale Uncertainty, Complexity and Innovation' (2011) 48 Common Market Law Review 357, 373–75.

[310]   It is important to underline that scientific uncertainty is generally considered to be much greater in relation to (the safety of) CO_2 storage than in relation to CO_2 transport.

[311]   In this regard, see also the following section on the state action defence. See Joined Cases C-379/08 and C-380/08 *ERG-II* [2010] ECR I-02007. The undertaking in question can in that case successfully rely on the state action defence.

tradition of deference in the judicial assessment of regulatory interventions made by the Member States to avert uncertain threats.[312] The ECJ has in a number of early cases such as *Eyssen*[313] and *Albert Heijn*[314] demonstrated that it is prepared to respect a Member State's preference for a precautionary approach in the case of scientific uncertainty.[315]

However, Member States will have to provide scientific backing for the perceived environmental/health risk. In the 2010 case of *Commission v. France*, the ECJ summarised the standards that Member States are required to meet in this regard.[316] By virtue of *Commission v. France*, a Member State seeking to successfully rely on the precautionary principle must identify potentially negative consequences for public health or the environment, assess those risks on the basis of the most reliable scientific data and most recent results of international research available (more than pure hypothesis) and make sure that the restrictive measures adopted are non-discriminatory and objective.[317]

For a Member State to successfully rely on the precautionary principle, the most difficult part will probably be to assess whether the scientific evidence used is reliable and recent enough. The ECJ did not further specify this requirement in *Commission v. France*. By analogy, a dominant CO_2 transport or storage operator bringing forward a precautionary defence for a refusal to grant access to a facility will probably have to meet the same standards.

6.4.2 THE 'STATE ACTION DEFENCE'

An important escape route under Article 102 is the 'state action defence'. A company that can demonstrate the requisite degree of state involvement in its actions can avoid the application of Article 102.[318] In this regard, the central question is whether the anti-competitive effect results from the autonomous conduct of the company concerned.[319] According to the Courts' case law,

---

[312] Heyvaert (n 304) 200.

[313] Case 53/80 *Eyssen* [1981] ECR 00409.

[314] Case 94/83 *Albert Heijn* [1984] ECR 03263.

[315] Stephen Weatherill, 'Recent Case Law Concerning the Free Movement of Goods: Mapping the Frontiers of Market Deregulation' (1999) 36 Common Market Law Review 51, 69; Alberto Alemanno, 'The Shaping of the Precautionary Principle by the European Courts: From Scientific Uncertainty to Legal Certainty' (2007) 3 <http://papers.ssrn.com/sol3/papers. cfm?abstract_id=1007404> accessed 24 May 2010, and Peter Oliver, *Free Movement of Goods in the European Community* (4th edn, Sweet & Maxwell 2003) 255–56.

[316] Case C-333/08 *European Commission v. French Republic* [2010] ECR I-00757.

[317] Ibid paras 92–93.

[318] O'Donoghue and Padilla (n 27) 28.

[319] Jones and Suffrin (n 36) 183–84.

undertakings cannot be found to infringe Article 102 if the entities engaging in the restrictive conduct are acting in the public interest, the undertakings concerned were compelled to participate by a state measure, or state regulation eliminated the possibility of competitive activity.[320] Undertakings participating in a state-implemented scheme that restricts competition can raise any of these points as a defence for their participation.[321] The Courts, however, apply the state action defence restrictively.[322]

Whether state regulation eliminates the possibility of competitive activity (third option) depends on a detailed analysis of the facts of the case and will be assumed under exceptional circumstances only.[323] If a national law merely encourages, or makes it easier for undertakings to engage in autonomous anti-competitive conduct, those undertakings remain subject to Article 102.[324] In the case of *Consorzio Industrie Fiammiferi (CIF)*, AG Jacobs opined that the crucial question is whether the restrictive effects on competition originate solely in the national law or, at least to an extent, in the applicant's conduct.[325] Under the regulatory framework at issue, the entity concerned has to enjoy 'sufficient autonomy' to restrict competition further than was already done by national legislation.[326] The fact that anti-competitive arrangements are communicated to and subject to authorisation by a public authority is not necessarily a decisive factor.[327] Of great importance to the scope of the third variant of the station action defence under Article 102 (c – state regulatory action defence) are the *Deutsche Telekom* decision and subsequent cases before the Courts.[328]

6.4.2.1   *Deutsche Telekom: decision 2003/707/EC and cases T-271/03 and C-280/08 P*

In *Deutsche Telekom*, the approval of anti-competitive conduct by a public authority played a prominent role. The case concerned the prices Deutsche Telekom, the incumbent German telecommunications operator, charged competitors and end-users for access to its local telephone network. The Commission had received complaints from competitors of Deutsche Telekom,

---

[320]   O'Donoghue and Padilla (n 27) 28.
[321]   Ibid.
[322]   See e.g. case T-513/93 *Consiglio Nazionale degli Spedizionieri Doganali* [2000] ECR II-08107, para 60.
[323]   O'Donoghue and Padilla (n 27) 30.
[324]   Case C-198/01, *Consorzio Industrie Fiammiferi (CIF)* [2003] ECR I-08055, para 56.
[325]   Case C-198/01 *Consorzio Industrie Fiammiferi*, Opinion of AG Jacobs, para 65.
[326]   Ibid para 66.
[327]   Ibid para 71.
[328]   *Deutsche Telekom AG* (Case COMP/C-1/37.451, 37.578, 37.579) Commission Decision 2003/707/EC [2003] OJ L263/9, Case T-271/03 *Deutsche Telekom AG* [2008] ECR II-00477 and Case C-280/08 P. *Deutsche Telekom AG* [2010] ECR I-09555.

who claimed that the margin between the two sets of prices was not sufficient to enable them to compete with Deutsche Telekom to provide end-user access over local networks.

In defence, Deutsche Telekom had argued that if there were any infringement of EU law, the Commission should not be acting against an undertaking whose charges were regulated (by the German telecommunications regulator), but against the German state under Article 258 TFEU.[329] This was, however, rejected by the Commission, which contended that the Courts had consistently held that the competition rules may apply where the sector-specific regulation does not preclude the relevant undertakings from engaging in autonomous conduct that prevents, restricts or distorts competition.[330] The Commission found Deutsche Telekom to have breached Article 102(a) TFEU (unfair pricing) as it had operated abusive prices by charging its competitors prices for access to its local network that were higher than the prices it charged final consumers for access to the same network.

In its application for annulment[331] of Commission Decision 2003/707/EC, Deutsche Telekom advanced a number of grounds of appeal.[332] In the essential first part of its first plea in law concerning the infringement of Article 102, Deutsche Telekom claimed that it had not abused its dominant position as it did not have sufficient scope to avoid the allegedly abusive behaviour.[333] According to Deutsche Telekom, it neither had the scope to fix charges for access to its local network by competitors (wholesale access), nor to do so for final consumers (retail access).[334]

In response, the CFI, referring to the cases of *van Landewijck*,[335] *Italy v. Commission*;[336] *Stichting Sigarettenindustrie*[337] and *CIF*,[338] provided that the state regulatory action defence had only been partially accepted by the ECJ.[339] According to the CFI, for the national legal framework to have the effect of

---

[329] Para 53. Article 258 TFEU allows the Commission to bring a case before the ECJ if it considers a Member State to have failed to fulfil an obligation under the Treaties.

[330] Para 54.

[331] Article 263 TFEU provides any natural or legal person the possibility to institute proceedings against an act (among which Commission decisions) addressed to that person.

[332] Case T-271/03 *Deutsche Telekom AG* (n 328).

[333] Ibid paras 70–83.

[334] Ibid paras 70 and 71.

[335] Joined Cases 209/78 to 215/78 and 218/78 *Van Landewijck and Others v. Commission* [1908] ECR 3125, paras 130–34.

[336] Case 41/83 *Italy v. Commission* [1985] ECR 873, para 19.

[337] Joined Cases 240/82 to 242/82, 261/82, 262/82, 268/82 and 269/82 *Stichting Sigarettenindustrie and Others v. Commission* [1985] ECR 3831, paras 27–29.

[338] Case C-198/01 *CIF* [2003] ECR I-8055, para 67.

[339] Case T-271/03 *Deutsche Telekom AG* (n 328) para 86.

making Article 102 inapplicable, the restrictive effects on competition had to originate 'solely in the national law'.[340] Article 102 may apply if it is found that the national legislation leaves open the possibility of preventing, restricting or distorting competition by, for instance, merely encouraging or making it easier for undertakings to engage in autonomous anticompetitive conduct.[341]

After considering the relevant German legal framework, the CFI found that Deutsche Telekom had sufficient discretion for its pricing policy to fall within the scope of Article 102.[342] It crucially noted that the fact that Deutsche Telekom's charges had to be approved by the national regulatory authority did not absolve it from responsibility under Article 102.[343] Since Deutsche Telekom influenced the level of its charges through applications to the national regulatory authority, the restrictive effects did not originate solely in the applicable national legal framework.[344] In the context of Deutsche Telekom's 'special responsibility' as an undertaking in a dominant position, it was obliged to submit applications for adjustment of its charges at a time when those charges had the effect of impairing genuine undistorted competition.[345] The CFI's judgment rejected all pleas advanced by Deutsche Telekom and upheld the decision of the Commission.[346]

Following the judgment of the CFI, Deutsche Telekom brought an appeal before the ECJ.[347] One of the three grounds of appeal again challenged the liability of Deutsche Telekom in relation to the alleged infringement on the ground of a lack of scope to adjust retail prices for end-user access services and the regulation of prices for telecommunications services by national regulatory authorities.[348] Deutsche Telekom largely reiterated the arguments put forward before the CFI and claimed that the latter had erred in law by adopting a legally incorrect test in respect of the liability of the infringement of Article 102.[349]

---

[340] Case T-271/03 *Deutsche Telekom AG* (n 328) para 87.
[341] Ibid paras 88 and 89.
[342] Ibid para 124.
[343] Ibid para 107.
[344] Ibid para 107.
[345] Ibid para 122.
[346] The CFI's judgment has been criticised for increasing the burden on regulated undertakings and not taking into account fundamental principles of EU law. See Robert O'Donoghue, 'Regulating the Regulated: Deutsche Telekom v. European Commission' (2008) 1 GCP Magazine 3, and P Alexiadis, 'Informative and Interesting: The CFI Rules in Deutsche Telekom v. European Commission' (2008) 1 GCP Magazine 10.
[347] Case C-280/08 P *Deutsche Telekom AG* (n 328).
[348] Ibid para 30. For a short summary of the Court's judgment on the two other grounds of appeal, see Emma-Jean Hinchy, 'Abuse of Dominant Position – Telecommunications' (2011) 34 European Competition Law Review 63.
[349] Case C-280/08 P *Deutsche Telekom AG* (n 328) para 77.

The ECJ, however, rejected Deutsche Telekom's claim.[350] Following the line of reasoning of the CFI, the ECJ provided that the mere fact that Deutsche Telekom was encouraged by the intervention of a national regulatory authority to maintain the pricing practices which led to the abuse, could not, as such, absolve it in any way from responsibility under Article 102.[351] According to it, the CFI was entitled to find, on the sole ground that Deutsche Telekom had scope to adjust its retail prices for end-user services, that the pricing practices were attributable to the undertaking.[352] Like the CFI, the ECJ dismissed Deutsche Telekom's appeal in its entirety, upholding the approach of the Commission in Decision 2003/707/EC.

One of the important outcomes of *Deutsche Telekom* is that it defined the scope of application of the EU competition rules in sectors subject to ex-ante (sector-specific) regulation, such as the telecommunications and electricity and gas sectors.[353] The approach of the Commission, which was upheld by the CFI and the ECJ, has been said to have very significant consequences for dominant operators.[354] It implies that, when a national regulatory authority adopts rules that fail to sufficiently protect the conditions of competition, a dominant operator could also itself be held responsible for violating competition rules if it nonetheless had the commercial freedom to adapt its tariff structure in such a way as to prevent the violation from occurring.[355]

Considering the *Deutsche Telekom* case, there is no reason to assume that this duty is limited to price setting. Central to the Commission's approach was the ability of Deutsche Telekom to autonomously engage in anti-competitive behaviour. Price is not the only regulated access condition and undertakings could also have a certain degree of autonomy in relation to the application of other access conditions such as, for instance, technical standards. What is more, the 'special responsibility' not to impair competition, which was attributed to the undertaking in a dominant position by the ECJ in the case of *Michelin*,[356] is not limited to a certain type of anti-competitive behaviour. The dominant undertaking has a broad responsibility 'not to allow its conduct to impair genuine undistorted competition on the common market'.[357]

---

[350]   Case C-280/08 P *Deutsche Telekom AG* (n 328) para 80.
[351]   Ibid para 84.
[352]   Ibid para 85.
[353]   Buigues and Klotz (n 20) 2.
[354]   O'Donoghue and Padilla (n 27) 31.
[355]   Ibid. As the dominant operator has a special responsibility, it must ensure that it uses the discretion allowed by the relevant regulation to avoid infringing Article 102.
[356]   Case 322/81 *Michelin* (n 30) para 12.
[357]   Ibid para 12.

Article 12(3) of the CCS Directive determines that the storage operator may only inject a CO_2 stream if a risk assessment has been carried out and if that risk assessment has shown that the contamination levels are in line with the criteria in Article 12(1).[358] In its guidelines on CO_2 stream composition, the Commission sheds some more light on how this may work in practice. A possible approach is to have the storage operator propose a certain CO_2 stream composition, which is then either approved or disapproved by the competent authority on the basis of its compatibility with requirements on pipeline safety, public health and environmental safety, storage integrity and well integrity.[359]

This would resemble the way in which access conditions were determined in the *Deutsche Telekom* case, in which the undertaking, as we have seen, proposed tariffs to the national regulatory authority for approval by the latter. By virtue of *Deutsche Telekom*, a dominant storage operator can, in case of an alleged breach of Article 102, probably not escape liability under Article 102 by claiming that the required CO_2 stream composition was approved by the competent authority. If the imposition of certain CO_2 stream purity criteria by either the storage or the pipeline operator is found to be anti-competitive, both can be held responsible for violating Article 102 if they had the commercial freedom to adapt the CO_2 stream requirements in such a way as to prevent the violation from occurring. Based on the text of Article 12(3) of the CCS Directive and the Commission's guidelines on CO_2 stream composition, it seems that at least the storage operator would likely have such freedom.

The availability of the state regulatory action defence obviously depends on the characteristics of the specific case, among which are the specifics of the applicable regulatory framework. O'Donoghue and Padilla have argued that the more prescriptive and detailed the regulatory framework is, the less likely it is that competition law has any residual role.[360] Nevertheless, it is important to underline that the test developed by the Commission and confirmed by the CFI and ECJ, is a binary one: the competition provisions are applicable, unless the anticompetitive behaviour originates *solely* in national law. No matter how prescriptive and detailed sector-specific regulation is, if it leaves the undertaking a choice,[361] competition law is applicable.

---

[358] On Article 12 and the CO_2 stream-purity criteria, see Chapter III on CO_2 stream-purity and Member States' scope to impose stricter norms.

[359] Commission, 'Guidance Document 2' (n 64) 60.

[360] O'Donoghue and Padilla (n 27) 31.

[361] See to this extent also Case C-52/09 *Teliasonera*, para 49, in which the ECJ states that 'it must be borne in mind that Article 102 TFEU applies only to anti-competitive conduct engaged in by undertakings on their own initiative. If anti-competitive conduct is required of undertakings by national legislation or if the latter creates a legal framework which itself eliminates *any* possibility of competitive activity on their part, Article 102 TFEU does not apply' (emphasis added).

6.4.3   ARTICLE 106(2) TFEU

Alongside the state action defence, there is another possibility for undertakings to escape liability under Article 102. Article 106(2) TFEU allows for an exemption from the rules in the Treaties, in particular those on competition. In order for Article 106(2) to apply, (1) the relevant undertaking has to be entrusted with the operation of services of general economic interest or have the character of a revenue-producing monopoly, (2) the application of the provisions of the Treaties would obstruct the performance of these services and (3) the development of trade must not be affected to an extent contrary to the interests of the Union.

The basic assumption underlying this provision is that since the market does not create sufficient incentives for the provision of certain public goods, it is necessary to regulate their supply.[362] As Article 106(2) is a provision which permits, under certain circumstances, derogation from the provisions of the Treaties, the category of eligible undertakings is strictly defined.[363] It may include private undertakings, as long as such undertakings have been entrusted with the operation of services of general economic interest by an act of the public authority.[364] The three cumulative conditions have been strictly construed by the Courts and are applicable to a relatively narrow range of bodies.[365]

The first criterion's crucial term 'services of general economic interest' is not defined in the Treaties.[366] The Commission has defined services of general economic interest as 'market services which the Member States subject to specific public service obligations by virtue of a general interest criterion', including, for instance, transport networks, energy and communications.[367] According to it, Member States are primarily responsible for defining services of general economic interest, their definition only being subject to control for manifest error.[368] In the *BUPA* case[369] the CFI confirmed that this is a matter over which

---

[362]   Amedeo Arena, 'The Relationship between Antitrust and Regulation in the US and the EU: An Institutional Assessment' (2011) Institute for International Law and Justice Emerging Scholar Papers 19, 31 <www.iilj.org/publications/documents/ESP19–2011Arena. pdf> accessed 1 April 2011.

[363]   Case 127/73 *BRT v. Sabam* [1974] ECR 00313, para 19.

[364]   Ibid para 20.

[365]   Goyder (n 41) 486.

[366]   In relation to services of general economic interest, see also Chapter VIII on the centralising of CO_2 storage site selection under EU law.

[367]   Commission, 'Communication from the Commission – Services of General Interest in Europe' 2001/C17/04, 23.

[368]   Ibid 8. See also Protocol No 26 to the TFEU on Services of General Interest, which provides that 'the shared values of the Union in respect of services of general economic interest (…) include in particular the essential role and the wide discretion of national, regional and local authorities in providing, commissioning and organising services of general economic interest as closely as possible to the needs of the users'.

[369]   Case T-289/03 *British United Provident Association Ltd (BUPA) v. Commission* [2008] ECR II-00081.

Member States have considerable latitude, while at the same time suggesting that there is an emerging set of EU-wide criteria, including the universal and compulsory nature of the service.[370]

As for the second criterion, in its early case law the ECJ held that this condition was met only if the relevant provisions were *incompatible* with the performance of the tasks assigned to the undertaking.[371] Later jurisprudence seems to suggest that there is no need to show that there is no other way to perform the task, only that the way in which the market is organised allows the provider of the relevant services to operate under economically acceptable conditions.[372] The common element is that the undertaking granted exclusivity has universal service obligations requiring it to perform some tasks that are not in themselves profitable.[373] The only way that it can do this is to have exclusive rights over those parts of the service that are profitable.[374]

Finally, while national courts are allowed to apply the first two criteria, the determination of whether the development of trade has been affected by the practices to an extent contrary to EU interests, is for the Commission alone to decide.[375] If the national court decides that protection from EU (competition) provisions is required, then it must, by virtue of the cases of *Commission v. Netherlands* and *Commission v. France*,[376] pass to the Commission the responsibility for determining if the derogation is likely to sufficiently affect trade between Member States in the relevant market.[377] The Commission will in that case assess the proportionality of the act entrusting the undertaking with the operation of the services of general economic interest.[378]

It could be argued that both the transport and the storage of CO_2 could, at least in the initial phase, be characterised as services of general economic interest as they are indispensable for capturers to dispose of the captured CO_2 and it is

---

[370] Damian Chalmers, Gareth Davies and Giorgio Monti, European Union Law (2nd edn, Oxford University Press 2010) 1030 and 1032. See also Harm Schepel and Wolf Sauter, *State and Market in European Law* (Cambridge University Press 2009) 174–79.

[371] Craig and de Búrca (n 26) 1132.

[372] Ibid, and Chalmers (n 370) 1034.

[373] Craig and de Búrca (n 26) 1133.

[374] Ibid.

[375] Goyder (n 41) 489.

[376] Cases C-157/94 *Commission v. Netherlands* [1997] ECR I-5699 and C-159/94 *Commission v. France* [1997] ECR I-5815.

[377] Goyder (n 41) 490.

[378] See e.g. Gert-Jan Koopman, 'Exemptions from the Application of Competition Law for State Owned Companies: An EU Perspective' (presentation held at the Global Competition Law and Economics Series Conference: Competition Law and the State, Hong Kong, 18 March 2011) <www.ucl.ac.uk/laws/global-competition/hongkong-2011/secure/docs/4_koopman.pdf > accessed 1 April 2011.

doubtful whether such services can in the short term be profitably offered in an emerging market (first and second criteria met).[379] Without the availability of CO_2 transport and storage services, CCS can obviously not take place. As indicated earlier, for certain heavy industries CCS might be the only means of realising deep emission cuts. Reducing greenhouse gas emissions from heavy industry can arguably be seen as a (environmental) public good. As the CCS business environment is likely to be highly uncertain for a number of years to come,[380] there might be scope for exempting transport and storage operators from the TFEU competition provisions. Member States seeking to do so will, however, have to come up with a thorough justification and will need to demonstrate that the service obligation requires the operators to perform a task that is not in itself profitable.

6.5 LESSONS FOR CO_2 TRANSPORT AND STORAGE OPERATORS

By virtue of the principle of precedence of primary EU (competition) law over both secondary (sector-specific) EU legislation and national (sector-specific) rules, the grounds for refusal of third-party access to transport networks or storage sites mentioned in Article 21(2) of the CCS Directive would not appear to allow an operator to refuse access to the relevant infrastructure if such refusal would constitute an abuse of a dominant position under Article 102 TFEU. For there to be an abuse of a dominant position, Article 102 requires the entity concerned to (1) be an undertaking within the meaning of EU competition law that is (2) dominant in a market (3) covering a substantial part of the internal market and (4) has abused its dominant position, (5) affecting the trade between Member States. In the enforcement of Article 102 TFEU, conditions (2) and (4)

---

[379] Yet it is likely that between the two, CO_2 storage will more easily, as well as for a longer period, qualify as a service of general economic interest, considering inter alia the costs and uncertainties associated with the storage of CO_2. The industry generally sees the long-term liability for the closed CO_2 storage site (n 56 of the Introduction) as a significant obstacle to the deployment of CCS. This has to do with the scale of projected CCS activities, the long timeframes over which the risks may present themselves, and the uncertainties of the geophysical system. See Mark de Figueiredo and others, 'The Liability of Carbon Dioxide Storage' 1 <http://sequestration.mit.edu/pdf/GHGT8_deFigueiredo.pdf> accessed 23 June 2010.

[380] The business case for CCS to a large extent depends on CO_2 prices under the EU ETS, which should rise in the scheme's third period (as of 2013). Whether future CO_2 prices will be sufficient to enable the large-scale deployment of CCS remains to be seen in view of the over-allocation of allowances in the first and second phases of the EU ETS. On the incentivisation of the deployment of CCS technologies see, for instance, Marijn Holwerda, 'Subsidizing Carbon Capture and Storage Demonstration through the EU ETS New Entrants Reserve: A Proportionality Test' 4 Carbon and Climate Law Review (2010) 228.

usually lead to significant problems of definition, while the other three, in practice, are generally more easily met.

As part of the assessment of an undertaking's alleged dominance, the product and geographic dimensions of the relevant market have to be defined and an analysis of the competitive position of the undertaking on the relevant market has to be conducted. An assessment of the product dimensions of the (future) relevant product markets for CO_2 transport and storage operators reveals that the three main activities in the CCS value chain will likely all constitute separate product markets. In certain specific individual situations, the product markets for CO_2 transport and storage could, respectively, be narrower or broader.

An analysis of the likely geographic scope of the future markets for the transport and storage of CO_2 suggests that these might not be wider than, at most, a single Member State before at least 2030. Yet should the mainly offshore storage of CO_2 in the EU materialise in the coming years, then there is a fair chance of the geographic scope of CO_2 transport and storage markets being wider than a single Member State, somewhere between 2020 and 2030. In any event, both markets could in practice be characterised by several barriers to entry.

If a refusal to grant access to CO_2 transport or storage infrastructure infringes Article 102, it will likely constitute an (exclusionary) refusal to deal under Article 102(2). Based on the decision practice, case law and the Commission's guidance notice on its enforcement priorities in applying Article 102 to abusive exclusionary conduct, the relevant legal test for determining whether a refusal by a dominant undertaking to grant access to CO_2 transport or storage infrastructure constitutes an abuse under Article 102 is probably still that of the *Bronner* case.

The refusal must be likely to eliminate all competition in the downstream market from the entity requesting access, there must be no actual or potential substitute for the relevant CO_2 transport or storage infrastructure and the refusal cannot be objectively justified. As several potential technical, legal and economic barriers to entry to, primarily, the market for the storage and monitoring of CO_2 can be identified, there is a real possibility of CO_2 storage facilities being, in practice, indispensable.

There are a number of limited ways for CO_2 transport and storage operators to try and avoid liability under Article 102. First, the Courts have developed the concept of objective justification. Potentially unlawful conduct might be objectively justified on the basis of, for instance, concerns about quality, safety or security. Due to the limited EU experience with the large-scale application of CCS technologies and limited scientific understanding of the impacts of contaminants in the CO_2 stream on the safety and integrity of, primarily, storage reservoirs, there might be scope for a dominant CO_2 transport or storage operator to bring forward a precautionary defence. Case law on the concept of objective justification, however, learns that chances of an alleged objective justification being accepted by the Courts are slim.

The state action defence provides a second instrument to avoid liability under Article 102. A dominant undertaking that can demonstrate the requisite degree of state involvement in its actions can avoid the application of Article 102. Several lines of defence are possible under the state action defence. Of great importance is the 2010 *Deutsche Telekom* case, which has significantly narrowed the possibilities of dominant undertakings to rely on one of the state action lines of defence by requiring that the anti-competitive conduct originates solely in national law. If the imposition of certain CO_2 stream purity criteria by either the storage or the pipeline operator is found to be anti-competitive, both can be held responsible for violating Article 102 if they had the commercial freedom to adapt the CO_2 stream requirements in such a way as to prevent the violation from occurring. Based on the text of Article 12(3) of the CCS Directive and the Commission's guidelines on CO_2 stream composition, it seems that at least the storage operator would likely have such freedom.

Finally, there is the possibility for undertakings to escape liability under Article 102 through Article 106(2) TFEU, allowing undertakings to, exceptionally, be exempted from the rules in the Treaties, in particular those on competition. There seems to be room to argue that both the transport and the storage of CO_2 could, at least in the short term, be characterised as services of general economic interest as they allow capturers to dispose of the captured CO_2 and it is doubtful whether such services can in the short term be profitably offered in an emerging market. Member States seeking to exempt transport and storage operators from the TFEU competition provisions will, however, have to come up with a thorough justification and need to demonstrate that the service obligation requires the operators to perform a task that is not in itself profitable.

Based on the above, CO_2 transport and storage operators are advised to be careful when refusing access to their infrastructure on technical grounds. Their scope under Article 102 to do so does not appear to be broad. CO_2 storage facilities, in particular, are likely to be considered indispensable under Article 102. In order to avoid liability under the latter provision in case either CO_2 transport or storage facilities are indispensable, the operators of these facilities would be wise to think of a precautionary defence for refusing access to their infrastructure. Such a defence would have to have proper scientific backing by having identified potentially negative consequences for public health/environment, having assessed those risks on the basis of the most reliable

232

scientific data and most recent results of international research available and having made sure that the refusal is non-discriminatory and objective.

Taking as a point of departure for assessing the required composition of CO_2 streams for storage the model suggested by the Commission in its guidelines, CO_2 transport and storage operators are not advised to rely on the state action defence. In such a scheme, at least the storage operator would arguably still have the freedom to adapt the CO_2 stream requirements in such a way as to prevent the infringement of Article 102 from occurring.

CHAPTER VII

THE DEVELOPMENT AND MANAGEMENT OF CO_2 TRANSPORT INFRASTRUCTURE AND EU ANTITRUST LAW

> In order not to infringe Article 102 TFEU, CO_2 transport operators are advised to:
>
> – conduct open seasons to market test (potential) demand (for cross-border) CO_2 transport capacity
>
> – explore the possibility of increasing planned capacity in order to facilitate (cross-border) third-party access
>
> – explore the willingness of third parties (from other Member States) to commit financially to an expansion project and take into (serious) consideration any co-financing offers
>
> – invest in additional transport capacity if there is (cross-border) demand for extra capacity and such expansion can be profitably realised

7.1 INTRODUCTION

Article 21 of the CCS Directive on the access of third parties to CO_2 transport networks and storage sites requires Member States to (generally) ensure that potential users are able to have access to CO_2 transport and storage infrastructure. Yet the CCS Directive is remarkably silent on the development and management (capacity allocation and congestion management)[1]

---

[1] The term capacity allocation refers to the mechanisms that are used by the network operators to allocate new capacity on EU gas and electricity transport markets. For an overview of the types of capacity allocation mechanisms in EU gas and electricity transport markets, see Commission, 'Commission Staff Working Document on Capacity Allocation and Congestion

of CO_2 transport infrastructure in relation to potential future capacity requirements.[2]

In particular, the Directive does not give any guidance as to how to deal with potential future cross-border capacity requirements when developing CO_2 transport infrastructure. The Directive's chapter five on third-party access does not mention future cross-border transport capacity requirements. Nor does the Directive deal with the development of CO_2 transport infrastructure. Article 24 merely provides that in cases of cross-border transport of CO_2, cross-border storage sites or cross-border storage complexes, the competent authorities of the Member States concerned shall jointly meet the requirements of the Directive and of other relevant EU legislation.[3]

Nevertheless, the absence of a provision in the CCS Directive on the development of CO_2 transport infrastructure in relation to potential future cross-border capacity requirements does not mean that parties developing such infrastructure will be under no EU obligations in this regard. A number of EU competition cases decided by the Commission seem to suggest that there could be several requirements related to the capacity allocation and congestion management on CO_2 transport infrastructure.

In these cases, the Commission addressed the development and management of gas transport infrastructure under Article 102 of the Treaty on the Functioning of the EU (TFEU). The parallel with the gas sector is arguably useful in view of the similar nature of activities and players in this sector and the CCS sector.[4] It is

---

Management for Access to the Natural Gas Transmission Networks Regulated under Article 5 of Regulation (EC) No 1775/2005 on Conditions for Access to the Natural Gas Transmission Networks' SEC (2007) 822. Congestion management procedures are instruments used to allocate and re-allocate capacity allocated by the transmission system operator to market participants in a transparent, fair and non-discriminatory manner to prevent or remedy a situation of scarcity of network capacity. See European Regulators' Group for Electricity and Gas (ERGEG), 'ERGEG Principles: Capacity Allocation and Congestion Management in Natural Gas Transmission Networks' (2008) 17.

[2] There is no regulation such as European Parliament and Council Regulation 715/2009 of 13 July 2009 on conditions for access to the natural gas transmission networks [2009] OJ L211.

[3] In this regard the CCS Directive resembles the first two Electricity Directives: European Parliament and Council Directive 96/92/EC of 19 December 1996 concerning common rules for the internal market in electricity [1997] OJ L27 and European Parliament and Council Directive 2003/54/EC of 26 June 2003 concerning common rules for the internal market in electricity and repealing Directive 96/92/EC [2003] L176/37. These Directives were also relatively silent on cross-border issues, merely providing some general principles such as third-party access to interconnectors and dispute resolution mechanisms. See Adrien de Hauteclocque, 'Long-term Supply Contracts in European Decentralized Electricity Markets: An Antitrust Perspective' (PhD thesis, University of Manchester 2009) 170.

[4] Both the gas and the CCS sector transport a gaseous substance that is (sometimes) stored. What is more, some of the market players in EU gas markets will likely also be active on future CCS markets.

important to underline, however, that the lessons that can be drawn from these cases are relevant in so far as future CO_2 transport operators are in a position of dominance – within the meaning of Article 102 TFEU – in the relevant market.

In the following sections, I answer the question in what way the above-mentioned Commission cases (also known as the 'gas foreclosure cases') impose requirements as to capacity allocation and congestion management on CO_2 transport infrastructure. Before the relevant Commission cases are explored in chronological order, section 7.2 briefly compares CO_2 transport and natural gas transport (infrastructure). Section 7.3 explores the *GDF* case. Section 7.4 deals with the *RWE* case. Section 7.5 examines the *E.ON* case. Section 7.6 deals with the Italian and EU *ENI* cases respectively. Finally, section 7.7 discusses the lessons that can be drawn for the development and management of CO_2 transport infrastructure and potential cross-border capacity requirements.

7.2 COMPARING NATURAL GAS AND CO_2 TRANSPORT (INFRASTRUCTURE)

The cases explored in section 7.3 predominantly concern natural gas transmission network infrastructure (high-pressure gas pipelines). This raises the question as to how to classify CO_2 transport infrastructure when comparing it to gas transmission infrastructure and the latter's place in the natural gas chain.[5] The CCS Directive does not classify CO_2 transport infrastructure in the way in which gas transport infrastructure is classified and categorised in Directive 2009/73/EC (Gas Directive).[6] The answer to the above question partly determines the usefulness of the lessons from the gas foreclosure cases for the development and management of CO_2 transport infrastructure.

Roggenkamp has argued that the CCS chain, from initial CO_2 capturing to final underground storage, appears to be the reverse of the natural gas chain, where the commodity leaves the geological reservoir and is transported through the upstream and downstream pipeline system before it arrives at its final destination.[7] According to Roggenkamp, CCS should, therefore, be characterised as a reversed downstream activity, to which the upstream laws and regulations

---

[5]   On this question, see, for instance, Martha Roggenkamp, 'The Concept of Third Party Access Applied to CCS' in Martha Roggenkamp and Edwin Woerdman (eds), *Legal Design of Carbon Capture and Storage* (Intersentia 2009), 273.

[6]   European Parliament and Council Directive 2009/73/EC of 13 July 2009 concerning common rules for the internal market in natural gas and repealing Directive 2003/55/EC [2009] OJ L211/94.

[7]   Roggenkamp (n 5) 293.

apply.[8] Roggenkamp infers this from the CCS Directive's wording, which appears to be similar to the wording in the Gas Directive concerning access to upstream (production) gas pipelines. Roggenkamp, in other words, appears to suggest that CO_2 transport (infrastructure) is most comparable to upstream gas transport (infrastructure).

I would, however, argue that CO_2 transport is comparable to natural gas transport over (primarily) transmission and (also) distribution infrastructure. When CO_2 is transported by pipeline, it is transported as a processed product. The CO_2 stream is purified, dehydrated and compressed before it enters the pipeline so as to prevent complications and facilitate transport. When natural gas enters the upstream pipeline, it is (at most) a partly processed product. It is transported as a fully processed product only when it enters the transmission pipeline, after the raw gas has been cleaned in a natural gas processing plant.

The above would, in my view, be and argument in favour of the classification of CO_2 pipelines as transport pipelines rather than reversed upstream pipelines. All the more so since CO_2 pipelines, like gas distribution (private consumers) or transmission pipelines (industrial consumers with a direct connection), deliver the product to the end user, the storage operator. In this respect, CO_2 transport pipelines could perhaps be seen as distribution and transmission pipelines in one.

When taking into account the possible future development of EU CCS markets and CO_2 transport infrastructure, the following considerations are of relevance. As we have seen in Chapter VI, some expect CO_2 transport infrastructure to consist of local point-to-point and hub and spoke connections in the initial years of EU CCS deployment (roughly until 2030).[9] From a scale perspective, such infrastructure is arguably comparable with gas distribution infrastructure. This would suggest that the lessons from the cases below would not appear to be of great relevance in the initial years of EU CCS deployment.

Yet, as I have argued in Chapter VI, this picture may change as a consequence of the low level of public acceptance of onshore CO_2 storage in (predominantly) North-West Europe. Should captured CO_2 mainly be stored offshore, then longer-distance cross-border CO_2 pipeline infrastructure, as well as CO_2 transport by ship, comes into play. In that case, CO_2 transport infrastructure could be said to be comparable with gas transmission infrastructure. Accordingly, the lessons from the gas foreclosure cases would appear to be of greater relevance. This conclusion is reinforced by the likelihood of the

---

[8]   Ibid.
[9]   See section 6.3.1.3.

Commission attaching greater importance to non-discriminatory access to CO$_2$ transport infrastructure in a cross-border setting (than in a purely domestic setting).[10] In sum, the more EU CO$_2$ transport networks develop and the more cross-border they become, the more relevant the findings from the cases below are likely to be.

7.3   GDF (39.316)

The *GDF* case centred around behaviour by the incumbent French gas company GDF which, according to the Commission, had the potential to prevent or reduce competition on downstream supply markets for natural gas in France through, in particular, a combination of long-term reservation of transport capacity and a network of import agreements, as well as through underinvestment in import infrastructure capacity.[11] The Commission suspected GDF to have abused its dominant position on the relevant markets contrary to Article 102 TFEU by (vertically) foreclosing[12] for a long period access to gas import capacity in each of the balancing zones[13] of the GRTgaz (a GDF subsidiary) network, thereby restricting competition on the markets for the supply of gas in each of those balancing zones.[14] It argued that the foreclosure was a result of the long-term reservation of most of the import capacity in the balancing zones of the GRTgaz network, the determination of reception capacity and the procedures for allocating long-term capacity at a new LNG terminal ('Fos Cavaou'), and the strategic limitation of investment in additional import capacity at an existing LNG terminal ('Montoir de Bretagne').[15]

---

[10]   By a cross-border setting, I refer to the situation in which a capturer in one Member State tries to have its captured CO$_2$ transported and stored in another Member State.

[11]   Commission, 'Antitrust: Commission Opens Formal Proceedings against Gaz de France Concerning Suspected Gas Supply Restrictions' MEMO/08/328 <http://europa.eu/rapid/pressReleasesAction.do?reference=MEMO/08/328&format=HTML&aged=0&language=EN&guiLanguage=en> accessed 26 January 2012.

[12]   Vertical foreclosure refers to the use of vertical integration or other vertical restraints by an input supplier ('upstream') to achieve market power in the output (or 'downstream') market. See American Bar Association Section of Antitrust Law, *Telecom Antitrust Handbook* (ABA Publishing 2005) 19–20. In a brief paper on the lessons learnt from the 2007 EU energy sector inquiry, Commission officials describe vertical foreclosure (in relation to the EU gas and electricity sectors) as 'the obstacles to competition stemming from the vertical integration of companies active in the supply and network businesses'. See Philip Lowe and others, 'Effective Unbundling of Energy Transmission Networks: Lessons from the Energy Sector Inquiry' Competition Policy Newsletter (2007) 23, 23 <http://ec.europa.eu/competition/publications/cpn/2007_1_23.pdf> accessed 10 January 2012.

[13]   The term balancing zone refers to the area in which the network operator ensures that the off-take of an amount of natural gas from the network is offset by network users by means of the injection of the same amount of natural gas in the transmission network.

[14]   *Gaz de France* (Case COMP/39.316) Commission Decision of 3 December 2009, para 24.

[15]   *Gaz de France* (n 14) para 25.

In relation to the long-term reservation of import capacity in France, the Commission considered that GDF's capacity reservations 'accounted for a very substantial part of total firm[16] import capacity in each of the balancing zones of the GRTgaz network for a very long period of time'.[17] As a result, third-party shippers did not have access to this capacity under conditions that would allow them to compete effectively on the downstream gas supply markets in these zones.[18] According to the Commission, both the gas infrastructure and import capacity in France constituted an input that was essential to enter the downstream gas supply market and extremely difficult to reproduce.[19]

With regard to the alleged problems in relation to the Fos Cavaou LNG terminal, the Commission, inter alia, found that 'despite genuine requests from a number of third-party shippers to reserve capacity, GDF Suez did not conduct an open, transparent and non-discriminatory procedure, for example an open season procedure,[20] to allocate capacity at the new terminal on a long-term basis'.[21] Furthermore, GDF allegedly never took into consideration, and de facto rejected, proposals by third-party shippers to co-finance the construction of the Fos Cavaou terminal.[22] While the terminal was being built, GDF did not explore the possibility of increasing its reception capacity in order to facilitate third-party access to the infrastructure.[23]

Finally, the Commission argued that GDF had strategically limited investment in additional capacity at the Montoir de Bretagne terminal.[24] Following an open season procedure, GDF allegedly had decided not to develop any additional import capacity, even though financial analyses indicated that, given the firm

---

[16] Article 2(16) of Regulation 715/2009 on conditions for access to the natural gas transmission networks defines 'firm capacity' as 'gas transmission capacity contractually guaranteed as uninterruptible by the transmission system operator'. The opposite of firm capacity is 'interruptible capacity', which Article 2(13) of Regulation 715/2009 defines as 'gas transmission capacity that may be interrupted by the transmission system operator in accordance with the conditions stipulated in the transport contract'.

[17] *Gaz de France* (n 14) para 28.

[18] Ibid para 30.

[19] Ibid paras 26–27.

[20] In an open season procedure, the name of a tender process for pipeline transport capacity, the developer makes it possible for anyone to apply to join the project. Open seasons act as regulatory assurance that private investors will coordinate to provide an efficient level of capacity through new investment, and that no pipeline developer is unduly excluding any other player. See NERA Economic Consulting, 'Developing a Regulatory Framework for CCS Transportation Infrastructure (Vol. 1 of 2)' (2009) 33 <www.decc.gov.uk/assets/decc/what%20we%20do/uk%20energy%20supply/energy%20mix/carbon%20capture%20and%20storage/1_20090617131338_e_@@_ccsreg1.pdf> accessed 4 January 2012.

[21] *Gaz de France* (n 14) para 32.

[22] Ibid para 33.

[23] Ibid.

[24] *Gaz de France* (n 14) para 37.

240

capacity requests received in the open season procedure, extension of the capacity at the Montoir de Bretagne terminal would have been sufficiently profitable.[25] As a result, the Commission contended, GDF had not only prevented another shipper who had submitted a firm capacity request in the open season procedure from reserving capacity on a long-term basis, but it had also greatly restricted any possibility of third-party access to the infrastructure until 2023, thereby making it even more difficult for third parties to import gas into France.[26]

Formally disagreeing with the Commission's main findings, GDF proposed a number of commitments indicating, inter alia, its preparedness to release firm long-term capacity at several entry points (into the French gas supply network) and at the Montoir de Bretagne and Fos Cavaou LNG terminals, as well as limiting its reservations of total firm long-term capacity throughout the French territory.[27] The final commitments proposed by GDF[28] were accepted by the Commission as constituting a proportionate and necessary solution to the competition problems identified by it.[29] The Commission's commitment decision outlines the various commitments in relation to the different natural gas qualities, essentially amounting to the commitments originally proposed by GDF.[30]

In a 2010 paper on the GDF commitment decision, Commission officials argued that the decision clarified that there can be limits on the extent to which dominant companies can reserve infrastructure on a long-term basis.[31] Cardoso

---

25 Ibid paras 38–39.
26 Ibid para 38.
27 See GDF, 'Engagements Proposés Formellement par GDF Suez, GRTgaz et Elengy dans la Cadre de la Procédure COMP/B-1/39.316 – version non confidentielle' (June 2009) <http://ec.europa.eu/competition/antitrust/cases/dec_docs/39316/39316_1854_9.pdf> accessed 26 January 2012.
28 See GDF, 'Engagements Proposés Formellement par GDF SUEZ, GRTgaz et Elengy dans la Cadre de la Procédure COMP/B-1/39.316 – version non confidentielle' (October 2009) <http://ec.europa.eu/competition/antitrust/cases/dec_docs/39316/39316_2144_9.pdf> accessed 27 January 2012.
29 *Gaz de France* (n 14) para 65. Also see para 87, where the Commission notes that the proposed commitments will put an end to the long-term foreclosure of access to gas import capacities by GDF. See also Commission, 'Antitrust: Commission Accepts Commitments by GDF Suez to Boost Competition in French Gas Market' IP/09/1872 <http://europa.eu/rapid/pressReleasesAction.do?reference=IP/09/1872&format=HTML&aged=0&language=EN> accessed 30 January 2012.
30 *Gaz de France* (n 14) 14–17.
31 See Ricardo Cardoso and others, 'The Commission's *GDF and E.ON Gas* Decisions Concerning Long-term Capacity Bookings: Use of Own Infrastructure as Possible Abuse Under Article 102 TFEU' (2010) European Commission Competition Policy Newsletter 2010-3, 8 <http://ec.europa.eu/competition/publications/cpn/cpn_2010_3.html> accessed 30 January 2012.

et al. noted that, even though the concept of anti-competitive long-term contracts is well established in EU antitrust law, the main competition concern in *GDF* was markedly different from previous (non-energy) cases involving anti-competitive long-term contracts and other energy cases concerning long-term supply contracts.[32] While it is common ground in EU competition law that long-term supply contracts can be a means to foreclose competitors from their customer base, the GDF investigation did not concern such long-term supply contracts with downstream customers, but reservations of transport capacity by the integrated company on its own transmission infrastructure.[33] According to Cardoso et al., in *GDF* it was not the (downstream) customer base that was foreclosed, but the access of third parties to GDF's transport infrastructure.[34]

Even though the *VEMW* case did not centre around Article 102 TFEU and a possible foreclosure of downstream supply markets, the issue of anti-competitive long-term transport capacity reservations had already come to the fore in this case.[35] Despite the fact that the case concerned a possible breach of the non-discrimination principle in the first Electricity Directive[36] and accordingly focused on the discriminatory (priority access) character of the long-term bookings,[37] *VEMW* does provide an early example of long-term capacity reservations in the EU energy sector being contrary to EU law. The statement by Cardoso et al. that in *GDF* it was not the downstream customer base, but access of third parties to GDF's transport infrastructure that was foreclosed, seems awkward. As the Commission indicated in its commitment decision, the foreclosure of the gas transport capacity markets restricted competition on the gas supply markets in the different balancing zones by negatively influencing the ability of other shippers to gain access to French gas supply markets.[38] Therefore, the downstream customer base was arguably foreclosed in *GDF*, at least indirectly.

GDF was not the first long-term energy transport capacity reservations case, but it did contain a novelty. In the *GDF* case, the Commission introduced a new kind of abusive behaviour under Article 102 TFEU: the strategic underinvestment in transport infrastructure and capacity. The core idea behind this controversial[39]

---

32   Cardoso and others (n 31) 10.
33   Ibid 10–11.
34   Ibid 11.
35   See case 17/03 *VEMW* [2005] ECR I-04983.
36   European Parliament and Council Directive of 19 December 1996 96/92/EC concerning common rules for the internal market in electricity.
37   Unlike *GDF*, the *VEMW* case did not deal with transport capacity reservations on a dominant undertaking's *own* network.
38   *Gaz de France* (n 14) paras 24–25 and 87.
39   See e.g. John Ratliff, 'Major Events and Policy Issues in EU Competition Law, 2009–2010 (Parts 1 and 2)' (2011) *International Company and Commercial Law Review* 128 <www.

new[40] abuse under Article 102 appears to be that dominant (gas) transport operators should either share or expand their indispensable infrastructure when it is profitable to do so in terms of the transport service alone.[41] As we have seen, the Commission argued that GDF had abused its dominant position in the gas transport markets by not investing in capacity expansion in the face of proven demand for additional capacity at the Montoir de Bretagne LNG terminal.[42]

Traditionally, the focus in Article 102 foreclosure cases has been on ensuring access to existing infrastructure/facilities. In *GDF*, the Commission widened the duty to deal[43] under Article 102 by requiring dominant transport operators to expand available capacity of indispensable infrastructure when there is proven demand for additional capacity and such expansion can be profitably done. The Commission had already pointed to the (perceived) risk of vertical integration leading to a delay or lack of investment in new transport infrastructure in its 2007 energy sector inquiry.[44] Yet in *GDF*, it applied the concept to a factual situation for the first time.

In relation to the development of transport infrastructure, several lessons can be drawn from *GDF*. First, the Commission's commitment decision seems to suggest that dominant transport operators should, in the face of third-party capacity requests, conduct open, transparent and non-discriminatory capacity allocation procedures – such as an open season procedure – when developing

---

wilmerhale.com/files/Publication/b999ed92-f1b5-4e75-b3a5-112e2c0d8d7d/Presentation/ PublicationAttachment/44f5a842-d65a-4738-baf1-14e56853292e/Ratliff_offprint.pdf> accessed 30 January 2012. Below, I will further discuss the concept of strategic underinvestment in relation to *ENI* (Case 39.315).

[40]   See, for instance, Richard Whish, *Competition Law* (6th edn, Oxford University Press 2009) 697. In 2009, Whish wrote that 'it has never been decided by the Commission or the Community Courts whether the owner of an essential facility can be under a duty to *increase* capacity in order to enable a third party to have access to it'. It could, however, be argued that not meeting demand is a form of abuse of a dominant position under Article 102 TFEU and that strategic underinvestment is nothing more than wilfully not meeting demand, which, as such, does not seem to be a novel abuse.

[41]   See Céline Gauer and John Ratliff, 'EU Competition Law and Energy: Recent Cases and Issues' (presentation held at the 51st Lunch Talk of the Global Competition Law Centre, 18 March 2011 <www.coleurope.be> accessed 30 January 2012.

[42]   This to protect the alleged *supra*-competitive rents reaped by GDF on the gas supply markets.

[43]   See section 6.3.2.2 of Chapter VI.

[44]   Commission, 'DG Competition Report on Energy Sector Inquiry' SEC (2006) 1724, 7–8, 13, 55 and 58–59 <http://ec.europa.eu/competition/sectors/energy/inquiry/full_report_part1. pdf> accessed 7 November 2011. The Commission notes that, inter alia, 'lack of investment and delayed investments by transmission companies with vertically integrated supply companies are another serious source of concern'. In relation to the gas sector, the Commission contends that it has found indications of this kind of behaviour taking place (58).

transport infrastructure.[45] It even appears to suggest that dominant transport operators are under a general duty[46] to explore the possibility of increasing capacity in order to facilitate third-party access when developing transport infrastructure.[47] Second, the decision indicates that when third-party shippers propose to co-finance the construction of transport infrastructure, such proposals should be taken into (serious) consideration by the dominant operator developing transport infrastructure.[48]

7.4 RWE (39.402)

In April 2007, the Commission decided to open proceedings against German utility RWE for a suspected breach of ex Article 82 TEC (now Article 102 TFEU).[49] The proceedings concerned the foreclosure of German gas supply markets through the creation of unjustified obstacles to third-party access to RWE's gas transport network in North Rhine-Westphalia.[50] The Commission's key concern was that RWE had abused its dominant position in the regional markets for the transport and wholesale supply of gas in North Rhine-Westphalia by raising rivals' costs[51] and preventing new entrants from getting access to capacity on gas infrastructure in Germany (thereby protecting RWE's dominant natural gas supply business).[52] It suspected RWE of charging high prices for access to gas networks operated by RWE's transmission system operator (RWE Transportnetz), inflation of the costs of RWE Transportnetz,

---

[45] *Gaz de France* (n 14) para 32.

[46] Even when no future capacity has actually been requested by third-party shippers.

[47] *Gaz de France* (n 14) para 33.

[48] Ibid.

[49] See Commission, 'Opening of Proceedings' (2007) <http://ec.europa.eu/competition/ antitrust/cases/dec_docs/39402/39402_43_10.pdf> accessed 1 February 2012, and Commission, 'Antitrust: Commission Initiates Proceedings against RWE Group Concerning Suspected Foreclosure of German Gas Supply Markets' MEMO/07/186 <http://europa.eu/ rapid/pressReleasesAction.do?reference=MEMO/07/186&format=HTML&aged=0&language =EN&guiLanguage=en> accessed 1 February 2012.

[50] Ibid.

[51] In a 2009 presentation on competition and energy, a member of the cabinet of former Competition Commissioner Kroes mentions several ways in which a vertically integrated undertaking can raise rivals' costs among which are making usage of the network by competitors difficult (e.g. 'book it yourself on a long-term basis'), increasing network costs (e.g. 'make sure that imbalances are very high') and giving yourself (as a vertically integrated undertaking) rebates. See Claes Bengtsson, 'Competition and Energy' (presentation at the CRA International Competition Workshop 'Competition Policy in the EU Energy Sector', Brussels, 12 February 2009) <www.crai.com/ecp/assets/Claes_BengtssonFeb09.pdf> accessed 1 February 2012.

[52] Commission, 'Opening of Proceedings' and 'Commission initiates proceedings' (n 49).

maintenance of an artificial network fragmentation and failure to release transport capacity to allow customer switching.[53]

In its 2009 commitment decision, the Commission indicated that the alleged abuse of a dominant position consisted of a refusal to supply/deal[54] and a margin squeeze.[55, 56] In relation to the alleged refusal to deal, the Commission argued that RWE Transportnetz had pursued a capacity management strategy according to which it tried to systematically keep the transport capacities on its own network for itself.[57] It had booked almost the entire capacity on its transmission network on a long-term basis, while there was steady and significant demand of third-party customers for transmission capacities on RWE's network.[58]

What is more, RWE had understated the capacity that was technically available to third customers, leading to unjustified refusals and deterring transport customers from requesting transport capacities; the difference between indicated and actually used capacity was so significant that it allegedly pointed to a strategy aimed at foreclosing potential third transport customers by understating the maximum technical capacity offered to the market.[59] Furthermore, the Commission maintained that RWE had not implemented an effective congestion

---

53   Ibid.
54   On the refusal to deal, see section 6.3.2.2 of Chapter VI.
55   A margin squeeze is a form of unfair pricing prohibited under Article 102(a) TFEU, consisting of conduct by an undertaking dominant in an upstream market and also active in a downstream market. A margin squeeze may occur if a vertically integrated undertaking which is dominant on an upstream market charges a price for a downstream product or service that prevents even equally efficient competitors from achieving a margin which allows them to compete effectively on the downstream market. See Oliver Koch and others, 'The RWE Gas Foreclosure Case: Another Energy Network Divestiture to Address Foreclosure Concerns' (2009) European Commission Competition Policy Newsletter 2009–2, 32, 32 <http://ec.europa.eu/competition/publications/cpn/2009_2_7.pdf> accessed 1 February 2012. See also Niamh Dunne, 'Margin Squeeze: Theory, Practice, Policy: Part 1' (2012) 33 European Competition Law Review 29, 29. For an example of a case in which margin squeeze played a role, see Case T-271/03 *Deutsche Telekom AG* [2008] ECR II-00477 and Case C-280/08 P *Deutsche Telekom AG* [2010] ECR I-09555.
56   *RWE Gas Foreclosure* (COMP/39.402) Commission Decision of 18 March 2009, para 21.
57   In this regard, the Commission referred to, inter alia, its decisions in *Sealink* (Case IV.34.689) Commission Decision of 21 December 1993; *Port of Rødby* Commission Decision 94/119/EC [1994] OJ L55/52 and *Frankfurt Airport* (Case IV.34.801) Commission Decision 98/190/EC [1998] OJ L72/30. It argued that it is noteworthy that the mere fact that the current capacities are fully used by the essential facility holder is not sufficient to exclude an abuse under ex Article 82 TEC. According to the Commission, in such a situation, a dominant essential facility holder is under the obligation to take all possible measures to remove the constraints imposed by the lack of capacity and to organise its business in a manner that makes the maximum capacities of the essential facility available. See *RWE Gas Foreclosure* (n 56) footnote 25.
58   *RWE Gas Foreclosure* (n 56) para 24.
59   *RWE Gas Foreclosure* (n 56) para 26.

management system to manage the scarce capacities on its network, which could have avoided many of the refused and delayed capacity requests.[60]

As for the alleged margin squeeze, the Commission contended that there was evidence that RWE had intentionally set its transmission tariffs at an artificially high level in order to squeeze RWE's competitors' margins.[61] RWE allegedly had embarked upon a strategy of raising its own network costs in order to charge higher network tariffs.[62] According to the Commission, there was also evidence that the network tariffs charged by RWE Transportnetz, which were already generally high, in practice may have been even higher for RWE's competitors than for its own trading company (RWE Energy).[63] This was exacerbated by RWE's rebate policy for long-term transmission contracts as it was almost impossible for new competitors to obtain long-term transport capacity.[64] Furthermore, RWE's balancing also had an 'asymmetrical negative impact on new entrants'.[65] While RWE Energy was exempted from paying balancing costs, other transport customers had to pay high penalty fees within RWE Transportnetz's network.[66]

While formally disagreeing with the Commission's allegations, RWE submitted several commitments which in essence consisted of RWE divesting its entire high-pressure gas transmission network (with the exception of some small regional network parts).[67] The Commission considered the commitments offered by RWE sufficient to effectively remove the competition concerns expressed by it.[68] In this regard, the Commission noted that 'these forms of behaviour (…) derive from an inherent conflict of interest within RWE as a vertically integrated gas company which controls both transmission and supply of gas'.[69] The

---

[60] Ibid para 27. The Commission argued that evidence indicated that RWE's intention might rather have been to protect RWE Energy from new competitors than to attract new transport customers for RWE Transportnetz.

[61] *RWE Gas Foreclosure* (n 56) para 30.

[62] Ibid para 32.

[63] Ibid 10.

[64] Ibid para 33.

[65] Ibid para 35.

[66] Ibid.

[67] See RWE, 'Fall COMP/B-1/39.402 – Deutscher Gasmarkt – Zusagen an Die Europäische Kommission' (2008) <http://ec.europa.eu/competition/antitrust/cases/dec_docs/39402/39402_437_11.pdf> accessed 1 February 2012 and *RWE Gas Foreclosure* (n 56) 11–12.

[68] *RWE Gas Foreclosure* (n 56) para 45. See also Commission, 'Summary of Commission Decision of 18 March 2009 Relating to a Proceeding under Article 82 of the EC Treaty and Article 53 of the EEA Agreement Case COMP/B-1/39.402 RWE Gas Foreclosure' [2009] OJ C133/08. At para 8, the Commission states that 'the sale of RWE's transmission business will ensure that RWE has no control over the gas transmission network and that RWE cannot engage in anticompetitive practices relating to the access to its network anymore'.

[69] *RWE Gas Foreclosure* (n 56) para 50. See also Commission, 'Antitrust: Commission Opens German Gas Market to Competition by Accepting Commitments from RWE to Divest

Commission considered that absent a remedy changing the structure of the RWE group, the incentives to further engage in such behaviour would not have been removed as effectively, resulting in a risk of a lasting or repeated infringement.[70]

There are several things to take away from the *RWE* case. First, there seem to be limits to the flexibility enjoyed by dominant (gas) network operators when calculating the maximum technical capacity offered to customers. A 'significant'[71] difference between indicated and actually used capacity may be interpreted by the Commission as revealing a strategy aimed at foreclosure of the relevant network. Second, a dominant network operator seems to be under the obligation to implement an effective congestion management system, which appears to require it to use 'all available means to make capacities available to its customers'.[72] Third, 'high'[73] network tariffs are suspect as they may indicate a strategy of raising own network costs in order to charge higher network tariffs and squeeze competitors' margins.[74] Finally, rebate schemes and exemptions from balancing fees for the own trading company are likewise practices which the Commission might consider to be indicative of a strategy aimed at squeezing competitors' margins.

7.5   E.ON (39.317)

Like the *GDF* and *RWE* cases, the *E.ON* case concerned long-term capacity bookings on the vertically integrated incumbent's own gas transmission network. The Commission suspected German utility E.ON to have infringed Article 102 TFEU by means of a refusal to supply gas transport capacity on its own transmission network.[75] E.ON had for a number of years in advance booked large parts of the available firm and freely allocable[76] entry capacity on

---

Transmission Network' IP/09/410 <http://europa.eu/rapid/pressReleasesAction.do?reference =IP/09/410&format=HTML&aged=0&language=EN&guiLanguage=en> accessed 1 February 2012. In the press release former Competition Commissioner Kroes is quoted stating that the divestment committed to by RWE ensures 'that RWE will no longer be able to use the control of its network to favour its own gas supply affiliate over its competitors'.

70   *RWE Gas Foreclosure* (n 56) para 50.
71   Unfortunately, in its commitment decision the Commission does not indicate what it considers to be a significant difference.
72   *RWE Gas Foreclosure* (n 56) para 27.
73   Again, the Commission did not state what it considered to be high network tariffs in *RWE*.
74   It should be underlined that in the gas and electricity sectors, transport tariffs are regulated by the energy regulators of the Member States.
75   See Commission, 'Case COMP/39.317 E.ON Gas Initiation of Proceedings' (2010) <http://ec.europa.eu/competition/antitrust/cases/dec_docs/39317/39317_1713_8.pdf> accessed 3 February 2012.
76   Since 2007, German law requires the application of an entry-exit system for gas transport capacity booking. Accordingly, network operators have to offer capacities which can be flexibly allocated ('freely allocable capacities') and which allow a shipper with a booking at an entry

its gas transmission grid.[77] Until 2019, it had booked at least 65–75% of all available firm freely allocable capacity on its high-calorific gas (H-gas) transmission network and 90–100% of all available freely allocable capacity on its low-calorific gas (L-gas) transmission network.[78]

The Commission argued that, as a consequence, little or no free capacity was available to competitors seeking to transport gas into E.ON's network.[79] The latter, allegedly, 'faced a permanent capacity bottleneck, severely limiting them to transport their gas to their actual or potential customers'.[80] According to the Commission, the tight capacity situation contrasted with steady and significant demand by transport customers for firm and freely allocable transmission capacities.[81] Referring to the *GDF* case, the Commission held that such long-term capacity bookings can be regarded as a refusal to supply/deal under Article 102 TFEU.[82]

Like GDF and RWE, E.ON offered commitments, while formally disagreeing with the Commission's allegations.[83] E.ON proposed to release firm freely allocable entry capacity into its gas transmission grid by late 2010 and to further reduce its overall share in the bookings of firm freely allocable entry capacity to 50% by late 2015.[84] E.ON would still be able to book interruptible capacities[85] and, in the first two years, short-term capacities that were not booked by third parties one month before the beginning of the gas year for which the capacity was released.[86] Following a market test by the Commission, E.ON proposed to adjust the commitments it offered by, inter alia, re-adjusting the capacity offered at the different entry points by 2010 and also offering exit capacity adjacent to an entry point covered by the entry capacity release insofar as E.ON held firm

---

point to choose any exit point within the operator's network. See *E.ON Gas* (Case COMP/39.317) Commission Decision of 4 May 2010, para 9. Under an entry-exit system, the booking of capacity is split into two parts: entry capacity, to transport gas from the injection point to a 'common or virtual balancing point', and exit capacity, to transport gas from the virtual balancing point to the different exit points in the system. This method of defining capacity is unrelated to the underlying physical characteristics of the network. See NERA (n 14) 30.

[77] *E.ON Gas* (n 76) para 2.
[78] Ibid para 37.
[79] *E.ON Gas* (n 76) para 38.
[80] Ibid.
[81] Ibid para 39.
[82] Ibid para 40.
[83] See E.ON, 'Anlage A zum Schreiben vom 07.01.2010 COMP/B-1/39.317 – E.ON Gas – Zusagen an die Europäische Kommission' (2010) <http://ec.europa.eu/competition/antitrust/cases/dec_docs/39317/39317_1729_6.pdf> accessed 6 February 2012.
[84] *E.ON Gas* (n 76) paras 44–47.
[85] See n 16.
[86] *E.ON Gas* (n 76) para 48. What is more, E.ON could also, as of October 2011, book long-term capacities under the condition that E.ON's overall booking share decreased over time until reaching the 2015 threshold.

bookings of the relevant exit capacity.[87] These capacities would be offered to interested third parties via the secondary capacity market.[88] The Commission considered these final commitments to be sufficient to effectively remove the competition concerns expressed by it.[89]

The *E.ON* case to a large extent resembles the *GDF* case.[90] In both cases, the incumbent had booked large parts of the capacity on its own transmission network.[91] As in *GDF*, the Commission argued that ownership unbundling[92] would not have resolved the competition problem since E.ON's long-term reservations would still close off competitors, even if the grid would have been sold to another operator.[93] A further parallel with *GDF* appears to lie in the linking of the anti-competitive behaviour at hand to a lack of investment in new transport infrastructure (capacity). In its FAQ on the *E.ON* commitments decision, the Commission has argued that E.ON's commitments will contribute to further infrastructure investments, thereby suggesting that such investments

---

87   *E.ON Gas* (n 76) para 59.
88   Ibid. Article 2(6) of Regulation (EC) 715/2009 defines the secondary capacity market as 'the market of the capacity traded otherwise than on the primary market'. Under Article 2(22) of the same Regulation, the primary market is defined as 'the market of the capacity traded directly by the transmission system operators'. In other words, secondary capacity is the capacity which is traded among market players.
89   *E.ON Gas* (n 76) para 60.
90   In a press release on E.ON's initial proposed commitments, the Commission refers to the GDF case as a parallel case. See Commission, 'Antitrust: Commission Welcomes E.ON Proposals to Increase Competition in German Gas Market' MEMO/09/567 <http://europa. eu/rapid/pressReleasesAction.do?reference=MEMO/09/567&format=HTML&aged=0&langu age=EN&guiLanguage=en> accessed 6 February 2012.
91   The unavailability of gas transmission capacity allegedly is a general problem in German gas markets. See Commission, 'Antitrust: E.ON's Commitments Open up German Gas Market to Competitors' IP/10/494 <http://europa.eu/rapid/pressReleasesAction.do?reference=IP/10/494 &format=HTML&aged=0&language=EN&guiLanguage=en> accessed 6 February 2012. In the press release, Competition Commissioner Almunia is quoted stating that 'the notorious lack of transport capacity is currently one of the major obstacles to gas competition in Germany'.
92   The term ownership unbundling refers to the separation of companies' production and sale operations from their transmission networks. Article 9(1) of European Parliament and Council Directive 2009/73/EC of 13 July 2009 concerning common rules for the internal market in natural gas and repealing Directive 2003/55/EC provides that the same person is entitled neither directly or indirectly to exercise control over an undertaking performing any of the functions of production or supply, and directly or indirectly to exercise control or exercise any right over a transmission system operator or over a transmission system, nor directly or indirectly to exercise control over a transmission system operator or over a transmission system, and directly or indirectly to exercise control or exercise any right over an undertaking performing any of the functions of production or supply.
93   See Commission, 'Antitrust: Commission's Commitment Decision Opens German Gas Pipelines to Competitors – Frequently Asked Questions' MEMO/10/164 <http://europa.eu/ rapid/pressReleasesAction.do?reference=MEMO/10/164&format=HTML&aged=0&language =EN&guiLanguage=en> accessed 6 February 2012.

did not (sufficiently) take place formerly.[94] Finally, the *E.ON* case confirms the *GDF* and *RWE* cases in declaring large long-term capacity bookings by dominant energy undertakings to, in principal, constitute a refusal to deal under Article 102 TFEU.[95]

7.6   *ENI* (CASES A358 AND 39.315)

Before discussing the EU *ENI* case (case 39.315), I first turn to a related (preceding) national competition case, the *ENI* case decided by the Italian Competition Authority (Case A358).

7.6.1   THE ITALIAN *ENI* CASE (A358)

In 2005, the Autorità Garante della Concorrenza e del Mercato (the Italian Competition Authority) commenced an investigation into an alleged abuse of a dominant position by Italian oil and gas incumbent ENI.[96] The Italian Competition Authority suspected ENI to have abused its dominant position on the Italian wholesale gas supply market by foreclosing access to that market.[97] In 2006, ENI received a €290 million fine[98] for having infringed ex Article 82 of the Treaty Establishing the European Community (TEC – now Article 102 TFEU) and was ordered to desist from its anticompetitive conduct and give third-party access to 6.5 billion cubic metres of gas transport capacity on the international gas pipeline operated by its subsidiary (the Trans-Tunisian Pipeline Company – TTPC), the TTPC pipeline.[99]

---

94   Commission, 'Antitrust' (n 93).

95   In this regard, see Also Kim Talus, 'Long-term Natural Gas Contracts and Antitrust Law in the European Union and the United States' (2011) 4 Journal of World Energy law and Business 260, 279. Talus, in my view rightly, argues that the *GDF* and *E.ON* cases indicate that the key issue is the volumes that are covered by these contracts. It is not the long-term nature as such, but the large volumes these contracts cover that are the problem.

96   Italian Competition Authority, 'ENI/Trans Tunisian Pipeline' (2005) <www.agcm.it/en/newsroom/press-releases/1456-enitrans-tunisian-pipeline.html> accessed 6 February 2012.

97   Ibid.

98   On administrative appeal, the fine was reduced to €20 million. See Carlo Baldini, 'Italian Competition Authority Case A358 – *International Transport of Gas*, ENI – TTPC (2005– 2006)' (presentation at the International Competition Network Teleseminar on Abuse of Dominance in the EU Energy Sector, 8 November 2011) <www.internationalcompetition network.org/uploads/library/doc768.pdf> accessed 8 February 2012.

99   Italian Competition Authority, 'ENI Fined € 290M for Abuse of Dominant Market Position in Wholesale Supply of Natural Gas' (2006) <www.agcm.it/en/newsroom/press-releases/1093-eni-trans-tunisian-pipeline.html> accessed 6 February 2012.

The Italian Competition Authority found ENI to have abused its dominant position in the national market for the wholesale supply of gas by having engaged in practices leading its subsidiary, TTPC, to abandon the (initially) planned expansion of capacity on the TTPC pipeline.[100] In 2002, TTPC had planned to increase the capacity of the – at that time fully booked[101] – TTPC pipeline.[102] In 2003, following the decision to increase the pipeline's capacity, TTPC assigned the additional capacity pro rata[103] on the basis of the requests received, entering into transport contracts with a number of operators.[104] According to the Italian Competition Authority, ENI subsequently, through its subsidiary TTPC, delayed the fulfilment of the contracts by, inter alia, announcing its intention to postpone the planned capacity expansion and to rescind the transport contracts signed.[105] Allegedly, ENI feared that the capacity expansion would lead to an oversupply of gas in the wholesale supply market, thereby threatening its position in that market.[106]

According to the Italian Competition Authority, ENI's dominant position in the national market for the wholesale supply of gas required it not to engage in any practice liable to influence, to the detriment of its (potential) competitors, the

---

[100] See Italian Competition Authority, 'Annual Report 2005' <www.agcm.it/en/annual-report/1804-annual-report-2005.html> accessed 6 February 2012, and Italian Competition Authority, *ENI-Trans Tunisian Pipeline* (Case A358) Decision n. 15174, published in Bulletin n. 5/2006 <www.google.nl/url?sa=t&rct=j&q=%22a358%20-%20eni-trans%20tunesian%20pipeline%22&source=web&cd=9&sqi=2&ved=0CHEQFjAI&url=http%3A%2F%2F62.149.237.43%2Fcomponent%2Fdomino%2Fdownload%2F41256297003874BD%2F95EADED91F188F9AC125711B0056A918.html%3Fa%3Dp15174.pdf&ei=CewvT5joEMiW8gOPp5SQDw&usg=AFQjCNF0KZYTO9J1X1XCwQrsvpcKLfYa_A&cad=rja> accessed 6 February 2012. Interestingly, in the past, ENI has, allegedly, been accused by the Italian regulator of not investing sufficiently in new underground gas storage facilities. See International Energy Agency, 'Energy Policies of IEA Countries: Italy 2009 Review' (2010) 120 <www.iea.org/textbase/nppdf/free/2009/italy2009.pdf> accessed 8 February 2012.
[101] By ENI and Italian utility ENEL.
[102] Italian Competition Authority, 'Annual report 2005' (n 100).
[103] Pro rata allocation of available capacity refers to capacity being allocated according to the proportion of each network user's requested capacity in relation to the total requested capacity.
[104] Italian Competition Authority, 'Annual report 2005' (n 100).
[105] International Competition Network, 'Case Annex to ICN Unilateral Conduct Working Group: Report on the Analysis of Refusal to Deal with a Rival under Unilateral Conduct Laws' (2010) 30 <www.internationalcompetitionnetwork.org/uploads/library/doc611.pdf> accessed 6 February 2012. See also the International Competition Network, '2009 Refusal to Deal Questionnaire: Italy' (2009) <www.internationalcompetitionnetwork.org/uploads/questionnaires/uc%20refusals/italy.pdf> accessed 6 February 2012. According to an employee of the Italian Competition Authority, during the investigation evidence of ENI's plans to delay the planned capacity expansion was found. In particular, letters were found, inter alia, revealing the influence of ENI over the decisions of its subsidiary and the willingness of ENI to stop the construction of the new capacity in order to preserve its downstream position in a context of stagnating demand. See Baldini (n 98).
[106] *ENI-Trans Tunisian Pipeline* (n 100).

conduct of its subsidiary TTPC, owner of the TTPC pipeline.[107] Had TTPC been acting as an independent operator, it would, considering the pre-existing transport contracts, have had every interest to proceed with the expansion of TTPC transport capacity.[108] As the investments required for the expansion of the pipeline would have been entirely recouped by the contracts signed with third-party shippers, there allegedly was no objective justification for the conduct delaying the entry of competitors into the market in which ENI held a dominant position (national wholesale gas supply market).[109]

Vasques and Nobili have argued that the Italian Competition Authority should have applied the 'intent test doctrine', apparently used in the US to ascertain whether a refusal to deal (or decision not to implement infrastructures that are essential to allow competitors to enter into the market) breaches the Sherman Act.[110] The US courts allegedly consider seemingly anti-competitive behaviour to infringe the Sherman Act if its *sole* objective is to monopolise the market in question.[111] According to Vasques and Nobili, such conduct is not considered to be anti-competitive if the undertaking concerned has a legitimate business reason justifying its decisions.[112] They have argued that ENI probably had legitimate business reasons justifying its decision not to invest millions to increase pipeline capacity for third parties.[113] Nevertheless, as indicated in section 6.3.2.3 of Chapter VI, the ECJ does not consider anti-competitive intent a prerequisite for a finding of an abuse of a dominant position. Under EU competition law, an undertaking can breach Article 102 TFEU, even though it did not intend to do so. The approach suggested by Vasques and Nobili would therefore be incompatible with standing EU competition case law.

According to Salerno, the Italian *ENI* case shows that a newcomer could expect the incumbent to refrain from stopping works for the expansion of a pipeline once the project has been launched.[114] Accordingly, an incumbent may be coerced by a newcomer to perform an activity that is beneficial to

---

[107]   *ENI-Trans Tunisian Pipeline* (n 100).
[108]   Ibid.
[109]   International Competition Network, 'Case Annex' (n 105) 30–31.
[110]   Luciano Vasques and Silvio Nobili, 'The Italian Competition Authority Fines ENI with the Highest Fine Ever Imposed to a Single Company in Italy for Abuse of a Dominant Position in the Wholesale Supply of Natural Gas on the Basis of Art. 82 EC (Trans Tunisian Pipeline Company-ENI)' (2006) e-Competitions No 501 <www.concurrences.com/article.php3?id_article=501&lang=fr> accessed 7 February 2012.
[111]   Vasques and Nobili (n 110).
[112]   Ibid.
[113]   Ibid.
[114]   Francesco Maria Salerno, 'The Competition Law-ization of Enforcement: The Way Forward for Making the Energy Market Work?' (2008) European University Institute Working Papers RSCAS 2008/07, 18 <http://cadmus.eui.eu/handle/1814/8108> accessed 8 February 2012.

competition.[115] Nevertheless, Salerno has argued that even if the Italian *ENI* case represents a leading precedent which is no doubt going to be observed closely by other network operators and supply undertakings in other markets, the fact remains that the case is an isolated one. In relation to EU energy regulation, the process of bringing to the fore applicable rights and obligations allegedly entails all the complex activities that are connected with a finding of a breach of competition law.[116] Yet, in view of the *GDF* and EU *ENI* cases (the latter is now discussed), it can be questioned whether the Italian *ENI* case indeed is an isolated one.

7.6.2   THE EU *ENI* CASE (39.315)

7.6.2.1   *Case 39.315*

In the same year that the Italian Competition Authority adopted decision A358, declaring ENI to have abused its dominant position in the Italian wholesale market for gas supply, the Commission carried out inspections on ENI premises and premises of ENI subsidiaries in Italy, Austria and Germany.[117] Based on the information gathered during these inspections, the Commission decided to initiate proceedings against ENI for a suspected infringement of ex Article 82 TEC (now Article 102 TFEU).[118] The Commission's concerns were related to the management and the operation of ENI's international gas transmission pipelines used for the import of natural gas from delivery points in Austria (TAG pipeline) and Germany (TENP/Transitgas pipelines) into Italy.[119] According to the Commission, ENI held a dominant position on the market for gas transmission

---

[115]   Salerno (n 114) 19. Importantly, Salerno has noted that the obligation was qualified by two factual circumstances, i.e. that work had been under way for some time and that onerous ship-or-pay contracts had already been signed. Ship-or-pay contracts are long-term contracts requiring payment regardless of whether any gas is actually shipped. See Goldman Sachs, 'Pipeline Financing Discussion' (presentation held at the Wyoming Natural Gas Pipeline Authority, 25 August 2003) <www.wyopipeline.com/information/presentations/2003/Goldman%20Sachs_files/frame.htm> accessed 8 February 2012.

[116]   Salerno (n 114) 19.

[117]   See Commission, 'Antitrust: Commission Initiates Proceedings against the ENI Group Concerning Suspected Foreclosure of Italian Gas Supply Markets' MEMO/07/187 <http://europa.eu/rapid/pressReleasesAction.do?reference=MEMO/07/187&format=HTML&aged=1&language=EN&guiLanguage=en> last accessed 6 February 2012.

[118]   Commission, 'Commission Initiates Proceedings' (n 117).

[119]   Commission, 'Antitrust: Commission Confirms Sending Statement of Objections to ENI Concerning the Italian Gas Market' MEMO/09/120 <http://europa.eu/rapid/pressReleasesAction.do?reference=MEMO/09/120&format=HTML&aged=1&language=EN&guiLanguage=en> accessed 6 February 2012.

to Italy[120] and on the gas supply markets[121] in Italy.[122] It suspected ENI of having managed and operated its gas transmission pipelines with the aim of limiting third-party access to available and new capacity.[123] This strategy allegedly was implemented by way of refusing to grant competitors access to capacity available on the transport network ('capacity hoarding'), granting access in an impractical manner ('capacity degradation') and strategically limiting investment ('strategic underinvestment') in ENI's international transmission pipelines.[124]

ENI either solely or jointly controlled all three international gas transmission pipelines, holding what the Commission considered to be significant transport rights on these infrastructures and having booked 'a substantial amount of the available capacity on a long-term basis'.[125] On the basis of the corporate structure of the relevant pipelines and the shareholding agreements, the Commission considered ENI to be likely to have the necessary information and powers to effectively influence decision making with regard to these infrastructures, namely the day-by-day management and operation of transport capacity as well as decisions with respect to new investment projects.[126]

In relation to the alleged *capacity hoarding*, the Commission argued that there was evidence that ENI had implemented a strategy to systematically reduce access to capacity for third parties on its gas transport infrastructure to Italy.[127] According to it, ENI had refused to offer existing available or unused capacity to other shippers on the three pipelines concerned, failed to increase the efficiency of its capacity management and understated the capacity technically available to third-party shippers.[128] What is more, ENI had allegedly secured significant primary capacity rights and corresponding long-term booking contracts on these pipelines.[129] In the eyes of the Commission, these practices

---

[120] By means of its ability to effectively control and influence the usage of all viable international pipelines for shipping gas into Italy as well as the Panigaglia LNG Terminal. See *ENI* (Case COMP/39.315) para 30.

[121] According to the Commission, the downstream gas supply market structure was 'still characterised by a quasi-monopoly of the vertically integrated incumbent ENI'. It held that ENI had a dominant position on the wholesale supply market in Italy as a whole and in particular on the market for supplies to gas fired power plants and on the market for supplies to large industrial customers. See *ENI* (n 120) paras 33 and 35.

[122] Ibid para 2.

[123] Ibid. Thereby abusing its dominant position on the gas transport market to Italy (contrary to the Italian *ENI* case, which concerned an abuse by ENI of its dominant position in the Italian wholesale gas supply market). See *ENI* (n 120) para 15.

[124] Ibid para 2.

[125] Ibid para 18.

[126] Ibid para 22.

[127] Ibid para 45.

[128] Ibid paras 46–48.

[129] Ibid para 49.

---

could be characterised as a constructive refusal to supply[130] capacity to third parties.[131]

As for the suspected *capacity degradation*, the Commission maintained that even when capacity on the three pipelines was offered by ENI, its purchase was made more difficult and less valuable to third parties by various means. ENI allegedly delayed the allocation of newly available capacity or offered it on a short-term basis through subsequent organised sales when it could have offered the capacity on a longer-term basis, designed allocation procedures in such a way that they would result in separate and uncoordinated capacity sales on complementary pipelines (e.g. TENP and Transitgas) and offered interruptible instead of firm capacity.[132]

Finally, the Commission argued that there had been significant and credible long-term capacity demand by third-party shippers on ENI's international pipelines.[133] Nevertheless, ENI's decisions to enhance transport capacity were allegedly mainly based on ENI's own needs resulting from new long-term contract commitments and the goal of keeping gas supply tight, while refusing to consider and eventually carry out expansions of capacity that would have allowed responding to third-party requests.[134] According to the Commission, ENI did not even gauge capacity demand from third-party shippers, for instance via open-season procedures,[135] and also did not explore the willingness of third parties to commit financially to an expansion project nor specific co-financing offers made by some shippers.[136] As a vertically integrated and dominant undertaking controlling the gas infrastructures to import gas into Italy, ENI had allegedly embarked upon a strategy of deliberately avoiding capacity expansions in order to ultimately limit third-party access to capacity and thereby prevent competition and lower prices on downstream markets (*strategic limitation of investments/underinvestment*).[137]

---

[130] The term 'constructive refusal to supply' refers to the dealing on unreasonable (or discriminatory) terms only. See *Deutsche Post* (Case COMP/C-1/36.915) Commission Decision 2001/892/EC [2001] OJ L331/40, para 141. See also *Sea Containers* (Case IV/34.689) Commission Decision 94/19/EEC [1994] OJ L15/8.

[131] *ENI* (n 120) 50.

[132] Ibid 13–14.

[133] Ibid para 56.

[134] Ibid para 57.

[135] See n 20.

[136] *ENI* (n 120) para 57.

[137] Ibid para 60. According to the Commission, ENI's limitation of investment was not driven by the lack of profitability of an increased transport activity for ENI as a TSO. Rather, such a strategy was the result of ENI's conflict of interest stemming from, on the one hand, the negative repercussions on the profitability of its own gas supply business from supplying capacity to downstream competitors in the Italian market and the relatively modest increase in profits due to expansions on the level of its transport network business on the other hand. See *ENI* (n 120) para 58. In a 2011 paper on the *ENI* commitment decision, Commission

ENI formally disagreed with the Commission's findings, but nonetheless committed to divest its shareholdings in the transmission system operators, and the companies holding the respective TSO's shareholding, of the three international gas transmission pipelines (TENP, Transitgas and TAG).[138] ENI's proposed commitments were market tested by the Commission, after which they were somewhat revised (without essentially affecting the proposed divestitures).[139] The Commission considered the final commitments sufficient to remove the competition concerns.[140] It argued that ENI would no longer be subject to the inherent conflict of interests it faced operating both as a transmission system operator and as an undertaking active on the Italian wholesale market.[141] Given the fact that ENI would lose control over the three gas transmission pipelines, it allegedly would no longer be in a position to refuse to grant access to these transport infrastructures, grant access in a less attractive manner, and limit investments in new capacity to transport gas into Italy.[142]

Even though the EU *ENI* case and the Italian *ENI* case both concerned alleged underinvestment in gas transmission infrastructure, there is an important difference between the two cases. Where the Commission accused ENI of having abused its dominant position in the (upstream) market for gas *transmission* to Italy,[143] the Italian Competition Authority based its case on ENI's dominant

---

officials note that '*concrete evidence* substantiated the Commission's concern that the absence of additional investments in transportation capacity was not driven by a lack of profitability' (emphasis added). See Frank Maier-Rigaud, Federica Manca and Ulrich von Koppenfels, 'Strategic Underinvestment and Gas network Foreclosure: The ENI Case' (2011) European Commission Competition Policy Newsletter 2011–1, 18, 21 <http://ec.europa.eu/competition/publications/cpn/2011_1_4_en.pdf> accessed 9 February 2012.

[138]   *ENI* (n 120) paras 63–64. See also Commission, 'Antitrust: Commission Welcomes ENI's Structural Remedies Proposal to Increase Competition in the Italian Gas Market' MEMO/10/29 <http://ec.europa.eu/rapid/pressReleasesAction.do?reference=MEMO/10/29&format=HTML&aged=0&language=EN&guiLanguage=en> accessed 8 February 2012; and ENI, 'Case COMP/B-1/39.315 ENI: Commitments Submitted to the European Commission' (2010) <http://ec.europa.eu/competition/antitrust/cases/dec_docs/39315/39315_3022_6.pdf> accessed 8 February 2012.

[139]   *ENI* (n 120) paras 73–81.

[140]   Ibid para 87. Even though the Commission always emphasises the voluntary character of commitments offered by dominant undertakings, it acknowledged in its commitment decision (para 88) that the divestment offered by ENI was 'a structural remedy of the type envisaged in the Statement of Objectives'. The process whereby commitments are offered by dominant undertakings and subsequently either rejected or accepted by the Commission is arguably one that leaves little scope for the undertakings to act in a voluntary manner.

[141]   *ENI* (n 120) para 89, See also Commission, 'Antitrust/ENI Case: Commission Opens up Access to Italy's Natural Gas Market' IP/10/1197 <http://europa.eu/rapid/press ReleasesAction.do?reference=IP/10/1197&format=HTML&aged=0&language=EN&guiLang uage=en> accessed 8 February 2012.

[142]   *ENI* (n 120) para 89.

[143]   The Commission's commitment decision is a bit confusing in this regard. In paras 1 and 2, the Commission states that the decision concerns 'ENI's conduct on the gas transportation

position in the (downstream) market for the wholesale *supply* of gas.[144] The Italian Competition Authority did not work from the premise that the transmission infrastructure controlled by ENI was indispensable. Rather, it held that, as an undertaking with a dominant position in the Italian market for the wholesale supply of gas, ENI was under the obligation not to induce its subsidiary to change an autonomously taken decision with the sole aim of preserving ENI's dominant position in that market.[145] Accordingly, the EU *ENI* case – as well as the *GDF* case – seems to go further than the Italian *ENI* case by requiring an undertaking dominant in the (gas) transport market which controls infrastructure that is indispensable to enter a downstream market to expand the infrastructure's capacity if there is proven demand for extra capacity and such expansion can profitably[146] be done.

7.6.2.2 *The concept of strategic underinvestment*

a)   Debating the concept of strategic underinvestment

The Commission's strategic underinvestment thesis has caused a fair deal of controversy among EU antitrust lawyers. Based on the Commission's commitment decision in *ENI*, Weitbrecht and Kallaugher have argued that the Commission appears to favour a very expansive interpretation of Article 102 TFEU, according to which the obligations of an owner of essential facilities[147] under certain circumstances are not limited to the granting of access to existing facilities but also require the dominant undertaking to enlarge the essential infrastructure facilities.[148] Similarly, Talus has noted that the Commission's cases on refusal to grant access to gas pipelines show its clear tendency to apply a

---

market to Italy and on the gas supply markets in Italy' and that 'ENI holds a dominant position on the market for transport of gas to Italy and on the gas supply markets in Italy'. Nevertheless, para 15 provides that 'the Commission has gathered evidence that ENI may have abused its dominant position on the gas *transportation* market into Italy to the detriment of competitors, competition and ultimately consumers on the downstream supply markets' (emphasis added).

[144]   See Baldini (n 98) and Ulrich Scholz and Stephan Purps, 'The Application of EC Competition Law in the Energy Sector' (2010) 1 Journal of European Competition Law and Practice 37, 47–48.

[145]   Baldini (n 98). Interestingly, in a 2009 International Competition Network questionnaire on refusal to deal, the Italian Competition Authority does characterise the *ENI* case (A358) as an 'access to essential facilities' case. See the International Competition Network, '2009 Refusal to Deal Questionnaire' (n 105) 3 and 5.

[146]   From the point of view of the operation of that infrastructure.

[147]   An essential facility can be described as a network or other type of infrastructure to which access is necessary to compete on a given market. See Cardoso and others (n 31) 3.

[148]   See John Kallaugher and Andreas Weitbrecht, 'Developments under Articles 101 and 102 TFEU in 2010' (2011) 32 European Competition Law Review 333, 336. See also Piotr Szlagowski, 'The Abuse of a Dominant Position through Strategic Underinvestment of Energy Transmission Network Interconnectors' (2010) 6 International Energy Law Review 201.

refusal to deal or an essential facility type of approach even beyond the original scope.[149] Instead of limiting the application to access refusals, the Commission has gone further by demanding capacity expansions or construction of new capacity on the back of Article 102.[150]

According to Scholz and Purps, the Commission is apparently demanding that an undertaking not only grant access to existing infrastructure but actually provide additional financial resources for expansion to adapt a given infrastructure to the actual demands.[151] They argue that it is questionable whether Article 102 is an adequate tool for competition authorities to establish and enforce investment obligations, since it would mean that these authorities would take entrepreneurial decisions with long-term effects.[152] According to Scholz and Purps, the expansion of existing infrastructure does not belong to the tasks of the Commission under EU competition law, regardless of its relevance for the competitive situation in upstream or downstream markets.[153] Rather, the setting of standards for investments into infrastructure assets should be subject to sector-specific legislation.[154]

Siragusa has argued that the Commission's commitment decision in *ENI* marks the culmination of a development started by the Commission in other energy antitrust cases (e.g. *E.ON*), as a consequence of which the traditionally negative principle of non-discriminatory third-party access to energy infrastructure has been transformed into a positive obligation for the essential facility holder to manage the infrastructure in a way that is beneficiary to its competitors, thereby restricting the dominant undertaking's ownership rights to its own infrastructure.[155]

According to Siragusa, the (abusive) concept of strategic underinvestment represents an absolute novelty in EU antitrust law, which goes counter to standing decision practice.[156] The latter allegedly dictates that lack of capacity constitutes an objective justification for refusing access.[157] Siragusa has noted

---

[149] See Kim Talus, *Vertical Natural Gas Transportation Capacity, Upstream Commodity Contracts and EU Competition Law* (Kluwer Law International 2011) 214.

[150] Ibid.

[151] Scholz and Purps (n 144) 48.

[152] Ibid.

[153] Ibid.

[154] Ibid.

[155] Mario Siragusa, 'Gli Obblighi di non Discriminazione nella Regolazione Settoriale e nella Disciplina Antitrust' (contribution to the second Energy Law Conference, Rome, 6–7 April 2011) 9.

[156] Ibid 11.

[157] Ibid. See also Article 35(1) of the Gas Directive. See also Article 32(2) of European Parliament and Council Directive 2009/72/EC of 13 July 2009 concerning common rules for the internal

that the rationale behind the Commission's strategic underinvestment approach appears to be based on the likely investment behaviour of an independent profit-maximising transmission system operator.[158] According to Siragusa, this approach is controversial, difficult to apply in practice and has never been explained or applied by the Commission in a systematic way.[159] Like Scholz and Purps, Siragusa has contended that incentives for capacity expansion should be given through sector-specific regulation.[160]

As argued in section 7.3, the concept of strategic underinvestment can be traced to the Commission's 2007 energy sector inquiry (sector inquiry). In its sector inquiry, the Commission mentioned vertical foreclosure as one of its eight key areas of attention.[161] It noted that vertical integration had negative repercussions on incentives to invest in networks, causing investment decisions not to be taken in the interest of network/infrastructure operations, but on the basis of supply interests of the integrated undertaking.[162] The Commission argued that it had found indications of discriminatory behaviour with regard to investment decisions taken by vertically integrated gas companies.[163] Allegedly, certain investment decisions on network extensions of the transport system operator had to be approved by an investment committee of the transmission system operator's parent company.[164] According to the Commission, in a number of cases, companies only invested in capacity expansions if their related supply arms had previously confirmed their interest for the bulk of the extra capacity.[165] By contrast, the investment did not take place if the interest in extra capacity stemmed from competitors.[166] The Commission held that ownership unbundling was the most efficient way to ensure that network operators do not have investment incentives that are distorted by supply interests (and *do* invest in further expansions).[167]

---

market in electricity and repealing Directive 2003/54/EC [2009] L211/55 (Electricity Directive).

[158]   Siragusa (n 155) 11.

[159]   Ibid 11–12.

[160]   Ibid 12.

[161]   Commission, 'Energy Sector Inquiry' (n 44) 4–5. According to the Commission, foreclosure of transport/storage infrastructure could particularly arise in cases where cross-border access was concerned.

[162]   Commission, 'Energy Sector Inquiry' (n 44) 7–8. In this regard, see also Lowe and others (n 12) 27. Lowe has argued that 'since the vertically integrated incumbents normally have very strong market positions as a supplier in the area where they control the network, it is often in their interest not to invest in infrastructure that would bring additional competition to this area: the interest in protecting the market power and the profitability of their supply business trumps their interest in increasing their (regulated) network business'.

[163]   Commission, 'Energy Sector Inquiry' (n 44) 58.

[164]   Ibid.

[165]   Ibid.

[166]   Ibid. In this regard, the Commission referred to the Italian *ENI* case.

[167]   Commission, 'Energy Sector Inquiry' (n 44) 62.

b)   The (economic) validity of the concept of strategic underinvestment

In order to better be able to assess the (economic) validity of the concept of strategic underinvestment introduced by the Commission in the *GDF* and *ENI* cases – as well as the criticism it received – I briefly turn to the economic literature on vertical foreclosure. The latter examines whether, and under what circumstances, foreclosure strategies such as strategic underinvestment would constitute rational, profit maximising behaviour and would survive in equilibrium.[168, 169] Vertical foreclosure theory, in other words, studies the (theoretical) likelihood of foreclosure strategies being pursued in practice.

In economic theory, the refusal to deal an upstream product (the input) to the market, as alleged in the *GDF* and *ENI* cases, is known as 'input foreclosure'.[170] Rey and Tirole have argued that for all its prominence in competition law, the notion of foreclosure until recently had poor intellectual foundations.[171] The theory of foreclosure is said to be complex as there are many different market structures in which foreclosure might arise, various strategies exist by which an undertaking can deny or limit access to its input for competitors and the practices that lead to foreclosure can have welfare enhancing as well as welfare reducing effects.[172] Furthermore, these effects can occur simultaneously and have the same origin.[173] In addition to the theory being complex and fragmented, empirical work on vertical foreclosure has generally lagged behind.[174]

Whether strategies that may lead to foreclosure, such as vertical restraints[175] or vertical integration, actually harm welfare is ambiguous.[176] The plausibility of claims that vertical integration leads to anti-competitive behaviour has been a

---

[168]   In economics, the term equilibrium refers to demand for and supply of a certain product being in balance.

[169]   American Bar Association (n 12) 21.

[170]   Ibid 20. See also Michiel Bijlsma and others, 'Vertical Foreclosure: a Policy Framework' (2008) CPB Netherlands Bureau for Economic Policy Analysis Document No 157 27 <www.cpb.nl/en/publication/vertical-foreclosure-policy-framework> accessed 13 February 2012.

[171]   See Patrick Rey and Jean Tirole, 'A Primer on Foreclosure' (2006) <http://idei.fr/doc/by/tirole/primer.pdf> accessed 13 February 2012.

[172]   Bijlsma and others (n 170) 16–18.

[173]   Ibid 18.

[174]   Ibid 14. See also Rey and Tirole (n 171) 27.

[175]   Vertical restraints can be defined as 'agreements or concerted practices entered into between two or more companies each of which operates, for the purposes of the agreement, at a different level of the production or distribution chain, and relating to the conditions under which the parties may purchase, sell or resell certain goods or services'. See Europa, 'Summaries of EU legislation: Guidelines on Vertical Restraints' <http://europa.eu/legislation_summaries/other/l26061_en.htm> accessed 13 February 2012.

[176]   Bijlsma and others (n 170) 14.

source of considerable controversy among (mainly US) economists.[177] The structure-conduct-performance perspective of the 1950s and 1960s ('Cambridge School') viewed vertical integration suspiciously, holding that it frequently embodied an anti-competitive strategy whereby a monopolist in an input (upstream) market would attempt to leverage its monopoly power into a retail (downstream) market for a product that relied on its input.[178]

In the 1960s, this view was challenged by the 'Chicago School', arguing that vertical integration and other restrictive vertical arrangements generally do not result in or facilitate anticompetitive outcomes because anti-competitive vertical arrangements would harm the upstream monopolist by imposing costs that equal or exceed the benefits of the strategy, or by reducing demand for the final product.[179] Rather, the Chicago School argued that vertical restraints can generally be explained on (economic) efficiency grounds.[180] Transaction cost economics[181] of the 1970s and 1980s staked a middle ground, identifying new efficiency rationales for vertical integration, while warning that undertakings with market power may have strategic goals poorly aligned with consumer welfare.[182] Most recently, new literature on vertical foreclosure (a.k.a. 'Post-Chicago Economics') has applied game theoretic tools[183] to develop new theories of strategic vertical integration and identify circumstances in which vertical integration alters industry conduct to the detriment of competitors and consumers.[184]

Of particular importance to network industries like the gas and electricity industries is the 'Bell Doctrine', developed by Baxter in the 1980s. This theory holds that incumbent, vertically integrated, regulated monopolies controlling

---

[177]   See Paul L Joskow and Roger G Noll, 'The Bell Doctrine: Applications in Telecommunications, Electricity, and other Network Industries' (1999) 51 Stanford Law Review 1249, 1254.

[178]   Michael H Riordan, 'Competitive Effects of Vertical Integration' (2005) Columbia University Department of Economics Discussion Papers 0506–11, 1 <www.columbia.edu/~mhr21/Vertical-Integration-Nov-11-2005.pdf> accessed 13 February 2012, and American Bar Association (n 12) 22.

[179]   American Bar Association (n 12) 22.

[180]   Ibid.

[181]   Transaction cost economics studies how trading partners protect themselves from the hazards associated with exchange relationships. See Howard A Shelanksi and Peter G Klein, 'Empirical Research in Transaction Cost Economics: A Review and Assessment' (1995) 11 Journal of Law, Economics and Organization 335, 336 <http://organizationsandmarkets.files.wordpress.com/2010/03/sk_jleo_1995.pdf> accessed 13 February 2012.

[182]   Riordan (n 178) 1.

[183]   Game theoretic concepts apply whenever the actions of several agents are interdependent. The concepts of game theory provide a language to formulate, structure, analyse and understand strategic scenarios. See Theodore L Turocy and Bernhard von Stengel, 'Game Theory' (2001) CDAM Research Report LSE-CDAM-2001–09, 4 <www.cdam.lse.ac.uk/Reports/Files/cdam-2001-09.pdf> accessed 13 February 2012.

[184]   Riordan (n 178) 1.

both monopoly segments and potentially competitive segments (like, historically, the incumbent EU gas undertakings) have the incentive and opportunity to monopolise related markets in which their monopolised service is an input.[185] The most effective solution to this problem, allegedly, is to 'quarantine' the regulated monopoly segment of the industry by separating its ownership and control from the ownership and control of undertakings that operate in potentially competitive segments (ownership unbundling).[186]

Alternatively, competition could be introduced in the monopolised segment and the regulation that applies to it.[187] In a 2008 paper on vertical foreclosure, Bijlsma et al. argued that in the presence of price regulation, foreclosure may remain a risk if there is vertical integration.[188] Allegedly, this is due to the existence of asymmetric information, in particular in assessing available capacity and in deciding on new investments.[189] According to Bijlsma et al., vertical separation can accommodate such risks.[190] The question in this regard, allegedly, is how large the costs of such separation are (e.g. loss of synergies),[191] and how they compare to the risks of foreclosure.[192]

As a consequence of advances in post-Chicago (game-theoretic) economic theory, it is now generally accepted that vertical restraints may be motivated by strategic objectives designed to change or resist changes to the structure of the industry.[193] More specifically, *refusals to interconnect* could be part of an anticompetitive strategy of vertically foreclosing competition.[194] Yet it is important to underline that in economic literature vertical foreclosure is not

---

[185]   Joskow and Noll (n 177) 1249–50. Joskow and Noll argue (p. 1313) that 'it is widely recognised that incumbent monopolies have the incentive and ability to use their control of monopoly network facilities to restrict competition in the potentially competitive segments where competitors require access to these networks'. See also Giulio Frederico, 'The Economic Analysis of Energy Mergers in Europe and in Spain' (2011) 00 Journal of Competition Law and Economics 1, 9.

[186]   Ibid 1250.

[187]   Ibid 1253. These, arguably, are the two solutions adopted by the Commission in the *RWE* and *ENI* cases on the one hand (unbundling) and the *GDF* and *E.ON* cases on the other (primarily capacity releases and ban on long-term capacity reservations).

[188]   Bijlsma and others (n 170) 77.

[189]   Ibid.

[190]   Ibid.

[191]   Synergy can be described as the tendency to unify the power of two or more elements and the perception that the whole is greater than the sum of the parts that constitute it. See Mihalis Giannakis, Simon Croom and Nigel Slack, 'Supply Chain Paradigms' 4 <http://fds.oup.com/www.oup.co.uk/pdf/0-19-925932-1.pdf> accessed 14 February 2012.

[192]   Bijlsma and others (n 170) 77. See also Joskow and Noll (n 177) 1253.

[193]   See Miguel Moura e Silva, 'EC Competition Law and the Market for Exclusionary Rights' <http://homepage.mac.com/mmsilva/documents/Lumextwp5.doc> accessed 14 February 2012. See also Bijlsma and others (n 164) 14 and American Bar Association (n 6) 32–33.

[194]   Adam Candeub, '*Trinko* and Re-grounding the Refusal to Deal Doctrine' (2005) 66 University of Pittsburgh Law Review 821, 825 and 848.

regarded as anti-competitive per se as it does not automatically harm welfare.[195] All potential pro- and anti-competitive effects have to be assessed and the latter must outweigh the former.[196] Bijlsma et al. have argued that foreclosure strategies lead to welfare loss when the main motivation for restricting access to an input is to put efficient rivals at a competitive disadvantage, or even to force them to leave the market (or not enter in the first place).[197] Riordan has noted that a convincing input foreclosure theory of harmful vertical integration has two crucial elements.[198] First, equally cost-effective substitute inputs are unavailable.[199] Second, a vertically integrated firm has an incentive to withdraw from the input market or raise the price of an input.[200]

Based on the above, the Commission's underlying hypothesis in *GDF* and *ENI* (as well as in the other gas foreclosure cases) of (anti-competitive) vertical foreclosure seems to have a (firm) foundation in economic literature. Yet whether the same goes for the alleged concept of strategic underinvestment can be doubted. In economic literature there appear to be few, if any, references to the concept of strategic underinvestment, as used by the Commission in the *GDF* and *ENI* cases. In a 2000 paper, Brennan, referring to a 1979 work by Flexner, mentioned the regulatory economics concept of 'undersizing'.[201] Accordingly, if a regulated transport facility (a pipeline, electric transmission line) is owned by one or more firms that compete at the end of the link, the owner(s) may have an incentive to reduce the capacity of the line. As the capacity of that link falls, the overall volume of delivery may fall, raising the price at which they can sell their output at the end of the line above competitive levels. This, arguably, is the kind of reasoning applied by the Commission in *ENI*, where it, *a contrario*, argued that ENI (as a supply business) had no interest in enlarging transmission capacity since such expansion would lead to lower downstream (supply) market prices.

---

[195]   See, for instance, Bijlsma and others (n 170) 43 and 61 and American Bar Association (n 12) 21. Vertical foreclosure can, as argued by the Chicago School, lead to economic efficiencies; it can, for instance, decrease supply by less efficient firms. See, for instance, Riordan (n 178) 48. For a critique of the Commission's dealing with vertical foreclosure and likely harm to welfare/consumers, see Philip Marsden, 'Some Outstanding Issues From the European Commission's Guidance on Article 102 TFEU: Not-so-faint Echoes of Ordoliberalism' in Federico Etro and Ioannis Kokkoris (eds), *Competition Law and the Enforcement of Article 102* (Oxford University Press 2010) 53.

[196]   Bijlsma and others (n 170) 43.

[197]   Ibid.

[198]   Riordan (n 178) 50.

[199]   Ibid.

[200]   Ibid.

[201]   See Timothy J Brennan, 'The Economics of Competition Policy: Recent Developments and Cautionary Notes in Antitrust and Regulation' (2000) Resources for the Future Discussion Paper 00–07, 17 <www.rff.org/documents/RFF-DP-00-07.pdf> accessed 14 February 2012.

In a 2011 study on the incentives of access-regulated undertakings to invest in network infrastructure they must share with competitors, Klumpp and Su concluded that their results lent considerable support to the hypothesis that access-regulated utilities do take strategic effects into account when making transmission investments.[202] According to Klumpp and Su, there is evidence that investments in (US) transmission capacity are made strategically; ceteris paribus, utilities are less likely to invest, and investment levels are lower, when competitors occupy a larger share of the market.[203] However, the results of this study are not necessarily characteristic of the investment incentives in other network industries (like the EU gas industry).[204] Furthermore, it is not certain whether the Klumpp and Su study lends support for the thesis that vertically integrated incumbents are less likely to invest in transport infrastructure (expansion) if the infrastructure is likely to be used by competitors.[205]

The concept of strategic underinvestment in transport infrastructure as used by the Commission in the *GDF* and *ENI* cases does not seem to figure prominently in economic theories of vertical foreclosure. However, this does not mean that the concept is invalid per se. As stated above, vertical foreclosure theory studies the likelihood of foreclosure strategies being pursued in practice. Even though certain strategies are unlikely to be pursued from a theoretical point of view – for instance due to those strategies harming the economic interests of those pursuing them – they can still be followed in practice. According to the Commission, it found evidence of such strategies being pursued in the EU gas sector in general[206] and in the *GDF* and *ENI* cases in specific.

The criticism that the Commission has (unjustly) broadened the scope of refusal to deal/supply under Article 102[207] does not, in my opinion, invalidate the concept of strategic underinvestment as such. Decisions on investment in network capacity are arguably just as much part of network management (and affect third-party access to infrastructure) as are decisions on the way in which capacity is allocated or congestion is managed. Admittedly, the finding of underinvestment in infrastructure can be difficult to make, since it involves the assessment of a great deal of detailed market and business information.

---

[202] Tilman Klumpp and Xuejuan Su, 'Strategic Investments under Open Access: Theory and Empirical Evidence' (2011) 19 <http://userwww.service.emory.edu/~tklumpp/docs/energy.pdf> accessed 14 February 2012.
[203] Ibid i.
[204] Ibid 20.
[205] Ibid 3.
[206] See the paragraph on the sector inquiry above.
[207] See the comments by Weitbrecht and Kallaugher, Talus, Scholz and Purps, Siragusa, and Szlagowski, discussed above.

However, there seem to be no grounds for arguing that the Commission generally is not competent to do so. When investigating possible strategic underinvestment, the Commission does not, as argued by Scholz and Purps, take entrepreneurial decisions. Rather, it evaluates whether entrepreneurial decisions (not to invest) have not been taken on grounds indicating a strategy to foreclose related (downstream) markets for (potential) competitors. The Commission does not (structurally) determine the investments to be made by network operators, in which case the argument for doing so under sector-specific regulation (Scholz and Purps, Siragusa) would make sense. It corrects instances of market failure[208] and appears to be fully competent to do so under Article 102.

Nevertheless, in economic theory, foreclosure is not assumed to automatically lead to loss of welfare. The Commission has long been (and still is) criticised for equating foreclosure to anti-competitiveness and welfare loss.[209] It has been criticised for applying a form-based instead of an effects-based (economic) approach.[210] The Commission has seemingly taken this criticism to heart in its 2009 guidance document on enforcement priorities in applying Article 102. Nevertheless, considering the enduring criticism[211] and the Commission's own desire to apply a more effects-based approach in its Article 102 decisions, it would have been helpful if the Commission had given some (more) insight into the economic reasoning underlying its application of the concept of strategic underinvestment. As both cases were settled and the Commission's commitment decisions are largely silent on the underlying economic reasoning, it is hard to judge the Commission's weighing of pro- and anti-competitive effects. In view of the novelty and apparent controversy of the concept of strategic underinvestment, the Commission would arguably have been wise not to introduce the concept by commitment decision.[212] By choosing to solve matters

---

[208]   In economics, the term market failure refers to a market failing to allocate resources efficiently, leading to a lower level of societal welfare.

[209]   See, for instance, Marsden (n 195).

[210]   See section 6.3.1.1 of the previous chapter.

[211]   Again, see, for instance, Marsden (n 195) 54. Marsden has argued that '[C]ommission policy is still focused on proving that dominant firm conduct may harm rivals – this in turn is assumed to reduce the competitive constraints on the dominant firm and thereby to allow it to harm consumers'.

[212]   See also Lars Kjølbye, 'The Commission's Evolving Commitment Practice – Its Impact and the Issues that it Raises' (presentation at Florence School of Regulation EU Energy Law & Policy Workshop on Priority Access for Renewable Energy into the Grid, Florence, 28 May 2010) <www.florence-school.eu/portal/page/portal/FSR_HOME/ENERGY/Policy_Events/ Workshops/2010/EU_Energy_Law_Policy/L.Kjolbye.pdf> accessed 15 February 2012. In relation to the commitment decisions in the EU energy sector, Kjølbye has argued that it is not an issue that the Commission runs challenging cases and pushes the law into new areas. Yet what may raise concern is that commitment decisions are not tested before the EU courts and that policy is developed by commitments and not by reasoned decisions, which reduces transparency and rigour. Likewise, Willis and Hughes have argued that the Commission does have the power to order full unbundling as a remedy in specific cases, provided that it has first

through a commitment decision, the Commission has created a considerable margin of discretion for itself.

In conclusion, the EU *ENI* case can be said to be one of the most important EU cases on access to energy transport infrastructure. In *ENI*, the Commission confirmed the *GDF*, *RWE* and *E.ON* cases in holding large/substantial long-term capacity bookings by dominant energy undertakings to, in principal, constitute a refusal to deal under Article 102 TFEU. It also indicated that a failure to increase the efficiency of capacity management and the understatement of capacity technically available to third-party shippers (see *RWE*) may constitute a constructive refusal to supply capacity to third parties. What is more, the Commission determined that dominant transport network operators may not make the purchase of capacity more difficult for third parties by, for instance, delaying the allocation of newly available capacity and offering interruptible instead of firm capacity.

Finally, in *ENI*, the Commission confirmed the concept of strategic underinvestment, as introduced in the *GDF* case. The Commission's use of the concept of strategic underinvestment seems to suggest that when dominant network operators controlling indispensable infrastructure are aware of third-party demand for extra transport capacity and such expansion can profitably be done (from the point of view of the transport business), the decision not to invest in capacity expansion can constitute abusive strategic underinvestment. The *ENI* case further indicates that a dominant network operator is under the obligation to at least gauge capacity demand from third-party shippers (e.g. open season conduct), explore the willingness of third parties to commit financially to an expansion project and take into consideration specific co-financing offers made by third-party shippers.

The review of the above cases allows for several conclusions to be drawn. First, large/substantial long-term capacity bookings by dominant vertically integrated gas (and possibly electricity) incumbents on their own transport infrastructure will likely constitute a refusal to deal under Article 102 TFEU.[213] In *GDF*, *RWE*, *E.ON* and *ENI*, the Commission found large long-term capacity reservations to be anti-competitive.

---

conducted a thorough economic analysis. However, in cases which have been settled, it is, in my opinion, not possible to assess whether the latter has indeed been done. See Peter Willis and Paul Hughes, 'Structural Remedies in Article 82 Energy Cases' (2008) 4 The Competition Law Review 147, 153 <www.clasf.org/CompLRev/Issues/Vol4Iss2Art3WillisHughes.pdf> accessed 15 February 2012.

[213] This also goes for capacity bookings made at the time of construction of the relevant facility (see *ENI* (n 120) paras 49–50).

The *RWE* and *ENI* cases indicate that dominant transport operators are under the obligation to have an efficient and effective congestion management system and to use all available means to make capacity available to third-party shippers. Both cases suggest that there are limits to the flexibility enjoyed by dominant (gas) network operators when calculating the maximum technical capacity offered to customers. The understatement of capacity technically available to third-party shippers may constitute a constructive refusal to supply. A 'significant' difference between indicated and actually used capacity may be interpreted by the Commission as revealing a strategy aimed at foreclosure of the relevant network.

The *RWE* case further shows that 'high' network tariffs are suspect as they may reveal a strategy of raising own network costs in order to charge higher network tariffs and squeeze competitors' margins. Rebate schemes and exemptions from balancing fees for the own trading company are likewise practices which the Commission might consider to be indicative of a strategy aimed at squeezing competitors' margins. What is more, in *ENI* the Commission determined that dominant transport operators may not make the purchase of capacity more difficult for third parties by, for instance, delaying the allocation of newly available capacity or offering interruptible instead of firm capacity (when the latter is available).

In relation to the development of energy transport infrastructure, the *GDF* and *ENI* cases are particularly instructive. The commitment decisions in both cases appear to suggest that dominant transport operators have to conduct open, transparent and non-discriminatory capacity allocation procedures – such as an open season procedure – when developing transport infrastructure. What is more, based on the *GDF* case, dominant transport operators seem to have the general duty to explore the possibility of increasing planned capacity in order to facilitate third-party access when developing transport infrastructure. Furthermore, dominant transport operators appear to be under the obligation to explore the willingness of third parties to commit financially to an expansion project and to take into (serious) consideration specific co-financing offers made by third-party shippers.

Finally, the *GDF* case introduced, and the *ENI* case confirmed, the vertical foreclosure concept of strategic underinvestment. The Commission's use of this novel antitrust concept seems to suggest that when dominant network operators are aware of third-party demand for extra transport capacity and such expansion can profitably be done (from the point of view of the transport business), the decision not to invest in capacity expansion can constitute abuse of a dominant position by refusal to deal. The concept of strategic underinvestment in transport infrastructure as used by the Commission in the *GDF* and *ENI* cases does not seem to figure prominently in economic theories of vertical foreclosure. Yet, this does not invalidate the concept as such, nor does the criticism that the Commission has (unjustly) broadened the scope of refusal to deal/supply under Article 102.

The incumbent gas undertakings being vertically integrated arguably played an important role in the four gas foreclosure cases. Even though the Commission explicitly referred to the anti-competitive conduct deriving from an inherent conflict of interests caused by vertical integration in the *RWE* and *ENI* cases only, it is argued here that the vertically integrated structure of the dominant undertakings played an equally important role in the *GDF* and *E.ON* cases.[214] The undertakings concerned had very solid positions (developed throughout time) in both upstream (network) and downstream (supply) markets. It is questionable whether the Commission would have scrutinised the dominant undertakings' conduct as stringently, had the undertakings concerned not had such strong positions on both upstream and downstream markets.

7.7 LESSONS FOR THE DEVELOPMENT AND MANAGEMENT OF CO_2 TRANSPORT INFRASTRUCTURE

In section 7.2, I argued that the more developed and the more cross-border EU CO_2 transport networks become, the more relevant the findings from the four gas foreclosure cases will likely be. Scale and the stage of development of EU CO_2 transport networks are arguably important factors for determining the relevance of the lessons from the above cases for the development and management of CO_2 transport infrastructure.

In addition, the four gas foreclosure cases and the lessons that can be drawn from these cases should be seen in the crucial context of the vertically integrated structure of the relevant dominant gas transport undertakings. GDF, RWE, E.ON and ENI were all vertically integrated incumbents which, as argued above, had very solid positions (developed over time) in both upstream (network) and downstream (supply) markets. In Chapter VI,[215] I indicated that future EU CCS markets could likewise be characterised by vertical integration.

---

[214] See also Adrien de Hauteclocque and Vincent Rious, 'Reconsidering the Regulation of Merchant Transmission Investment in the Light of the Third Energy Package: The Role of Dominant Generators' (2009) Reflexive Governance in the Public Interest Working Paper Series REFGOV-IFM-67, 3.

[215] See section 6.3.1.4.

However, should the Commission investigate the competitiveness of CO$_2$ transport markets, it can be questioned whether it would conduct such investigation with the same level of scrutiny as it has done in the gas sector. Unlike EU gas markets, EU CCS markets are yet to develop. The Commission's fierce antitrust stance as regards the former markets can be explained by the solid position of incumbents in these historically shielded markets. What is more, in the case of the CCS markets, the strong underlying right of gas end-users to freely choose their supplier of gas seems to be lacking.

Nevertheless, the *GDF, RWE, E.ON* and *ENI* cases all concerned the anti-competitive conduct of vertically integrated dominant undertakings controlling gas transmission infrastructure. As we have seen in Chapter VI, undertakings active in future EU CO$_2$ transport markets could have a dominant position in these markets until at least 2030. Regardless of the particular antitrust approach adopted by the Commission towards the CCS sector, Article 102 TFEU will, in that case, come into play. As illustrated by recital 38 of the preamble to the CCS Directive[216] – access to CO$_2$ transport/storage sites could become a condition for entry into/competitive operation within EU electricity markets – there might be incentives for vertically integrated CO$_2$ transport operators (transport and capture) to discriminate against (potential) rivals in the internal electricity and heat market. What is more, in its decision practice on Article 102, the Commission has created a considerable margin of discretion for itself. In order to prevent CO$_2$ transport operators from running into problems with this provision, it might therefore be wise to take the lessons from the four gas foreclosure cases to heart.

Based on these lessons, CO$_2$ transport operators are advised to conduct open seasons to market test (potential) demand for CO$_2$ transport capacity, including (potential) demand for cross-border CO$_2$ transport capacity. They would also be wise to explore the possibility of increasing planned capacity in order to facilitate third-party access, as well as the willingness of third parties (from other Member States) to commit financially to an expansion project, and to take into (serious) consideration co-financing offers made by third-party shippers. Finally, in order to prevent allegations of strategic underinvestment, CO$_2$ transport operators are advised to invest in additional transport capacity if there is demand for extra capacity and such expansion can be profitably realised. In this regard, it is worthwhile mentioning that in case of point-to-point transport infrastructure, the capacity of that infrastructure can obviously not grow beyond the maximum capacity of the linked storage reservoir.

---

[216] In this regard, see Chapter V on refusing access to CCS infrastructure and the general EU law principle of loyalty.

Even though the four gas foreclosure cases primarily focused on pipeline infrastructure, these lessons are not only applicable to CO_2 pipeline infrastructure. Shipping terminals or CO_2 hubs used to load captured CO_2 onto tankers for transport to offshore storage locations can be seen as the reverse of LNG terminals. Should access to such infrastructure be indispensable for access to offshore storage locations, the above lessons will arguably likewise be of relevance.

CHAPTER VIII

CENTRALISING CO_2 STORAGE SITE SELECTION UNDER EU LAW

- The Commission would seem to be able to base a possible proposal for amendment of Article 4(1) CCS Directive, allowing it to force Member States to accept CO_2 storage in (parts of) their territory, on Article 192(1) TFEU

- Such amendment does not appear to have to be based on Article 192(2) TFEU (adoption by unanimity)

- Nor would Article 194(2) TFEU seem to provide a proper legal basis for such amendment

- An amended Article 4(1) allowing the Commission to force Member States to accept CO_2 storage in (parts of) their territory, would probably infringe Article 345 TFEU, as it would likely impair the very substance of the right of ownership of the relevant storage reservoir

8.1   INTRODUCTION

As a consequence of the low level of public acceptance of predominantly onshore CO_2 storage in several Member States, little CO_2 storage capacity has become available in the first years of EU CCS demonstration (2009–2012).[1] The poor availability of CO_2 storage capacity sharply contrasts with the role attributed to CCS technologies in EU efforts to reduce greenhouse gas emissions and the European ambitions with regard to the early demonstration of CCS technologies. According to the European Commission (Commission), CO_2 emissions avoided

---

[1]   In this regard, see also the introduction to Chapter III. See also European Commission, 'Communication from the Commission to the European Parliament, the Council, the European Economic and Social Committee and the Committee of the Regions on the Future of Carbon Capture and Storage in Europe' COM (2013) 180 final 18. According to the Commission, 'several Member States are banning or restricting storage of CO_2 on their territories'.

through CCS could account for some 15% of the reductions required in 2030.[2] Furthermore, the EU aims to have up to 12 large-scale CCS demonstration projects running by 2015.[3] Considering the interests at stake, it is not unthinkable that the Commission will try and obtain a more central role in the allocation of CO_2 storage locations in the EU. It could possibly attempt to centralise the allocation of CO_2 storage locations by forcing Member States with storage capacity to allow CO_2 storage in suitable potential storage reservoirs.

At the time of writing, the CCS Directive does not allow the Commission to force Member States to accept CO_2 storage in (parts of) their territory.[4] Recital 19 of the preamble to the Directive provides that Member States should retain the right to determine the areas within their territory from which storage sites may be selected. This includes the right not to allow any storage in parts or on the whole of their territory, or to give priority to any other use of the underground.

What is more, these rights are explicitly anchored in (the binding) Article 4(1) of the Directive, which reiterates the essential wording of recital 19. Nevertheless, as part of the review of the CCS Directive in 2015, the Commission could try to amend Article 4(1) in order for the provision to allow it to force Member States to accept CO_2 storage in (parts of) their territory.[5] Chances are that the Commission

---

[2]   See the European Commission's website on carbon capture and geological storage at <http://ec.europa.eu/clima/policies/lowcarbon/ccs/index_en.htm> accessed 20 July 2012.

[3]   The idea to have up to 12 large-scale demonstrations of sustainable fossil fuel technologies in commercial power generation running by 2015 was first mentioned in the 2007 Commission Communication, 'Sustainable Power Generation from Fossil Fuels: Aiming for Near-zero Emissions from Coal after 2020' COM(2006) 843 final, and endorsed two months later by heads of state and government during the 2007 Spring European Council; see European Council, 'Brussels European Council 8/9 March 2007: Presidency Conclusions' 7224/1/07, 22.

[4]   In this regard, see also Global CCS Institute, 'Strategic Analysis of the Global Status of Carbon Capture and Storage Report 3: Countries Studies – The European Union' (2009) 7 <http://cdn.globalccsinstitute.com/sites/default/files/publications/8517/strategic-analysis-global-status-ccs-country-study-european-union.pdf> accessed 27 September 2012. According to the Global CCS Institute, '[t]he CCS Directive leaves significant discretion for Member States to resist the development of CCS facilities in their jurisdiction'.

[5]   Recent developments in EU nuclear law show that the poor availability of final disposal capacity – in the case of CCS CO_2 storage capacity – could lead to the adoption of legislation obliging Member States to realise final disposal facilities. While leaving some flexibility as to the dates a disposal facility is put into operation, Articles 1(1), 2, 4(3)(c), 11 and 12 of Directive 2011/70/Euratom of 19 July 2011 on the responsible and safe management of spent fuel and radioactive waste [2011] OJ L199/48, oblige the Member States to initialise without undue delays the process towards the planning and the realisation of disposal of spent fuel and radioactive waste. Article 4(4) of the Directive indicates that, even though the export of radioactive waste or spent fuel to another Member State or third country is not excluded, the basic principle is that radioactive waste shall be disposed of in the Member State in which it was generated. As the Commission's impact assessment of its proposal for Directive 2011/70/Euratom indicates, the proposal for the new Directive was predominantly motivated by the poor state of development of final disposal repositories for radioactive waste and spent fuel in

will base such an amendment on Article 192(1) of the Treaty on the Functioning of the EU (TFEU) (ex Article 175(1) of the Treaty establishing the European Community (TEC)), the current legal basis of the CCS Directive, or on Article 194 TFEU. The latter is the new energy title in the TFEU, which requires Parliament and the Council to take measures that, inter alia, promote energy efficiency and energy saving and the development of new and renewable forms of energy.[6]

In the following, I sketch the EU legal framework with which an amendment of Article 4(1) will have to comply. More specifically, I answer the questions of to what extent Articles 192 and 194 TFEU would provide proper legal bases for such an amendment of Article 4(1) and to what extent an amended Article 4(1) would be compatible with Article 345 TFEU. Section 8.2 discusses Articles 192 and 194 TFEU, the environmental and energy legal bases in the TFEU, with a special focus on Articles 192(2) and 194(2). Section 8.3 addresses Article 345 TFEU, which provides that the Treaties shall in no way prejudice the rules in Member States governing the system of property ownership.[7] Finally, section 8.4 gives an overview of the legal framework with which an amendment of Article 4(1) will have to comply.

I do not address the international legal framework of an amendment of Article 4(1). Such an analysis would fall outside the scope of this thesis, as well as

---

the EU. See Council Directive 2011/70/Euratom of 19 July 2011 establishing a Community framework for the responsible and safe management of spent fuel and radioactive waste [2011] L199/48; Ute Blohm-Hieber, 'The Radioactive Waste Directive: A Necessary Step in the Management of Spent Fuel and Radioactive Waste in the European Union' (2011) 87 Nuclear Law Bulletin 21, 34; Euratom Supply Agency, 'Annual Report 2011' 9 <http://ec.europa.eu/euratom/ar/last.pdf> accessed 8 August 2012: and Commission, 'Commission Staff Working Document: Accompanying Document to the Revised Proposal for a Council Directive (Euratom) on the management of spent fuel and radioactive waste' SEC (2010) 1289 final 10. See also Commission, 'Questions & Answers: Safety Standards for Nuclear Waste Disposal' MEMO/10/540 <http://europa.eu/rapid/pressReleasesAction.do?reference=MEMO/10/540&format=HTML&aged=0&language=EN> accessed 7 August 2012: and Commission, 'Commission Staff Working Document: Summary of the Impact Assessment Accompanying Document to the Revised Proposal for a Council Directive (Euratom) on the Management of Spent Fuel and Radioactive Waste' SEC (2010) 1290 final 3 <http://eur-lex.europa.eu/LexUriServ/LexUriServ.do?uri=SEC:2010:1290:FIN:EN:PDF> accessed 7 August 2012.

6   Article 194(1)(c) TFEU.
7   For the argument that Article 345 TFEU can function as a barrier for the exercise of competences by the EU, see Johann-Christian Pielow, Gert Brunekreeft and Eckhart Ehlers, 'Legal and Economic Aspects of Ownership Unbundling in the EU' (2009) 2 Journal of World Energy Law and Business 1, 10; Johann-Christian Pielow and Eckhart Ehlers, 'Ownership Unbundling and Constitutional Conflict: A Typical German Debate?' (2008) 2 European Review of Energy Markets 1, 13 and Nicole Ahner, 'Final Report' (Florence School of Regulation workshop on unbundling of energy undertakings in relation to corporate governance principles, Berlin, 25 September 2009) 8 <www.florence-school.eu/portal/page/portal/LDP_HOME/Events/Workshops_Conferences/2009/Unbundling/CorporateGovernance_AX-Programme.pdf> accessed 6 August 2012.

my expertise. Suffice it to say that United Nations General Assembly Resolution 1803 (XVII) acknowledges the right of nations to permanent sovereignty over their natural wealth and resources[8] and that Article 18 of the Energy Charter Treaty recognises state sovereignty and sovereign rights over energy resources.[9] All 27 EU Member States, as well as the European Community (Community) and the European Atomic Energy Community (Euratom), are signatories of the Energy Charter Treaty. What is more, the international legal principle of permanent sovereignty contained in United Nations General Assembly Resolution 1803 arguably has legal value.[10] This raises interesting questions with regard to the relationship between these international legal instruments and the potential possibility for the Commission to force Member States to allow CO_2 storage in their territory; these are questions which will, however, not be answered in this work.[11]

---

[8] On the question of whether CO_2 storage capacity can be seen as a natural resource see, for instance, Marcel Brus, 'Challenging complexities of CCS in public international law' in Martha M Roggenkamp and Edwin Woerdman (eds), *Legal Design of Carbon Capture and Storage* (Intersentia 2009), 19, 33. Brus states that 'a general definition of a natural resource does not exist, but natural storage capacity that can be used for purposes of allowing further generation of human wealth and/or reducing degradation of the human environment, could be considered a natural resource without stretching our imagination too far'.

[9] United Nations General Assembly Resolution 1803 (XVII) of 14 December 1962, 'Permanent Sovereignty over Natural Resources' <www2.ohchr.org/english/law/resources.htm> accessed 23 July 2012: and Article 18 of the Energy Charter Treaty <www.encharter.org/fileadmin/user_upload/document/EN.pdf> accessed 23 July 2012.

[10] Nico Schrijver, *Sovereignty over Natural Resources – Balancing Rights and Duties* (Cambridge University Press 1997) 375 and 377. Schrijver has argued that 'as far as doctrine is concerned, hardly any contemporary international lawyer would deny the principle of permanent sovereignty a legal value' and that 'despite its complicated genesis the principle of permanent sovereignty over natural resources has achieved a firm status in international law and is now a widely accepted and recognised principle of international law'.

[11] One of these is the question whether these international legal instruments would be covered by Article 351 TFEU. Article 351(1) provides that the rights and agreements arising from agreements concluded before 1 January 1958 or, for acceding states, before the date of their accession to the EU, between a Member State and a third country shall not be affected by the provisions of the Treaties (TFEU and TEU). To the extent that such agreements are not compatible with the Treaties, Article 351(2) obliges the Member State concerned to take all appropriate steps to eliminate the incompatibilities established. At first sight, both legal instruments would not appear to fall under Article 351 since the ECJ has confirmed from the outset that the first paragraph of this provision covers the rights of *third countries* and not those of *Member States* (only their obligations). Article 351 does not enable Member States to exercise rights conferred upon them by international agreements in violation of EU law. See Piet Eeckhout, *EU External Relations Law* (2nd edn, Oxford University Press 2011) 424, Panos Koutrakos, *EU International Relations Law* (Hart Publishing 2006) 302, and Rass Holdgaard, *External Relations Law of the European Community* (Kluwer Law International 2008) 138.

8.2 ARTICLES 192 AND 194 TFEU

In the following sections, I explore Articles 192 and 194 TFEU, in order to examine their relevance as a legal basis for a possible amendment of Article 4(1) of the CCS Directive. In relation to Article 192 TFEU, I answer the question to what extent Article 192(2) (adoption by unanimity) might constitute a proper legal basis for a possible amendment of Article 4(1).

8.2.1 ARTICLE 192 TFEU

Article 192 contains the environmental legal basis in the TFEU. Article 192(1) states that:

> [T]he European Parliament and the Council, acting in accordance with the ordinary legislative procedure and after consulting the Economic and Social Committee and the Committee of the Regions, shall decide what action is to be taken by the Union in order to achieve the objectives referred to in Article 191.

Article 192(1) is the successor to Article 175(1) TEC, which is the legal basis of the CCS Directive. The ordinary procedure referred to in Article 192(1) is the legislative procedure mentioned in Articles 289 and 294 TFEU.[12] The objectives referred to in Article 191 are the EU's objectives in the field of environmental protection, including, among other things, the protection of human health and the prudent and rational utilisation of natural resources.

The second paragraph of Article 192 provides that:

> [B]y way of derogation from the decision-making procedure provided for in paragraph 1 and without prejudice to Article 114, the Council acting unanimously in accordance with a special legislative procedure and after consulting the European Parliament, the Economic and Social Committee and the Committee of the Regions, shall adopt:
> a) Provisions primarily of a fiscal nature;
> b) Measures affecting:
>   – Town and country planning;
>   – Quantitative management of water resources or affecting, directly or indirectly, the availability of those resources;
>   – Land use, with the exception of waste management;

---

12   Article 289(1) TFEU provides that the ordinary legislative procedure shall consist in the joint adoption by the European Parliament and the Council of a Regulation, Directive or Decision on a proposal from the Commission. This procedure is defined in Article 294 TFEU.

    c)   Measures significantly affecting a Member State's choice between different energy sources and the general structure of its energy supply;

The Council, acting unanimously on a proposal from the Commission and after consulting the European Parliament, the Economic and Social Committee and the Committee of the Regions, may make the ordinary legislative procedure applicable to the matters referred to in the first subparagraph.

As an exception to Article 192(1) TFEU, Article 192(2) generally has to be interpreted narrowly.[13] In *Spain v. Council*, discussed in the next section, the European Court of Justice (ECJ) indicated that it will likely, at least in relation to the second ground under Article 192(2)(b),[14] interpret the exceptions in Article 192(2) restrictively.[15] Although Article 192(2) TFEU is hardly used in practice, its interpretation generates considerable problems.[16] In the light of a possible amendment of Article 4(1) of the CCS Directive, the second and third grounds in subparagraph (b) and the ground in subparagraph (c) arguably are of most relevance. In the following, we will therefore look at the interpretation of these provisions.

8.2.1.1   Measures affecting quantitative management of water resources

The second ground in Article 192(2)(b) refers to measures affecting quantitative management of water resources or affecting, directly or indirectly, the availability of those resources. On the face of it, this ground does not seem to be of relevance for determining the correct legal basis for a possible amendment of Article 4(1) of the CCS Directive. A measure allowing the Commission to force Member States to accept CO_2 storage in (parts of) their territory would not immediately appear to affect Member States' quantitative management of water resources or the availability of those resources. Nevertheless, as illustrated by the seminal case in this regard, *Spain v. Council*, there might be a link with a possible amendment of Article 4(1).

In *Spain v. Council*, Spain challenged the validity of Decision 97/825/EEC,[17] which had as a legal basis ex Article 130s(1) TEC (now Article 192(1) TFEU –

---

[13]   Ludwig Krämer, *EU Environmental Law* (7th edn, Sweet & Maxwell 2011) 81–82.

[14]   Measures affecting the quantitative management of water resources or affecting, directly or indirectly, the availability of those resources.

[15]   Case C-36/98 *Spain v. Council* [2001] ECR I-00779. On the ECJ's restrictive interpretation of the second ground under Article 192(2)(b), see also Han Somsen, 'Discretion in European Community Environmental Law: An Analysis of ECJ Case Law' (2003) 40 Common Market Law Review 1413, 1417.

[16]   Jan H Jans and Hans HB Vedder, *European Environmental Law* (4th edn, Europa Law Publishing 2012) 59.

[17]   Council Decision 97/825/EEC of 24 November 1997 concerning the conclusion of the Convention for the protection and sustainable use of the river Danube [1997] OJ L342.

adoption by qualified majority) in conjunction with the first sentence of ex Article 228(2) TEC and the first subparagraph of ex Article 228(3) TEC.[18] Spain argued that Decision 97/825/EEC should have been based on ex Article 130s(2) TEC (now Article 192(2) TFEU – adoption by unanimity) in conjunction with the second sentence of ex Article 228(2) TEC and the first subparagraph of ex Article 228(3) TEC.[19] In essence, Spain held that the Decision on the conclusion of the Danube Convention was to be characterised as a measure 'concerning the management of water resources'.[20]

After comparing the different language versions of ex Article 130s(2), the ECJ held that it was clear from the objectives of Community policy on the environment and from a reading of ex Article 130r TEC (now Article 191 TFEU) in conjunction with ex Article 130s(1) and (2) (now Article 192(1) and (2) TFEU) that the inclusion of the 'management of water resources' in the first subparagraph of ex Article 130s(2) TEC was not intended to exclude any measure dealing with the use of water by man from the application of ex Article 130s(1) TEC (the ordinary legislative procedure).[21]

The ECJ argued that, apart from the measures concerning the management of water resources, the second indent in ex Article 130s(2) TEC referred to measures relating to town and country planning and to land use with the exception of waste management and measures of a general nature.[22] According to it, these

---

[18] Ex Article 300 TEC has been replaced by Article 218 TFEU. The first sentence of ex Article 300(2) TEC provided that: 'subject to the powers vested in the Commission in this field [the conclusion of agreements between the Community and third states/international organisations], the agreements shall be concluded by the Council, acting by a qualified majority on a proposal from the Commission'. The first subparagraph of ex Article 300(3) TEC provided that: 'the Council shall conclude agreements after consulting the European Parliament, except for the agreements referred to in Article 113(3), including cases where the agreement covers a field for which the procedure referred to in Article 189b or that referred to in Article 189c is required for the adoption of internal rules. The European Parliament shall deliver its opinion within a time-limit which the Council may lay down according to the urgency of the matter. In the absence of an opinion within that time-limit, the Council may act'.

[19] The second sentence of ex Article 228(2) TEC reads that: 'the Council shall act unanimously when the agreement covers a field for which unanimity is required for the adoption of internal rules, and for the agreements referred to in Article 238'.

[20] See case C-36/98 *Spain v. Council* (n 15) paras 9–15. Ex Article 130s(2) TEC referred to 'measures concerning town and country planning, land use with the exception of waste management and measures of a general nature, and *management of water resources*'. The Treaty of Amsterdam changed the wording of ex Article 130s(2) TEC to that of ex Article 175(2) TEC, which is, in relation to the exception for measures affecting Member States' management of water resources, identical to that of the current Article 192(2)(b) TFEU ('quantitative management of water resources or affecting, directly or indirectly, the availability of those resources').

[21] Case C-36/98 *Spain v. Council* (n 15) para 50.

[22] Ibid para 51.

---

were measures regulating the use of the territory of the Member States.[23] The ECJ held that the territory and land of the Member States and their water resources were limited resources and that the second indent of the first subparagraph of ex Article 130s(2) TEC therefore referred to measures relating to the management of limited resources in its quantitative aspects and not to measures concerning the improvement and the protection of the quality of those resources.[24] According to the ECJ, a consideration of the various factors led to the conclusion that the concept of 'management of water resources' did not cover every measure concerned with water, but only measures concerning the regulation of the use of water and the management of water in its quantitative aspects.[25]

In *Spain v. Council*, the ECJ clearly indicated that the second ground in Article 192(2)(b) TFEU should be interpreted in a restrictive manner. In relation to a possible amendment of Article 4(1) of the CCS Directive, the importance of the ECJ's ruling lies in its characterisation of Article 192(2) TFEU as a provision referring to measures relating to the quantitative management of limited resources. This characterisation seems to suggest that there might be other measures, not mentioned in Article 192(2) TFEU, that could likewise provide a ground for following the legislative procedure referred to in that provision, as long as these measures affect the quantitative management of limited resources.[26]

An amended Article 4(1) of the CCS Directive providing the Commission with the possibility of forcing Member States to allow CO_2 storage in (parts of) their territory could arguably be such a measure. The measure would clearly have an environmental objective, i.e. the reduction of greenhouse gas emissions through the storage of captured CO_2. What is more, like the territory, land and water resources of Member States – and perhaps even more so than the latter – potential CO_2 storage formations can be said to be a limited resource. Even if, for instance, technological development would increase the possibilities to store captured CO_2 in new kinds of geological formations, a Member State would still only have a limited number of suitable formations in its territory. In addition, an amended Article 4(1) could clearly affect the quantitative management of these limited resources, since it could lead to the obligation for some Member States to make (part of) these resources available for CO_2 storage. As a consequence, these storage formations would not, or would to a lesser extent, be available for other (energy-related) applications such as oil and gas storage.

---

23 Case C-36/98 *Spain v. Council* (n 15).
24 Ibid para 52.
25 Ibid para 55.
26 At first sight, the ECJ's ruling gives the impression that the list of grounds in Article 192(2) TFEU might be illustrative instead of limitative.

Nevertheless, it can be doubted whether this would be a proper reading of the ECJ's ruling in *Spain v. Council*. First, a careful reading of the relevant paragraphs in the judgement suggests that the list of grounds in Article 192(2) TFEU *is* limitative. The crucial paragraph in the ECJ's ruling reads that:

> The territory and land of the Member States and their water resources are limited resources and the second indent of the first subparagraph of Article 130s(2) of the Treaty therefore refers to the measures which affect *them* as such, that is measures which regulate the quantitative aspects of the use *of those resources* or, in other words, measures relating to the management of limited resources in its quantitative aspects and not those concerning the improvement and the protection of the quality *of those resources*.[27]

Essential in this paragraph is the use of the phrases 'them' and 'of those resources'. Both phrases refer to the limited resources mentioned at the beginning of the sentence: territory, land and water resources. This means that, even though it lacks an explicit link to the limited resources mentioned at the beginning of the sentence, the clause 'measures relating to the management of limited resources in its quantitative aspects' should not be read to refer to an open category of limited resources, but relates to the three types of resources mentioned at the beginning of the sentence. Furthermore, the ECJ's restrictive interpretation of one of the grounds under Article 192(2) TFEU suggests that it would probably not be inclined to interpret this provision in such a (broad) manner as to allow for other grounds to be added to those listed in the provision. There is nothing in the wording of the provision to suggest that such would be a proper reading of Article 192(2) TFEU.[28]

Finally, it is worth briefly looking at the possibility of an amended Article 4(1) of the CCS Directive affecting Member States' quantitative management of water resources or (indirectly) affecting the availability of those resources. Even though such a provision would not immediately seem to affect the management and availability of water resources, it could nonetheless have an impact on such resources. Captured CO₂ – stored as a consequence of the Commission having forced a Member State to accept storage in a certain reservoir – leaking from the CO₂ storage reservoir could contaminate (near-surface) drinking water reservoirs.[29] The stored CO₂ stream could, for instance,

---

27   C-36/98 *Spain v. Council* (n 15) para 52 (emphasis added).
28   Typical phrases like 'such as', 'for example' or 'including' are missing.
29   See, for instance, Intergovernmental Panel on Climate Change (IPCC), 'IPCC Special Report on Carbon Dioxide Capture and Storage. Prepared by Working Group III of the Intergovernmental Panel on Climate Change' (Bert Metz and others (eds) Cambridge University Press 2005) 63, and 'Report Says CO2 Leaks Could Contaminate Drinking Water' (2010) Carbon Capture Journal <www.carboncapturejournal.com/displaynews.php?News ID=685> accessed 25 July 2012.

contain heavy metals, which could pollute underground sources of drinking water.[30]

Yet I doubt whether an amended Article 4(1) would therefore fall within the scope of the second exception in Article 192(2)(b) TFEU. Such a measure would not automatically affect Member States' management and availability of water resources, but could do so, depending on whether the stored CO_2 leaks from the reservoir and, consequently, reaches drinking water reservoirs. In my opinion, it would therefore go too far to hold that an amended Article 4(1) would affect Member States' management and availability of water resources. There arguably are too many ifs for that line of reasoning to hold.

Considering the above, there are no reasons to assume that an amended Article 4(1) of the CCS Directive will fall within the scope of the second exception under Article 192(2)(b) TFEU and will therefore have to be adopted by unanimity. In the following section, I examine whether an amended Article 4(1) could fall within the scope of the third exception under Article 192(2)(b), measures affecting land use, with the exception of waste management.

8.2.1.2 Measures affecting land use

In addition to measures affecting quantitative management of water resources, Article 192(2)(b) also refers to measures affecting land use, with the exception of waste management. According to Krämer, the objective of this exclusion is to ensure that Member States stay responsible for the decision (where and in which dimensions) to build.[31] At first sight, a provision granting the Commission the possibility to force Member States to accept CO_2 storage in (parts of) their territory would seem to fall within the scope of this exception. Such a provision appears to directly affect land use in the Member States. What is more, as captured CO_2 for storage has – by virtue of Articles 35 and 36 of the CCS Directive – been removed from the scope of the Waste Directive (Directive 2008/89/EC)[32] and the Regulation on the shipments of waste (Regulation (EC) No 1013/2006),[33] it no longer classifies as 'waste' under EU law. An amended Article 4(1) would therefore not appear to qualify for the exception to the exception in Article 192(2)(b).

---

[30] See Commission, 'Implementation of Directive 2009/31/EC on the Geological Storage of Carbon Dioxide Guidance Document 2: Characterisation of the Storage Complex, CO_2 Stream Composition, Monitoring and Corrective Measures' 80 <http://ec.europa.eu/clima/policies/lowcarbon/ccs/implementation/docs/gd2_en.pdf> accessed 25 July 2012.

[31] Krämer (n 13) 83.

[32] European Parliament and Council Directive 2008/98/EC of on waste and repealing certain Directives [2008] L312/3.

[33] European Parliament and Council Regulation (EC) 1013/2006 of 14 June 2006 on shipments of waste [2006] L190/1.

Jans and Vedder, however, have argued that the reference to measures affecting land use in Article 192(2)(b) does not intend to exclude any measures on land use from the application of Article 192(1) (the ordinary legislative procedure). In support, they refer to Directive 96/82/EC on the control of major-accident hazards involving dangerous substances (Seveso II Directive),[34] which, although clearly affecting land use, has ex Article 130s(1) TEC (now Article 192(1) TFEU) as a legal basis.[35] It could, nevertheless, be questioned whether the Council's interpretation of Article 192(2), as reflected in its choice of legal basis for the Seveso II Directive, should be seen as an objective and authoritative source for construing the meaning of this provision. By contrast, had the ECJ interpreted the provision, the authority and objectivity of its interpretation would have been beyond doubt.[36]

A more convincing argument against a potential amendment of Article 4(1) of the CCS Directive falling within the scope of the third ground in Article 192(2)(b), relates to the argument that CO$_2$ captured for storage does not formally qualify as waste under EU law. It can be doubted whether CO$_2$ captured for storage should not still been seen as waste under EU law. Notwithstanding the exclusion of CO$_2$ captured for storage from the scope of the Waste Directive and the Regulation on the shipments of waste, there is a strong analogy between waste for disposal and CO$_2$ captured for storage. Both are disposed of with the intention not to use the relevant substance again. Adopting that line of reasoning, a potential amendment of Article 4(1) would arguably be possible through the ordinary legislative procedure provided for by Article 192(1) TFEU.

In relation to the legal bases of Directives 2009/147/EC and 92/43/EC (Wild Bird and Habitats Directives),[37] Krämer has suggested another way to determine the applicability of the exception concerning measures affecting land use: by assessing the 'centre of gravity' of the relevant Directive.[38] This term refers to the primary objective of the Directive. I doubt, however, whether this is a correct interpretation of the wording of Article 192(2)(b). The provision literally refers to measures *affecting* land use. As such, it is not concerned with intention/motive, but with effect. A measure that does not intend to regulate land use, could

---

34 Council Directive 96/82/EC of 9 December 1996 on the control of major-accident hazards involving dangerous substances [1997] OJ L10.

35 Ex Article 130s(2) contained a provision similar to the current Article 192(2) TFEU.

36 Apart from any doubts about the content of the interpretation, which frequently arise in relation to judgements of the ECJ.

37 European Parliament and Council Directive 2009/147/EC of 30 November 2009 on the conservation of wild birds [2010] L20 and Council Directive 92/43/EEC of 21 May 1992 on the conservation of natural habitats and of wild fauna and flora [1992] L206.

38 Krämer (n 13) 83.

nonetheless affect the latter. I would therefore argue that this is not a reliable method for determining the proper environmental legal basis.

Considering the above, Article 192(2)(b)'s exception concerning measures affecting land use (and the related unanimity procedure) would not seem apply to a possible amendment of Article 4(1) of the CCS Directive. Even though such an amendment would likely affect the land use in Member States, it could probably still be adopted by the ordinary legislative procedure. In the next section, we will look at the final relevant exception in Article 192(2) TFEU, Article 192(2)(c).

8.2.1.3 Measures significantly affecting a Member State's energy source choice

Article 192(2)(c) refers to measures significantly affecting a Member State's choice between different energy supplies and the general structure of its energy supply. A first thing to note is that subparagraph (c) speaks of measures *significantly* affecting a Member State's energy source choice. The threshold 'significantly affecting' is clearly much higher than 'affecting'.[39] According to Jans and Vedder, the interpretation of this category of measures will, at least in theory, give rise to the necessary problems.[40]

Krämer has argued that measures which come under Article 192(2)(c) are, for instance, measures which lead to the abandonment, for environmental reasons, of nuclear energy or of lignite or coal that has too high a sulphur content.[41] In his Opinion in *Spain v. Council*, AG Léger contended that the exception in Article 192(2)(c) must be interpreted as meaning that only measures to protect the environment which substantially affect the choice made by a Member State between different sources of energy and the general structure of energy supply must be adopted by unanimity.[42] According to the AG, among these are measures which oblige Member States to construct hydroelectric dams to conserve non-renewable natural resources used to produce energy and, conversely, measures which prohibit the construction of such dams in order to safeguard water supplies for certain areas of the EU.[43]

The examples given by AG Léger suggest that the defining characteristic of a measure significantly/substantially affecting a Member State's energy source choice and energy supply structure is that it (directly) obliges the Member State to (generally) use or stop using a particular energy source. In that light, it is

---

[39]   Jans and Vedder (n 16) 62.
[40]   Ibid 63.
[41]   Krämer (n 13) 83.
[42]   See Case C-36/98 *Spain v. Council* [2001] ECR I-00779, Opinion of AG Léger, para 96.
[43]   Ibid para 100.

questionable whether an amended Article 4(1) of the CCS Directive, allowing the Commission to force Member States to accept CO$_2$ storage in (parts of) their territory, would *significantly* affect Member States' energy source choice/energy supply structure. It could certainly have an impact on the latter, as recognised by recital 19 of the preamble to the CCS Directive. Recital 19 refers to the possible interference of the use of potential storage sites for CO$_2$ storage with the use of the same structures for other energy-related options, 'including options which are strategic for the security of the Member State's energy supply or for the development of renewable sources of energy'. The recital gives a few examples of these options, such as the exploration, production and storage of hydrocarbons (oil and gas) and the geothermal use of aquifers. An amended Article 4(1) could have similar effects.

Nevertheless, such effects would arguably not be significant/substantial in the sense that a Member State would (directly) be obliged to (generally) use or stop using a particular energy source. While it is true that, in individual instances, the effect could be that a Member State will no longer be able to use a particular reservoir for a competing energy-related purpose, this does not mean that the relevant Member State will no longer be able to use that energy source in its entirety. The latter could perhaps be the case when the Commission would force a Member State to surrender *all* its potential CO$_2$ storage sites, leaving the Member State no geological formation in which to store gas or oil or to develop geothermal energy. But even then, the question is whether this would truly prevent a Member State from, for instance, having gas fired power production or developing geothermal energy.

By contrast, recital 19 indicates that the EU legislator – or at least the Member States – would probably be of the opinion that an amended Article 4(1) would have to be adopted unanimously. It is no coincidence that the recital that refers to the possibility of CO$_2$ storage interfering with the use of potential CO$_2$ storage reservoirs for other energy-related purposes explicitly acknowledges the right of Member States to determine the areas within their territory from which CO$_2$ storage sites may be selected, including the right not to allow any CO$_2$ storage in parts or on the whole of their territory. The same could be concluded from the superfluous – from the point of view of the principle of conferral[44] – superfluous presence of Article 4(1).[45]

---

[44]   The principle of conferral dictates that the EU can act only in those policy areas where the Member States have given it the power to act through the Treaty on European Union (TEU) and the TFEU (see Article 5(1) and (2) TEU). The EU has no competence to create competences ('*Kompetenz-kompetenz*'). As the Member States have never given the EU the competence to decide over the use of Member States' underground, a provision such as Article 4(1) is unnecessary. On the principle of conferral see, for instance, John Fairhurst, *Law of the European Union* (8th edn, Pearson Education Limited 2010) 58–59.

[45]   Nevertheless, this principle has not prevented the Member States, as co-legislator, from doing similar things in other areas of EU law.

Nevertheless, the intention of the EU legislator with regard to decision making concerning the allocation of CO_2 storage sites – a Member State's consent is required for storage in its territory – does not alter the likelihood of an amended Article 4(1) not *significantly* affecting a Member State's energy source choice/ energy supply structure and therefore likely not having to be adopted unanimously.

What is more, an interpretation of an amended Article 4(1) not significantly affecting a Member State's energy source choice/energy supply structure would appear to be in line with the ECJ's apparently generally strict interpretation of the exceptions in Article 192(2). The ECJ would likely not interpret the exceptions in Article 192(2) in such a broad manner as to have an amended Article 4(1) fall within the scope of Article 192(2)(c), if only for the likelihood that a vast number of measures based on Article 192 would then have to be adopted unanimously. Most of the measures aimed at the reduction of EU greenhouse gas emissions (such as the CCS Directive) would, for instance, very likely affect the Member States' energy source choice or energy supply structure in one way or the other.[46] Such an interpretation of Article 192(2)(c) would seriously undermine the effectiveness of the EU environmental policy making and therefore not likely be adopted by the ECJ.

Based on the above, it is not likely that an amended Article 4(1) of the CCS Directive would have to be adopted by unanimity since it would significantly affect the Member States' energy source choice or energy supply structure. Such a provision will likely affect the latter, but arguably not to such an extent that it obliges the Member States to use or stop using certain energy sources.

The above assessment shows that of the three potentially relevant grounds in Article 192(2), none is likely to influence the choice of legal basis of a possible amendment of Article 4(1) of the CCS Directive. This is partly caused by the ECJ's generally restrictive interpretation of the exceptions in Article 192(2), but also dictated by the logic of the provision. If the ECJ were to, for instance, interpret Article 192(2) in such a broad manner as for an amended Article 4(1) of the CCS Directive to fall within the scope of Article 192(2)(c), the effectiveness of EU environmental policy making could, arguably, be seriously undermined. In the next section, I examine the framework of Article 194, which contains a

---

[46] A good example is Directive 2009/28/EC (Renewable Energy Directive), which contains a list of binding targets for Member States' renewable energy consumption in 2020, but was based on ex Article 175(1) TEC (now Article 192(1) TFEU – adoption by qualified majority through the ordinary procedure). See European Parliament and Council Directive 2009/28/EC of 29 April 2009 on the promotion of the use of energy from renewable sources and amending and subsequently repealing Directives 2001/77/EC and 2003/30/EC [2009] OJ L140.

provision similar to that in Article 192(2)(c) TFEU, in order to assess whether the former provision would provide a proper legal basis for an amended Article 4(1) of the CCS Directive.

8.2.2 ARTICLE 194 TFEU

Article 194 TFEU contains the new energy legal basis introduced by the Treaty of Lisbon.[47] Its first paragraph lists the objectives to be pursued by EU energy policy:

> [I]n the context of the establishment and functioning of the internal market and with regard for the need to preserve and improve the environment, Union policy on energy shall aim, in a spirit of solidarity between Member States, to
> (a) Ensure the functioning of the energy market;
> (b) Ensure security of energy supply in the Union;
> (c) Promote energy efficiency and energy saving and the development of new and renewable forms of energy; and
> (d) Promote the interconnection of energy networks.

The second paragraph provides that:

> [W]ithout prejudice to the application of the other provisions of the Treaties, the European Parliament and the Council, acting in accordance with the ordinary legislative procedure, shall establish the measures necessary to achieve the objectives in paragraph 1. Such measures shall be adopted after consultation of the Economic and Social Committee and the Committee of the Regions.

> Such measures shall not affect a Member State's right to determine the conditions for exploiting its energy resources, its choice between different energy sources and the general structure of its energy supply, without prejudice to Article 192(2)(c).

Finally, Article 194(3) reads that:

> [B]y way of derogation from paragraph 2, the Council, acting in accordance with a special legislative procedure, shall unanimously and after consulting the European Parliament, establish the measures referred to therein when they are primarily of a fiscal nature.

---

[47]   Treaty of Lisbon amending the Treaty on European Union and the Treaty establishing the European Community [2007] OJ C306.

When reading Article 194, it immediately becomes clear that the provision is closely related to Article 192 TFEU.[48] For our analysis, the main relevance of Article 194 lies in the second part of the second paragraph, which contains a reference to Article 192(2)(c) TFEU. The phrase 'shall not affect a Member State's right (...) without prejudice to Article 192(2)(c)' is strongly reminiscent of the reference to 'a Member State's choice between different energy sources and the general structure of its energy supply' in Article 192(2)(c).

Yet there is an important difference between the two provisions. Whereas Article 192(2)(c) allows for the unanimous adoption of measures that significantly affect a Member State's energy source choice and energy supply structure, Article 194(2) does not allow for EU energy measures to affect the same matters, as well as a Member State's right to determine the conditions for exploiting its energy resources *at all*. While both limitations serve to protect national sovereignty with respect to the Member States' energy resources, the energy legal basis is far more rigorous: it forbids any EU measure affecting, whether significantly or not, this national reservation, and the prohibition cannot, it would seem, be overcome unanimously in the Council.[49]

As a consequence, the choice of legal basis can have significant consequences. Article 192 allows the EU legislator more room for manoeuvre than Article 194. The phrase 'without prejudice to Article 192(2)(c)' in Article 194(2) clarifies that the restrictions on the EU's policy on energy do not affect the ability of the EU to adopt measures implementing the EU's policy on the environment which affect the national energy mix and the general structure of the energy supply.[50]

The above implies that if the objective of an amended Article 4(1)[51] could be brought under one of the objectives in Article 194(1), for instance the promotion of energy efficiency and energy saving and the development of new and renewable forms of energy,[52] then Article 194(2) would arguably not be available as a legal basis for such an amendment. As we have seen in section 8.2.1.3 above, an amended Article 4(1) is likely to affect a Member State's energy source choice and energy supply structure. As a consequence of the more stringent Member State reservation in Article 194(2), the latter provision

---

[48]   Simm has argued that Article 194 could be seen as the lex specialis provision – compared with the lex generalis provision in Article 192 – for market liberalisation and environmental resource saving measures in the energy sector. See Marion Simm, 'Institutional Report – The Interface between Energy, Environment and Competition Rules of the European Union' 19 <www.fide2012.eu/index.php?doc_id=87> accessed 25 July 2012.

[49]   Ibid 22.

[50]   Ibid.

[51]   That is the reduction of greenhouse gas emissions through the storage of captured CO_2.

[52]   In this regard, see, for instance, section 5.7.2 of Chapter V.

will therefore likely not be available as a legal basis for an amended Article 4(1) of the CCS Directive.

8.3   ARTICLE 345 TFEU

Article 345 TFEU states that 'the Treaties shall in no way prejudice the rules in Member States governing the system of property ownership'. At first sight, this provision could create problems for an amendment of Article 4(1) of the CCS Directive which allows the Commission to force Member States to accept CO$_2$ storage in their territory. An amended Article 4(1) would not lead to the expropriation of the owners of the relevant potential CO$_2$ storage formations, but it could lead to these owners being forced to allow for a different use of their property. The question is whether such forced change in use of property would be allowed under Article 345. In the following, I answer this question by further examining the meaning and scope of Article 345. To this end, I first address the (limited) literature on Article 345 and then proceed to the European courts' case law on the provision.

8.3.1   LITERATURE

In a 2010 paper, Akkermans and Ramaekers have sought to clarify the exact meaning and scope of Article 345 through an analysis of the drafting of the provision as well as its use by the EU's institutions and Member States.[53] In the following, I will go through the most important findings of their thorough writing.

Akkermans and Ramaekers start by (linguistically) dissecting Article 345 into four different parts, which are consequently examined:

1) the meaning of the term 'prejudice';
2) the place of Article 345 in the TFEU;
3) the meaning of the term 'system of ownership'; and
4) the meaning of the term 'the Treaties';

For the interpretation of the central terms of Article 345, Akkermans and Ramaekers work from a comparison of the French and English language versions of the provision (supplemented with other language versions), since the first predecessor of Article 345 TFEU, Article 222 of the Treaty Establishing the

---

[53]   Bram Akkermans and Eveline Ramaekers, 'Article 345 TFEU (ex Article 295 EC), Its Meanings and Interpretations' (2010) 16 European Law Journal 292–314.

European Economic Community (TEEC), was drafted in French and the current Article 345 is in English.

With regard to the term 'prejudice', Akkermans and Ramaekers note that it is doubtful whether 'to prejudice' is a proper translation of the original French *'préjugent'*. The equivalents in the German, Dutch and Italian language versions – *'lassen unberührt'*, *'laten onverlet'* and *'lasciano impreguidicato'* – appear to have a stronger meaning (leaving something unaffected), which matches the original French expression.[54] In fact, Akkermans and Ramaekers argue, the French term *'préjuge'* is even stronger than this, in the sense that it implies that the TEEC did also not prejudge the systems of property ownership of the Member States as they existed when the TEEC was drafted; their compatibility with the TEEC at that moment was assured, as was the neutral stance of the then Community towards these systems after the entry into force of the TEEC.[55]

After having compared the French and English versions of Article 345 with those of Article 296 TFEU, Akkermans and Ramaekers conclude that the expression 'shall in no way prejudice' refers to the neutral stance that is taken by the EU towards the way in which Member States regulate their system of property ownership; the Treaties may apply, but may not cause harm to the Member State rules governing the system of ownership.[56]

In relation to the place of Article 345 in the TFEU, Akkermans and Ramaekers contend that the place of ex Article 222 TEEC in the final part of the TEEC (General and Final Provisions) implies that the drafters did not consider it a provision fundamental to the TEEC and intended to be directly connected to a legal basis in the TEEC; ex Article 222 TEC was not meant to provide a general limitation on the harmonising powers of the then Community.[57] Furthermore, they argue, being surrounded by closing provisions dealing with the application (and thus with the full content) of the TEEC, ex Article 222 TEEC must be interpreted as a general provision dealing with the full content of the TEEC. Accordingly, Akkermans and Ramaekers state, the phrase 'shall in no way prejudice the rules' is likely to mean that the Treaties will not apply to Member States' rules governing the ownership of property.[58]

As for the meaning of the term 'system of ownership', on the basis of a historical interpretation Akkermans and Ramaekers argue that, even though the original reference to undertakings (*'des enterprises'* – in relation to the term system of

---

54   Akkermans and Ramaekers (n 53) 297–98.
55   Ibid 298.
56   Ibid 298.
57   Ibid 299.
58   Ibid.

ownership ('*regime de propriété*')) was removed from the final version of ex Article 222 TEC, Article 345 relates to the neutrality of the Treaties in respect to the ownership, private or public, of undertakings only.[59] In support, they, inter alia, refer to the explanatory memorandums of the governments of the initial six Member States.[60] What is more, Akkermans and Ramaekers note that this conclusion is supported by property law; the use of the term 'system' suggests that we are not dealing with the right of ownership itself, but rather with the way in which the right of ownership can be held.[61] Article 345 does not concern the content of the right of ownership, nor the objects of a right of ownership.[62] Finally, Akkermans and Ramaekers state that the words 'the Treaties' in Article 345 refer to the Treaty on European Union (TEU) and the TFEU.[63]

Combining the above, Akkermans and Ramaekers come to the preliminary conclusion that Article 345 is a provision that limits, but does not prevent, the application of the Treaties in their entirety to the way in which Member State rules deal with the right of ownership of undertakings.[64] The provision only concerns the private or public ownership of undertakings, with which the EU shall not concern itself and which can thus be regulated by the Member States themselves.[65]

After having researched the textual, historical and contextual interpretation of Article 345, Akkermans and Ramaekers test their preliminary conclusion against the institutional and judicial use of the provision. In relation to the former, they conclude that in general Article 345 does not confer any exclusive powers to deal with property law to the EU or the Member States.[66] Instead, the institutions only use the provision to confirm the neutrality of the Treaties to questions of private or state ownership.[67]

---

[59]   Akkermans and Ramaekers (n 53) 302.

[60]   Ibid. The German government, for instance, commented that ex Article 222 TEEC was in the Treaty, because no other provision in the TEEC and no measure of the Community institutions was to be allowed to have the effect of nationalising or privatising undertakings. Likewise, Italy and France sought to have ex Article 222 included in the TFEU out of concern that the Community would interfere with their practice of nationalising certain industries.

[61]   Akkermans and Ramaekers (n 53) 303.

[62]   Ibid 304.

[63]   Ibid.

[64]   Ibid 304–05.

[65]   Ibid 305. For a similar reading, see Inigo del Guayo, Gunther Kühne and Martha Roggenkamp, 'Ownership Unbundling and Property Rights in the EU Energy Sector' in Aileen McHarg and others (eds), *Property and the Law in Energy and Natural Resources* (Oxford University Press 2010) 326, 337–38.

[66]   Akkermans and Ramaekers (n 53) 308.

[67]   Ibid. In relation to the Commission's application of Article 345 TFEU, see also Peter Willis and Paul Hughes, 'Structural Remedies in Article 82 Energy Cases' (2008) 4 The Competition Law Review 147, 150.

With regard to the judicial use of Article 345, Akkermans and Ramaekers refer to a number of cases before the ECJ.[68] From these cases,[69] Akkermans and Ramaekers conclude that the ECJ, at least in relation to intellectual property rights, draws a fundamental distinction between the existence and the exercise of property rights.[70] As a consequence of this distinction, Member States may, under the provisions of the Treaties, be free to legislate in the area of property law.[71] The existence of such rights remains untouched by the Treaties.[72] Akkermans and Ramaekers argue that, even though this distinction has only been applied in intellectual property cases, a number of cases appear to suggest that such reasoning would also apply to property law in the more classic meaning of the term.[73]

Akkermans and Ramaekers conclude that, although the provision is occasionally used in a way not consistent with this interpretation, Article 345 is most likely to embed the principle of neutrality, according to which the Treaties are neutral as to whether an undertaking is held in public or private ownership.[74] Yet, the case law of the ECJ indicates that the rules of the internal market remain applicable with regard to the exercise, by the Member States, of their competence to nationalise.[75]

Akkermans and Ramaeker's conclusion that the scope of Article 345 is limited to the Treaties not affecting the ownership of undertakings (neutrality principle) has been criticised by Fraga et al.[76] They argue that as the reason for removing the reference to undertakings from the final version of ex Article 222 TEEC is

---

[68] Akkermans and Ramaekers (n 53) 309–11.
[69] Akkermans and Ramaekers discuss, inter alia, Joined Cases 56 and 58/64 *Consten & Grundig* [1966] ECR 00429, Case 78/70 *Deutsche Grammaphon* [1971] ECR 004487, Case 182/83 *Fearon* [1984] ECR 03677, Case 24/67 *Parke Davis* [1986] ECR 00081, Case C-367/98 *Commission v. Portugal* [2002] ECR I-04731, Case C-483/99 *Commission v. France* [2002] ECRI-04781, Case C-503/99 *Commission v. Belgium* [2002] ECR I-04809. I address the case law on Article 345 in section 8.3.2.
[70] Akkermans and Ramaekers (n 53) 309–10. In this regard, see also Floris Becker, 'Market Regulation and the "Right to Property" in the European Economic Constitution' in *Yearbook of European Law* (Oxford University Press 2007) 255, 277.
[71] Akkermans and Ramaekers (n 53) 310.
[72] Ibid.
[73] Ibid 311.
[74] Ibid 314. In this regard, see also Erika Szyszczak, 'The Survival of the Market Economic Investor Principle in Liberalised Markets' (2011) 10 European State Aid Law Quarterly 35, 37; and Erika Szyszczak, 'Public Services in the New Economy' in Cosmo Graham and Fiona Smith (eds), *Competition, Regulation and the New Economy* (Hart Publishing 2004) 185, 187.
[75] Akkermans and Ramaekers (n 53) 314.
[76] Fernando Losada Fraga, Teemu Juutilainen, Katri Havu and Juha Vesala, 'Property and European Integration: Dimensions of Article 345 TFEU' (2012) University of Helsinki Faculty of Law Legal Studies Research Paper Series No 17 <http://papers.ssrn.com/sol3/papers.cfm?abstract_id=2012983> accessed 27 July 2012.

unknown,[77] it is possible that the phrase was removed with a view to broadening the scope of the article beyond mere ownership of undertakings.[78] According to Fraga et al., it seems unlikely that the Member States would have wanted to limit the provision safeguarding their competence to nationalise property to the ownership of undertakings (and not involve other property rights), since they were generally concerned with their competence to nationalise property.[79] The version of ex Article 222 TEEC before the version with the reference to undertakings contained a reference to 'means of production'.[80]

Fraga et al. contend that the reference to undertakings might have been removed so as to broaden the scope of the provision to what it would have been with the qualification concerning ownership of means of production, or even broader.[81] They conclude by stating that the study by Akkermans and Ramaekers shows that the question of public or private ownership of undertakings belongs to the core of the scope of Article 345, but that it is still unknown whether the scope is not broader than that.[82]

Ehlers has noted that the wording of Article 345 at first sight seems to support a broad applicability; the system of property ownership consists of all constitutional provisions concerning private ownership and in particular expropriation, socialisation and the regulation of the use of property.[83] Yet, Ehlers argues, such a wide interpretation would neglect the fact that the EU cannot establish a common market without having the competence to regulate the use of property.[84] As to the view that Article 345 refers to the ownership of undertakings only, Ehlers contends that the wording of Article 345 is clear in that it does not refer to a transfer of property from private to public ownership and vice versa.[85] As a result, the provision must have a primary meaning going beyond the mere prohibition of privatisations and socializations by the EU.[86] Ehlers suggests a third, intermediate, interpretation, according to which the EU cannot take decisions which immediately affect the system (or existence) of property ownership but can take decisions which affect the exercise of property

---

[77] The *travaux préparatoires* do not provide any guidance in this regard.
[78] Fraga and others (n 76) 10.
[79] Ibid.
[80] Ibid.
[81] Ibid 11.
[82] Ibid 13.
[83] Eckart Ehlers, 'Electricity and Gas Supply Network Unbundling in Germany, Great Britain and the Netherlands and the Law of the European Union: A Comparison' (PhD thesis, University of Tilburg 2009) 150.
[84] Ibid.
[85] Ibid 152.
[86] Ibid.

ownership and which are equivalent to a restriction of the allocation of property ownership.[87]

Ehlers concludes that Article 345 is designed to ensure that fundamental decisions of economic policy stay in the remit of the Member States and that the national systems of property ownership are left untouched.[88] The fundamental decisions whether to nationalise or socialise private sector market activity or to privatise public sector market activity fall within the competences of the Member States and not the EU.[89] Once property ownership has been allocated, the exercise of property ownership falls within the scope of the provisions of the Treaties.[90]

In addition, Ehlers argues that, given the ever growing fundamental rights protection on the EU level, Article 345 also disallows EU legislation which encroaches upon the essential content ('*Wesensgehalt*') of the fundamental right to property, which all Member States' constitutions have in common.[91] Referring to Article 17 of the Charter of Fundamental Rights of the European Union (CFREU – right to property),[92] Ehlers notes that the regulation of the use of property necessary for the benefit of the general public may not violate the essential content ('*Wesensgehalt*') of the right to property.[93] Accordingly, Ehlers argues, Article 345 prevents the EU from substituting ownership with something which does not deserve to be called 'ownership' any longer; the 'untouchable' core of the guarantee to own property may not be undermined.[94]

Ehlers's interpretation of Article 345 as a provision which prevents the EU 'from substituting ownership with something that does not deserve to be called

---

[87]   Ehlers (n 83) 154.
[88]   Ibid 155.
[89]   Ibid 155.
[90]   Ibid 156.
[91]   Ibid 157.
[92]   Charter of Fundamental Rights of the European Union [2010] OJ C83/389. Article 17 of the Charter provides that '(1) Everyone has the right to own, use, dispose of and bequeath his or her lawfully acquired possessions. No one may be deprived of his or her possessions, except in the public interest and in the cases and under the conditions provided for by law, subject to fair compensation being paid in good time for loss. The use of property may be regulated by law in so far as is necessary for the general interest. (2) Intellectual property shall be protected'.
[93]   Ehlers (n 83) 158.
[94]   Ibid 158. Ehlers refers to the case law of the German Federal Constitutional Court (the Bundesverfassungsgericht), by virtue of which the essential content of the right to property consists of the allocation of a particular property to a particular owner in order for it to be available for his economic use and benefit on the one hand, and, on the other hand, to give him the principal power to dispose of such property. The latter does not only include the positive power for the owner to do with his property whatever he pleases but also the (negative) power, not to have to dispose of it; the power to dispose thus reflects the control, or better, the sovereignty over the property. For a similar line of reasoning, see also Pielow, Brunekreeft and Ehlers (n 7) 12–13.

ownership any longer' seems to heavily lean on the case law of the German Constitutional Court. Yet, it can be questioned whether Article 17 of the CFREU can be used as a vehicle for imposing German constitutional standards upon the other Member States. The wording of the provision certainly does not lend support for such an interpretation.

The fact that all the constitutions of the EU Member States recognise the fundamental right to property does not mean that all Member States share the German interpretation of that right, as determined by the German Constitutional Court. What is more, there does not seem to be anything in the wording or in the legislative history to indicate that the scope of Article 345 would be that wide. I would therefore argue that Ehlers's interpretation of Article 345 in this regard strays too far from the wording and background of this provision.

8.3.2 CASE LAW

8.3.2.1 *General lessons*

Akkermans and Ramaekers have stated that there is hardly any case law on Article 345 TFEU.[95] Nevertheless, an examination of the European Courts' jurisprudence indicates that there are quite a few cases in which Article 345 TFEU or one of its predecessors (Articles 222 TEEC and 295 TEC) play a role.[96]

---

[95]   Akkermans and Ramaekers (n 53) 293.
[96]   For this section, I have examined the following cases and opinions: Joined Cases 56 and 58/64 *Consten and Grundig* [1966] ECR 00299; Case 78/70 *Deutsche Grammaphon* [1971] ECR 00487; Case 24/67 *Parke Davis* [1986] ECR 00081; Case 32/65 *Italy v. Council* [1966] ECR 00389; Joined Cases 188 to 190/80 *France, Italy and UK v. Commission* [1982] ECR 02545; Case 323/82 *Intermills v. Commission* [1984] ECR 03839; Case 182/83 *Robert Fearon* [1984] ECR 03677; Case 264/81 *Savma v. Commission* [1984] 03915; Case 41/83 *Italy v. Commission* [1985] 00873, Opinion of AG General Darmon; Case 41/83 *Italy v. Commission* [1985] ECR 00873; Case C-10/89 *SA CNL* [1990] ECR I-03711, Opinion of AG Jacobs; Case C-41/90 *Hoefner and Esler* [1991] ECR I-01979, Opinion of AG Jacobs; Case C-202/88 *France v. Commission* [1991] ECR I-01223, Opinion of AG Tesauro; Case C-305/89 *Italy v. Commission* [1991] ECR I-01603; Case C-235/89 *Commission v. Italy* [1992] ECR I-00777; Case C-30/90 *Commission v. UK* [1992] ECR I-0829; Case C-235/89 *Commission v. UK* [1992] ECR I-00777, Joined Opinion of AG van Gerven; Joined Cases C-92/92 and C-326/92 *Phil Collins* [1993]; Joined Cases C-92/92 and C-326/92 *Phil Collins* [1993] ECR I-05145, Opinion of AG Jacobs; Case C-350/92 *Spain v. Council* [1995] ECR I-01985, Opinion of AG Jacobs; Case C-350/92 *Spain v. Council* [1995] ECR I-01985; Case C-309/96 *Annibaldi* (1997) ECR I-07505; Case C-302/97 *Konle* [1999] ECJ I-03099, Opinion of AG La Pergola; Case C-302/97 *Konle* [1999] ECJ I-03099; Case C-44/98 *BASF* [1999] ECR I-06269, Opinion of AG La Pergola; Case C-38/98 *Renault* [2000] ECR I-02973, Opinion of AG Alber; Case C-344/98 *Masterfoods* [2000] ECR I-11369, Opinion of AG Cosmas; Case T-128/98 *Aeroport de Paris* [2000] ECR II-03929; Case C-36/98 *Spain v. Council*, Opinion of AG Leger (n 39); Case C-163/99 *Portugal v. Commission* [2001] ECR I-02613; Order of the president of the CFI of 26 October 2001 in

Not all of these cases provide an in-depth examination of Article 345. Yet, the whole of the Courts' case law on Article 345 paints a fairly good picture of the provision.

A first thing to note from the Courts' case law on Article 345 is that the provision is characterised differently in the various cases. There is support for Akkermans and Ramaekers's characterisation of Article 345 as a principle which is neutral to public or private ownership of a Member State's undertakings ('neutrality principle'),[97] but the provision has also been described as a principle of equal treatment of public and private undertakings.[98] At first sight, the characterisation of Article 345 as a principle of equal treatment of public and private undertakings appears to go further than that of the provision as a neutrality principle.

Yet Article 345 is unlikely to always guarantee the equal treatment of public and private undertakings under EU law. Private and public undertakings can be in different situations, for instance due to subsidies being granted to public undertakings, and can therefore not always be treated equally under, for

---

Case 184/01 R *IMS Health v. Commission* [2001] II-03193; Case C-367/98 *Commission v. Portugal* [2002] ECR I-04731; Case C-482/99 *France v. Commission* [2002] ECR I-04397, Opinion of AG Jacobs; Case C-367/98 *Commission v. Portugal* [2002] ECR I-04731; Cases C-367/98, C-483/99 and C-503/99 [2002] ECR I-04731, Joined Opinion of AG Colomer; Case C-483/99 *Commission v. France* [2002] ECR I-04781; Case C-503/99 *Commission v. Belgium* [2002] ECR I-4809; Case C-482/99 *France v. European Commission* [2002] ECR I-04397; Case C-491/01 *British American Tobacco* [2002] ECR I-11453, Opinion of AG Geelhoed; Case C-491/01 *British American Tobacco* [2002] I-11453; Cases C-463/00 and /C-98/01 *Commission v. Spain* [2003] ECR I-04581, Joined Opinion of AG Colomer; Case C-452/01 *Ospelt* [2003] ECR I-09743; Joined Cases T-228/99 and T-233/99 *Westdeutsche Landesbank* [2003] ECR II-00435; Case T-65/98 *Van den Bergh Foods* [2003] ECR II-04653; Case C-463/00 *Commission v. Spain* (2003) ECR I-04581; C-363/01 *Flughafen Hannover* (2003) ECR I-11893, Opinion of AG Mischo; Case C-300/01 *Doris Salzmann* [2003] ECR I-04899, Opinion of AG Leger; Case C-300/01 *Doris Salzmann* [2003] ECR I-04899; Case C-452/01 *Ospelt* [2003] ECR I-09743; Joined Cases T-116/01 and T-118/01 *P&O* [2003] II-02957; Case C-174/04 *Commission v. Italy* [2005] ECR I-04933, Opinion of AG Kokott; Case C-436/03 *European Parliament v. Council* [2006] ECR I-03733, Opinion of AG Stix-Hack; Case C-503/04 *Commission v. Germany* [2007] ECR I-06153; Case C-112/05 *Commission v. Germany* [2007] ECR I-08995, Opinion of AG Colomer; Case T-362/04 *Leonid Minin v. Commission* [2007] ECR II-002003; Case C-326/07 *Commission v. Italy* [2009] ECR I-02291, Opinion of AG Colomer; Case T-156/04 *EDF v. Commission* [2009] ECR II-04503; Case C-52/09 *Teliasonera* [2011] ECR I-00527, Opinion of AG Mazak; Case C-124/10 P *Commission v. EDF* [2012] not yet reported, Opinion of AG Mazak; and Case C-124/10 P *Commission v. EDF* [2012] not yet reported.

[97] See the Opinions of AG Jacobs in Case C-41/90 *Höfner and Elser* (para 43) and Case C-350/92 *Spain v. Council* (n 96) (para 29).

[98] See Joined Cases 188 to 190/80 *France, Italy, UK v. Commission* (n 96) paras 20 and 21, Case C-482/99 *France v. Commission*, Opinion of AG Jacobs (n 96) para 47 and Case C-124/10 P *Commission v. EDF*, Opinion of AG Mazák (n 96) para 1. In addition, AG Tesauro has described Article 345 as a provision that reflects the fundamental contradiction inherent in the TFEU between, on the one hand, the common market and system of free competition and, on the other, the freedom for Member States to take economic-policy decisions. See Case C-202/88 *France v. Commission*, Opinion of AG Tesauro (n 96) para 11.

instance, EU competition law.[99] So, even though Article 345 could perhaps be characterised as a principle of equal treatment of public and private undertakings, this characterisation cannot guarantee an equal treatment of public and private undertakings when primary EU law is applied. The EU is indifferent as to whether a Member State's undertakings are publicly or privately owned, but this neutrality does not mean that these undertakings will always be treated equally when primary EU law is applied.

What becomes most clear from the case law on Article 345 is that the provision cannot prevent/limit the application of primary EU law/the fundamental rules of the Treaties to the system of property ownership in the Member States. The ECJ's ruling in the early case of *Italy v. Commission* could be read to suggest that Article 345 *is* able to limit the application of primary EU law,[100] but a vast number of (predominantly) subsequent cases clearly indicate that this is not so.[101] These cases dealt with the application of the EU competition rules, free movement provisions and general principle of non-discrimination to Member States' systems of property ownership.

Even though the Courts usually do not (explicitly) make a distinction between the existence and the exercise of property rights (the 'existence/exercise dichotomy' mentioned by Akkermans and Ramaekers),[102] it seems that the *exercise* of property rights may (contrary to the existence of property rights), in the general interest, be restricted under EU law.[103] As to the scope of the

---

99　See Case C-482/99 *France v. Commission*, Opinion of AG Jacobs (n 96) para 47.

100　In Case 41/83 *Commission v. Italy* (n 96) para 22, the ECJ countered Italy's argument that by applying the EU competition rules to a statutory monopoly the Commission infringed Article 345, by stating that the relevant undertaking did not have a statutory monopoly in the relevant area, instead of arguing that the provision could not limit the application of EU competition rules to Member States' systems of ownership. This implicitly leaves open the possibility that Article 345 *can*, under different circumstances, limit the application of primary EU law.

101　See Joined Cases 56 and 58–64 *Consten and Grundig*, p. 345; Case 182/83 *Robert Fearon*, para 7; Case C-305/89 *Italy v. Commission*, para 24; Case C-235/89 *Commission v. Italy*, para 14; Case C-30/90 *Commission v. UK*, para 18; Joined Cases C-92/92 and C-326/92 *Phil Collins*, Opinion of AG Jacobs, para 22; Case C-350/92 *Spain v. Council*, para 18; Case C-302/97 *Konle*, Opinion of AG La Pergola, para 14; Case C-302/97 *Konle*, para 38; Case C-44/98 *BASF*, Opinion of AG La Pergola, para 14; Case C-344/98 *Masterfoods*, Opinion of AG Cosmas, paras 103–06; Case C-163/99 *Portugal v. Commission*, para 59; Case C-367/98 *Commission v. Portugal*, para 48; Case C-483/99 *Commission v. France*, para 44; Case C-503/99 *Commission v. Belgium*, para 44; Joined Cases T-288/99 and T-233/99 *Westdeutsche Landesbank Girozentrale*, para 192; Case C-452/01 *Ospelt*, para 24; Case C-463/00 *Commission v. Spain*, para 67; Case C-300/01 *Doris Salzmann*, Opinion of AG Léger, para 34; Case C-300/01 *Doris Salzmann*, para 39; Joined Cases T-116/01 and T-118/01 *P&O European Ferries*, para 151; and Case C-503/04 *Commission v. Germany*, para 37 (n 96).

102　Akkermans and Ramaekers (n 53) 309–10.

103　See, for instance, Joined Cases 56 and 58–64 *Consten and Grundig* (n 96) p. 345; and Case C-344/98 *Masterfoods*, Opinion of AG Cosmas (n 96) paras 105–06.

existence/exercise dichotomy, at first it appeared to apply in relation to rights to intangible (intellectual) property and the application of substantive Treaty rules to the exercise of those property rights only.[104] Yet, as suggested by Akkermans and Ramaekers, the existence/exercise dichotomy also seems to apply to rights to tangible property.[105] What is more, the dichotomy likewise appears to apply in relation to the EU competence to regulate a matter falling within the scope of Article 345.[106] In other words, secondary EU legislation may, under certain circumstances, restrict the exercise of property rights. I further address the scope of the existence/exercise dichotomy when discussing a number of Article 345 cases that are of particular relevance, below.

8.3.2.2 Selected Article 345 TFEU cases

In the following, I first explore the cases also known as the 'golden shares' cases.[107] These cases are worth examining if only for the fact that the Joined Opinion of AG Colomer in these cases arguably contains the most elaborate judicial discussion of Article 345 to date.[108] Subsequently, I examine two cases in which the ECJ held that Article 345 does not exclude any influence whatsoever of EU law on the exercise of national property rights and in which it made clear that the exercise of the right to property can be restricted, provided that the restriction corresponds to objectives of general economic interest pursued by the EU and that it does not constitute a disproportionate and intolerable interference, impairing the very substance of the rights guaranteed (*British American Tobacco* and *Van den Bergh Foods*).[109]

a) The golden shares cases

The golden shares cases dealt with the compatibility with primary EU law (free movement of capital, freedom of establishment) of national systems granting the executive certain prerogatives to intervene in the share structure and in the management of privatised undertakings in strategically important areas of the

---

[104] Cases like Joined Cases 56 and 58–64 *Consten and Grundig* (n 92), Case 78/70 *Deutsche Grammaphon* (n 96) and Case 24/67 *Parke Davis* (n 96) all related to the exercise of intellectual property rights. On the existence/exercise dichotomy initially applying to the exercise of substantive EU rules only (and not in relation to the question whether there is EU competence to regulate a matter falling within the scope of Article 345), see, for instance, Case C-350/92 *Spain v. Council*, Opinion of AG Jacobs (n 96) para 20.

[105] See Case C-463/03 *European Parliament v. Council*, Opinion of AG Stix-Hackl, para 70.

[106] See Cases C-491/01 *British American Tobacco*, Opinion of AG Geelhoed (n 96) para 256 and C-491/01 *British American Tobacco* (n 96) paras 147 and 148.

[107] These are Cases C-367/98 *Commission v. Portugal*, C-483/99 *Commission v. France* and C-503/99 *Commission v. Belgium* (n 96).

[108] Cases C-367/98 *Commission v. Portugal*, C-483/99 *Commission v. France* and C-503/99 *Commission v. Belgium*, Joined Opinion of AG Colomer (n 96).

[109] Cases C-491/01 *British American Tobacco* (n 92) and T-65/98 *Van den Bergh Foods* (n 96).

economy.[110] These prerogatives are generally known as golden shares. Since the Commission considered the golden shares schemes in Portugal, France and Belgium to infringe the freedom of establishment and the free movement of capital (ex Articles 52 and 56 TEC – now Articles 59 and 63 TFEU), it had brought three infringement actions under ex Article 226 TEC (now Article 258 TFEU).

Article 345 played a central role in AG Colomer's Opinion in the golden shares cases. The AG essentially argued that the Commission, in its actions against Portugal, France and Belgium, had sidestepped the fundamental importance of Article 345, which could be derived from the provision's place in the TFEU,[111] its forceful and unconditional wording,[112] its origin in the 1950 Schuman declaration[113] and the historical and teleological background of the provision.[114] According to the AG, the historical and teleological background of Article 345 indicated that the term system of ownership referred to any measure which, through intervention in the public sector (understood in the economic sense), allowed the state to contribute to the organisation of the nation's financial activity.[115] The fundamental importance of Article 345, so AG Colomer argued, gave measures such as the golden shares a presumption of validity when assessing their compatibility with the fundamental rules of the Treaties; such measures were to be seen as compatible with the Treaties, unless proven otherwise.[116]

According to the AG, the golden shares constituted rules governing public intervention in the activities of certain undertakings, which were on the same footing as forms of ownership of the undertakings whose organisation

---

[110] See the introduction of the Opinion of AG Colomer in the three cases (n 96).

[111] Cases C-367/98 *Commission v. Portugal*, C-483/99 *Commission v. France* and C-503/99 *Commission v. Belgium*, Joined Opinion of AG Colomer (n 96) para 43. The AG notes that the provision can be found in what is now part 7 of the TFEU (general and final provisions), which affects all the provisions in the TFEU.

[112] Cases C-367/98 *Commission v. Portugal*, C-483/99 *Commission v. France* and C-503/99 *Commission v. Belgium*, Joined Opinion of AG Colomer (n 96) para 44.

[113] Ibid para 45. The 1950 Schuman declaration proposed the creation of a European Coal and Steel Community. For more information on the Schuman declaration, see the EU's website on the declaration at <http://europa.eu/about-eu/basic-information/symbols/europe-day/schuman-declaration/index_en.htm> accessed 3 August 2012.

[114] Cases C-367/98 *Commission v. Portugal*, C-483/99 *Commission v. France* and C-503/99 *Commission v. Belgium*, Joined Opinion of AG Colomer (n 96) paras 49–56.

[115] Ibid para 56.

[116] Cases C-367/98 *Commission v. Portugal*, C-483/99 *Commission v. France* and C-503/99 *Commission v. Belgium*, Joined Opinion of AG Colomer (n 96) para 67. It is important to underline that AG Colomer did not, however, argue that primary EU law does not, as a consequence of Article 345, apply to national measures such as the ones at stake in the golden shares cases.

constituted a matter for the Member States by virtue of Article 345.[117] The existence of such rules was not in itself contrary to the fundamental freedoms established by the TFEU, but the specific manner in which they were applied could be.[118] Accordingly, part of the action against Portugal, as well as the actions against France and Belgium, had to be dismissed.[119]

By contrast, the ECJ did not place any significance on the role of Article 345 in its rulings in the three cases.[120] In *Commission v. Portugal*, the ECJ denied the relevance of Article 345 by stressing the primacy of the fundamental provisions in the Treaties:

> [A]rticle 222 of the Treaty is irrelevant in the present case. That article merely signifies that each Member State may organise as it thinks fit the system of ownership of undertakings whilst at the same time respecting the fundamental freedoms enshrined in the Treaty (...) However, those concerns cannot entitle Member States to plead their own systems of property ownership, referred to in Article 222 of the Treaty, by way of justification for obstacles, resulting from privileges attaching to their position as shareholder in a privatised undertaking, to the exercise of the freedoms provided for by the Treaty. As is apparent from the Court's case-law (*Konle*, cited above, paragraph 38), that article does not have the effect of exempting the Member States' systems of property ownership from the fundamental rules of the Treaty.[121]

Likewise, in *Commission v. France*, the ECJ dismissed Article 345 as irrelevant and again underlined that Article 345 could not have the effect of exempting the Member States' systems of property ownership from the fundamental rules of the Treaties.[122] Finally, in *Commission v. Belgium*, the ECJ stated that Article 345 was irrelevant since the national rules concerning the privatisation of companies in any event had to respect EU law.[123] Again, the ECJ pointed out that Article 345 could not have the effect of exempting the Member State's systems of property ownership from the fundamental rules of the Treaties.[124]

---

[117] Cases C-367/98 *Commission v. Portugal*, C-483/99 *Commission v. France* and C-503/99 *Commission v. Belgium*, Joined Opinion of AG Colomer (n 96) para 91.
[118] Ibid.
[119] Ibid para 92.
[120] Erika Szyszczak, 'Golden Shares and Market Governance' (2002) 29 Legal Issues of Economic Integration 255, 268.
[121] Case C-367/98 *Commission v. Portugal* (n 96) paras 28 and 48.
[122] Case C-483/99 *Commission v. France* (n 96) paras 23 and 44.
[123] Case C-503/99 *Commission v. Belgium* (n 96) para 22.
[124] Ibid para 44.

In his Opinion in two related golden shares cases,[125] AG Colomer discussed the ECJ's rulings in the three above-mentioned cases.[126] In essence, the AG reiterated the argumentation in his Opinion in the golden shares cases,[127] while criticising the ECJ's rulings in the three cases. AG Colomer's criticism boiled down to the three judgments having rendered Article 345 devoid of all its practical effect.[128] According to the AG, the three judgments marked the end of free state intervention in companies as it had up till then been understood.[129] Without stating why, the judgments had ignored the question of the application and scope of Article 345, which could not be done with impunity, since Article 345 was as important as the fundamental freedoms were.[130] According to the AG, both Spain and the UK were to benefit from the presumption of legality, finally resulting in the action against the former having to be dismissed and that against the latter having to be upheld.[131]

In its ruling in *Commission v. Spain*, the ECJ, perhaps unsurprisingly, stuck to its line of reasoning in the three golden shares cases. Without wasting too many words on the matter, the ECJ simply restated the formula that Article 345 could not exempt the Member States' systems of property ownership from the fundamental rules of the Treaties.[132] It upheld the Commission's action against Spain. In the *Commission v. UK* judgement, this standard formula was not even mentioned.[133]

---

[125]   These are Case C-463/00 *Commission v. Spain* [2003] ECR I-04581 and Case C-98/01 *Commission v. UK* [2003] ECR I-04641 (n 96).

[126]   Cases C-463–00 *Commission v. Spain* and C-98/01 *Commission v. UK*, Joined Opinion of AG Colomer (n 96).

[127]   In para 57 of his Opinion, AG Colomer argues that: 'the Treaty's observance, enshrined in Article 295 EC, of the system of property ownership in the Member States must extend to any measure which, through intervention in the public sector, understood in the economic sense, allows the State to contribute to the organisation of the nation's economic activity. It implies that those measures should not be considered *per se* as incompatible with the Treaty; therefore they are covered by the presumption of validity conferred on them by the legitimacy of Article 295 EC. For these purposes, it is particularly enlightening that the reservation in Article 295 EC is worded as a prohibition against prejudicing. If the Treaty in no way prejudices, this means, at the very least, that a national measure concerning the public sector system for adopting decisions must be judged compatible with the Treaty, *unless it is proved otherwise*. And prejudice is specifically what is involved when it is assumed that a measure which is in itself not discriminatory will be used in an unjustifiably discriminatory manner'.

[128]   Cases C-463–00 *Commission v. Spain* and C-98/01 *Commission v. UK*, Joined Opinion of AG Colomer (n 96) para 37.

[129]   Ibid.

[130]   Ibid.

[131]   Ibid para 58.

[132]   Case C-463/00 *Commission v. Spain* (n 96) para 67.

[133]   The only paragraph which referred to Article 345, was para 7, in which the content of ex Article 295 TEC was mentioned.

Nevertheless, as far as the AG was concerned, this was not the end of it. In his Opinions in *Commission v. Germany* and *Commission v. Italy*, AG Colomer again reiterated his view that the observance of the system of property ownership enshrined in Article 345 had to extend to any measure which, through intervention in the public sector (understood in the economic sense), allowed the state to contribute to the organisation of the nation's financial activity.[134] In other words, golden shares were to be seen as a form of ownership of undertakings, and thus constituted a matter for the Member States by virtue of Article 345.[135] In its rulings in both cases, the ECJ again did not respond to the AG's pleas.[136]

I cannot but feel sympathy for the AG's line of reasoning. When the interpretation of Article 345 as a neutrality principle is taken as a point of departure, the ECJ's reasoning in the golden shares cases indeed seems to be remarkably meagre. If the public ownership of undertakings constitutes a way for Member States to safeguard certain strategic public interests (and thus cannot be subject to the fundamental rules in the Treaties), then it does not make sense to simply draw the line at the existence of such rights. Simply owning such undertakings would arguably not guarantee the securing of the interests concerned. In order to secure those interests, the Member State would also need to be able to influence decision making in the relevant undertakings, for example through the holding of golden shares.

AG Colomer essentially held that the exercise of ownership rights through the holding of golden shares did not substantially differ from the decision to require an undertaking to be in public hands. As argued by the AG, this does not mean that the exercise of such ownership rights does not have to comply with the fundamental rules in the Treaties. However, the presumption of invalidity applied by the ECJ, without assessing the proportionality of the relevant restrictions, does not seem to be in line with the functional similarity between the ownership rights on the one hand and the exercise of those rights on the other. The ECJ appears to draw an artificial line in this regard.

Nevertheless, the golden shares cases indicate that the ECJ strictly distinguishes between the existence and the exercise of ownership rights. The former are not subject to the fundamental provisions in the Treaties, whereas the latter are. It is apparent from the above cases that the ECJ's interpretation of the exercise of ownership rights is broad. Even when the ownership rights as such would not

---

[134] See Case C-112/05 *Commission v. Germany*, Opinion of AG Colomer (n 92) para 49 and Case C-326/07 *Commission v. Italy*, Opinion of AG Colomer (n 96) para 37.

[135] Case C-326/07 *Commission v. Italy*, Opinion of AG Colomer (n 96) para 41.

[136] Case C-112/05 *Commission v. Germany* (n 96) and Case C-326/07 *Commission v. Italy* (n 96).

appear to suffice to guarantee the relevant strategic public interests, the exercise of these rights still needs to comply with primary EU law. This interpretation seems to be in line with the rest of the case law on Article 345. In the next section, I will discuss a number of cases that were already briefly touched upon in relation to the general lessons from the case law on Article 345.

b)   Cases C-491/01 *British American Tobacco* and T-65/98 *Van den Bergh Foods*

Before turning to ECJ's ruling in *British American Tobacco*, I briefly examine AG Geelhoed's Opinion in the case,[137] which appears to have influenced the ECJ's discussion of Article 345 to a great extent.

British American Tobacco concerned the validity of Directive 2001/37/EC on the manufacture, presentation and sale of tobacco products.[138] Article 5 of the Directive ('labelling') provides, inter alia, that each unit of tobacco products must carry the general warnings that 'smoking kills/can kill you' and that 'smoking seriously harms you and others around you'. The English High Court had referred to the ECJ the question, inter alia, of whether Directive 2001/37/EC could be invalid on the ground that it infringed the right to property. In national proceedings, British American Tobacco and Imperial Tobacco (supported by Japan Tobacco and JT International) had, among other things, claimed that Directive 2001/37/EC was invalid because of an infringement of Article 345 and the fundamental right to property.

In his Opinion in the case, AG Geelhoed argued that Article 345 was not relevant to the case.[139] According to the AG, the UK, France and Belgium had correctly pointed out that the provisions of Directive 2001/37/EC bore no relation to the systems of property ownership in the Member States within the meaning of Article 345.[140] The Directive did no more than to impose a restriction on the *exercise* of specified property rights by cigarette manufacturers.[141] Referring to *Fearon*,[142] the AG argued that Article 345 could not be invoked in order to set aside a restriction on the exercise of property rights which resulted from the application of EU provisions.[143]

---

137   Case C-491/01 *British American Tobacco*, Opinion of AG Geelhoed (n 96).
138   European Parliament and Council Directive 2001/37/EC of 5 June 2001 on the approximation of the laws, regulations and administrative provisions of the Member States concerning the manufacture, presentation and sale of tobacco products [2001] L194/26.
139   Case C-491/01 *British American Tobacco*, Opinion of AG Geelhoed (n 96) para 255.
140   Ibid 256.
141   Ibid.
142   Case 182/83 *Fearon* (n 96) para 7.
143   Ibid.

In relation to the right to property (recognised by the EU as a fundamental right through Article 6 TEU), the AG held that it was settled case law that the exercise of the right to property could be made subject to restrictions, provided that such restrictions in fact corresponded to objectives of general interest pursued by the EU and did not constitute, with regard to the aim pursued, disproportionate and unreasonable interference undermining the very substance of the right.

In its ruling in *British American Tobacco*, the ECJ followed AG Geelhoed's line of reasoning on Article 345 and the right to property. Referring to *Consten and Grundig*,[144] it held that Article 345 merely recognised the power of Member States to define the rules governing the system of property ownership and did not exclude any influence whatsoever of EU law on the exercise of national property rights.[145] The ECJ argued that Directive 2001/37/EC did not impinge in any way on the rules governing the system of property ownership in the Member States within the meaning of Article 345, which was irrelevant in relation to any effect produced by the Directive on the exercise by the manufacturers of tobacco products of their trademark rights over those products.[146]

With regard to the validity of Directive 2001/37/EC in respect of the right to property, the ECJ stated that it had consistently held that, while that right formed part of the general principles of EU law, it was not an absolute right and had to be viewed in relation to its social function.[147] Consequently, its exercise could be restricted, provided that those restrictions in fact corresponded to objectives of general interest pursued by the EU and did not constitute a disproportionate and intolerable interference, impairing the very substance of the rights guaranteed.[148] The ECJ held that the only effect produced by Article 5 of the Directive was to restrict the right of manufacturers of tobacco products to use the space on some sides of cigarette packets or unit packets of tobacco products to show their trademarks, without prejudicing the substance of their trade mark rights, the purpose being to ensure a high level of health protection once the obstacles created by national laws on labelling had been eliminated.[149] In the light of this analysis, the ECJ concluded that Article 5 constituted a proportionate restriction on the use of the right to property compatible with the protection afforded to that right by EU law.[150]

---

[144] Joined Cases 56 and 58–64 *Consten and Grundig* (n 96).
[145] Case C-491/01 *British American Tobacco* (n 96) para 147.
[146] Ibid para 148.
[147] Ibid para 149.
[148] Ibid.
[149] Ibid para 150.
[150] Ibid.

The case of *Van den Bergh Foods* concerned an action for annulment under ex Article 230(4) TEC (now Article 263(4) TFEU) of Commission Decision 98/531/EC, in which the Commission had held that several distribution agreements concluded by ice cream manufacturer Van den Bergh Foods in Ireland were incompatible with ex Articles 85 and 86 TEC (now Articles 101 and 102 TFEU). In its fifth plea, Van den Bergh Foods had argued that the application of the competition rules in the contested decision constituted an unjustified and disproportionate infringement of its right to property, as recognised in Article 345.[151]

As in *British American Tobacco*, the ECJ responded by arguing that even though the right to property formed part of the general principles of EU law, it was not an absolute right but had to be viewed in relation to its social function.[152] Consequently, the right to property could be restricted, provided that the restrictions in fact corresponded to objectives of general interest pursued by the EU and did not constitute a disproportionate and intolerable interference, impairing the very substance of the rights guaranteed.[153] The ECJ held that the contested decision did not affect Van den Bergh Foods's ownership of its assets, but merely regulated, in the public interest, one particular manner of exploiting them.[154] It did not, therefore, contain any undue limitation on the exercise of Van den Bergh Foods's right to property.[155]

The cases of *British American Tobacco* and *Van den Bergh Foods* are, for several reasons, of great relevance. First, in *British American Tobacco*, the ECJ clearly and explicitly acknowledged that secondary EU legislation may, in principle, restrict the exercise of property rights in the Member States. In many cases, the ECJ has firmly underlined the principle that Article 345 cannot exempt the exercise of property rights in the Member States from the provisions in the Treaties. In *British American Tobacco*, it went a bit further by holding that Article 345 likewise cannot prevent secondary EU legislation from restricting the exercise of property rights in the Member States.

The fact that *British American Tobacco* dealt with rights to intellectual (as opposed to tangible) property should not be problematic in this regard. As argued above, the existence/exercise dichotomy also appears to apply to rights to tangible property. Furthermore, the ECJ's reference to a case dealing with the application of primary EU law to the exercise of property rights[156] (and not with

---

151   Case T-65/98 *Van den Bergh Foods* (n 96) para 164.
152   Ibid para 170.
153   Ibid.
154   Ibid para 171.
155   Ibid.
156   The ECJ referred to Joined Cases 56 and 58-64 *Consten and Grundig* (n 96).

a restriction of that exercise through secondary EU law) does not make this conclusion less valid. In *British American Tobacco*, the ECJ clearly stated that Article 345 does not exclude any influence whatsoever of EU law (including secondary EU law such as a Directive) on the exercise of national property rights.

Second, in relation to the general principle of the right to property – which can be interpreted as being enshrined in Article 345 (as done by Van den Bergh Foods) – the ECJ has, in both cases, made it clear that the exercise of this right can be restricted, provided that the restriction corresponds to objectives of general interest pursued by the EU and that it does not constitute a disproportionate and intolerable interference, impairing the very substance of the rights guaranteed.[157] In *British American Tobacco*, the ECJ found, inter alia, the mandatory labelling of cigarette packages not to prejudice the substance of the producers' trade mark rights; it only restricted the right of producers to use *some* sides of the packaging to show their trademarks.

In *Van den Bergh Foods*, the ECJ held that the Commission decision did not deprive Van den Bergh Foods of its rights of property in its stock of freezer cabinets or prevent it from exploiting those assets by renting them out on commercial terms; the decision merely required that if Van den Bergh Foods decided to exploit its freezers by making them available without charge to retailers, it could not do so on the basis of an exclusivity clause,[158] so long as it held a dominant position on the relevant market.[159]

The above review of the literature and case law on Article 345 shows that the meaning and scope of the provision are still not entirely clear. Nevertheless, a number of conclusions can be drawn.

First, Article 345 seems to primarily constitute a neutrality principle, according to which the Treaties are neutral as to whether an undertaking is held in public or private ownership. Based on a historical interpretation, Fraga et al. have argued that the scope of Article 345 might possibly be broader, but the case law of the European Courts does not appear to lend support for this interpretation.

What becomes clear from the case law on Article 345 is that the provision cannot prevent or limit the application of primary EU law to the systems of property ownership in the Member States. Even though the Courts usually do

---

[157] See also Lars Henriksson, 'Structural Measures in EC Competition Law: A Bridge Too Far?' in Ulf Bernitz and others (eds), *General Principles of EC Law in a Process of Development* (Kluwer Law International 2008) 141, 151.
[158] The ice creamer retailers could use the free freezers supplied by Van den Bergh Foods for displaying the ice creams produced by the latter only.
[159] Case T-65/98 *Van den Bergh Foods* (n 96) para 171.

not (explicitly) make a distinction between the existence and the exercise of property rights (the 'existence/exercise dichotomy' mentioned by Akkermans and Ramaekers), it seems that the *exercise* of property rights may (contrary to the *existence* of property rights), in the general interest, be restricted under EU law.

As to the scope of the existence/exercise dichotomy, at first it appeared to apply in relation to rights to intangible (intellectual) property and the application of substantive Treaty rules to the exercise of those property rights only. However, as suggested by Akkermans and Ramaekers, the existence/exercise dichotomy also seems to apply to rights to tangible property; the exercise of rights to tangible property may also be restricted. What is more, the dichotomy likewise appears to apply in relation to the EU competence to regulate a matter falling within the scope of Article 345. In other words, secondary EU legislation may, under certain circumstances, restrict the exercise of property rights.

The golden shares cases indicate that the ECJ draws a strict line between the existence and the exercise of ownership rights. It is apparent from these cases that the ECJ's interpretation of the exercise of ownership rights is broad. Even when the ownership rights as such would not appear to suffice to guarantee the relevant strategic public interests, the exercise of these rights still needs to comply with primary EU law. This interpretation seems to be in line with the rest of the case law on Article 345.

c)   The implications of an amended Article 4(1) of the CCS Directive

Considering the above, it seems that Article 345 would not, in principle, stand in the way of an amended Article 4(1) of the CCS Directive giving the Commission the possibility to oblige Member States to accept CO₂ storage in (parts of) their territory. However, this provision will have to meet the requirements listed in both cases: (1) the resulting restriction of the exercise of the right of ownership of suitable storage formations should correspond to objectives of general interest pursued by the EU, and (2) it should not constitute a disproportionate and intolerable interference, impairing the vary substance of the rights guaranteed. As for the first requirement, this will likely be met. As argued in section 8.2.1.1, the objective of such a provision would be to reduce greenhouse gas emissions through the storage of captured CO₂. That objective is unmistakably related to one of the main EU environmental objectives, the combating of climate change.[160]

---

[160]   See Article 191(1) TFEU.

As for the requirement that the restriction resulting from the provision should not constitute a disproportionate and intolerable interference, impairing the vary substance of the ownership rights, things might be different. The ECJ's rulings in *British American Tobacco* and *Van den Bergh Foods* show that the ECJ will likely deem this requirement to have been met as long as there remains some substantial way for the owner to exercise his property right. In the former case, the ECJ essentially argued that producers could still use the sides of the cigarette packaging that were not labelled for advertising their brand. In the latter case, the ECJ essentially held that Van den Bergh Foods could still use its ice cream freezers commercially – for instance, by, renting them out – but that it could not make them available without charge to retailers and demand that they were used for the displaying of its ice creams only, as long as Van den Bergh Foods held a dominant position in the relevant market.

On the basis of the above, I would argue that as long as the owner of a potential CO_2 storage formation will remain able to use that formation for other purposes, for instance commercial gas storage, in a commercially viable manner, this requirement will be met. However, should the obligation to use a certain reservoir for CO_2 storage make the use of that reservoir for other purposes in a commercially viable manner impossible, then this requirement would arguably not be met. Article 345 would probably not allow for an amended Article 4(1) of the CCS Directive leading to owners of potential CO_2 storage reservoirs being able to use these reservoirs for CO_2 storage only. That would arguably impair the very substance of the right of ownership of the relevant reservoir.

Considering the requirements for CO_2 storage permits in the CCS Directive, the latter could very well be the case. Article 6(1) of the CCS Directive provides that no conflicting uses are permitted on the storage site. The CCS Directive does not further define the term conflicting use, but it would most likely cover activities such as commercial gas storage. This would mean that an amended Article 4(1) leading to the owners of a reservoir having to use that reservoir for CO_2 storage would probably not be compatible with Article 345; as Article 6(1) prohibits conflicting uses of that reservoir, it could then only be used for CO_2 storage.

The above conclusions hold true, independent from the answer to the question of *who* owns the relevant potential storage reservoir. Take, for instance, the scenario that the potential storage reservoir is a nearly depleted oil or gas field that is owned by the state (in the form of a state ownership of the subsoil and its natural resources),[161] but for which the right of use resides with a private party in the form of a mining permit. In this case, the question is whether the exercise

---

[161] A potential CO_2 storage reservoir would then have to be seen as a kind of natural resource. In this regard, see n 8, above.

of the right of use given to the private entity can be seen as the exercise of a property right within the meaning of the *British American Tobacco* and *Van den Bergh Foods* cases. If this question can be answered in the positive, an amended Article 4(1) would restrict the private entity's exercise of his property right in a disproportionate and intolerable manner.

In the in my opinion more likely case that this question must be answered in the negative,[162] it would be the state's exercise of its property right with which Article 4(1) would interfere in a disproportionate and intolerable manner. After all, as a consequence of Article 4(1), the state would no longer be able to use the reservoir for other purposes in a commercially viable manner, such as giving a right of use to a private entity. Similarly, when the ownership of the potential storage reservoir resides with a private party, an amended Article 4(1) would, arguably interfere with the entity's exercise of its property right in a disproportionate and intolerable manner. No matter who holds the right of property over the relevant storage reservoir, as a consequence of Article 6(1) of the CCS Directive, an amended Article 4(1) is likely to interfere with the exercise of the relevant property right in a disproportionate and intolerable manner and thus to breach Article 345 TFEU.

An exception would arguably be the situation in which the reservoir in question could not sensibly be used for any other commercial application than CO$_2$ storage. This would seem to be more likely in case of the potential storage reservoir being a saline aquifer than in case of it being a (nearly) depleted oil or gas field. For the latter, the use of the reservoir for mining or gas storage purposes is an obvious option, whereas for the former this is not, or is to a lesser extent, the case. Another exception would be the situation in which there is, according to the laws of the relevant Member State, no ownership of the potential storage reservoir. In such situation, there would, in my opinion, be no breach of Article 345 TFEU.

8.4 LESSONS FOR AN AMENDMENT OF ARTICLE 4(1) OF THE CCS DIRECTIVE

The above analysis provides several lessons for a possible amendment of Article 4(1) of the CCS Directive. First, even though the grounds in Article 192(2) (b) and (c) TFEU at first sight appear to be of great relevance, none of these grounds are likely to influence the choice of legal basis of a possible amendment

---

[162] The ownership of the reservoir, after all, still resides with the state.

of Article 4(1) in order to allow the Commission to force Member States to accept CO_2 storage in (parts of) their territory.

This is partly caused by the ECJ's generally restrictive interpretation of the exceptions in Article 192(2), but also dictated by the logic of the provision. If the ECJ were to, for instance, interpret Article 192(2) in such a broad manner as for an amended Article 4(1) of the CCS Directive to fall within the scope of Article 192(2)(c), the effectiveness of EU environmental policy making could arguably be seriously undermined. As a result, the grounds in Article 192(2)(b) and (c) would not appear to lead to an amendment of Article 4(1) of the CCS Directive having to be adopted by unanimity.

Likewise, if the objective of an amended Article 4(1) – the reduction of greenhouse gas emissions through CO_2 storage – could be brought under one of the objectives in Article 194(1) TFEU, for instance the promotion of energy efficiency and energy saving and the development of new and renewable forms of energy, then Article 194(2) would arguably not be available as a legal basis for such amendment. Article 194(2) provides that EU energy measures shall not (at all) affect a Member State's right to determine the conditions for exploiting its energy resources, its choice between different energy sources and the general structure of its energy supply. As indicated in section 8.2.1.3 above, an amended Article 4(1) is likely to affect, even though not significantly, a Member State's energy source choice and energy supply structure. As a consequence of the more stringent Member State reservation in Article 194(2), the latter provision will therefore likely not be available as a legal basis for an amended Article 4(1) of the CCS Directive.

As for Article 345 TFEU, the *British American Tobacco* and *Van den Bergh Foods* cases indicate that Article 345 would probably stand in the way of an amended Article 4(1) of the CCS Directive giving the Commission the possibility to oblige Member States to accept CO_2 storage in (parts of) their territory. Considering the requirements in Article 6(1) of the CCS Directive, an amended Article 4(1) will likely not be compatible with Article 345 as it would force the owner of a potential storage reservoir to use that reservoir for CO_2 storage only. Such a restriction would arguably impair the very substance of the right of ownership of the relevant reservoir. An exception is arguably the situation in which the reservoir in question could not sensibly be used for any other commercial application than CO_2 storage. This is more likely to be the case when the potential storage reservoir in question is a saline aquifer than when it is a (nearly) depleted oil or gas field. Another exception would be the situation in which there is, according to the laws of the relevant Member State, no ownership of the potential storage reservoir. Then there can, in my opinion, be no breach of Article 345 TFEU.

Based on the above, the Commission could base an amendment of Article 4(1) of the CCS Directive on the TFEU equivalent of the original legal basis of the CCS Directive (Article 175(1) TEC), Article 192(1) TFEU (ordinary legislative procedure). There do not appear to be any convincing grounds for such an amendment having to be adopted unanimously. Nor would Article 194(2) appear to provide a proper legal basis for such amendment. However, considering the requirements in Article 6(1) of the CCS Directive, the Commission is advised not to amend Article 4(1) in such a way that it would allow it to force Member States to accept CO_2 storage in (parts of) their territory, since a thus amended Article 4(1) could very well breach Article 345 TFEU.

CHAPTER IX

PUBLIC FUNDING OF CCS INFRASTRUCTURE AND EU STATE AID LAW

- To avoid public support for the development of CO_2 transport and storage infrastructure being classified as state aid under Art. 107(1) TFEU, Member States are advised to select the relevant transport/storage operator through an open, transparent and non-discriminatory public procurement procedure in line with applicable EU public procurement legislation

- Such procedure can be based on qualitative selection criteria – e.g., geographical scope or safety – as long as the final selection is based on the relevant infrastructure being developed at lowest societal cost

9.1 INTRODUCTION

I have argued before that CO_2 transport and storage infrastructure is likely to be constructed (partly) with public funds.[1] When Member States decide to support the development of CO_2 transport and storage infrastructure, EU rules on state aid come into play. Article 107(1) of the Treaty on the Functioning of the EU (TFEU) prohibits Member States from granting aid that distorts competition and affects trade between Member States. If public support meets the conditions of Article 107(1) TFEU and qualifies as state aid,[2] it will have to be notified to the Commission under Article 108(3) TFEU. Proposed national measures may not be put into effect until approved by the Commission or Council.[3]

---

[1]   See section 6.3.2.3 of Chapter VI. See also Hans Vedder, 'An Assessment of Carbon Capture and Storage under EC Competition Law' (2008) 29 European Competition Law Review 586, 586–89.

[2]   These conditions are briefly discussed in section 9.2, below.

[3]   Article 108(3) TFEU. This prohibition is effective during the whole of the assessment period and thus includes both the period for a preliminary assessment and, if applicable, the period of the formal investigation procedure under Article 108(3). See Conor Quigley, *European State Aid Law and Policy* (2nd edn, Hart Publishing 2009) 371.

In its revised Environmental State Aid Guidelines, the Commission notes that some of the means to support CCS envisaged by Member States could constitute state aid, but that it is too early to lay down guidelines relating to the authorisation of such aid.[4] It states that it will, under certain circumstances, have a generally positive attitude towards state aid for CCS projects.[5] Aid for such projects could then be approved under Article 107(3) TFEU, which allows the Commission to declare aid compatible with the common market. At the beginning of 2012, the Commission had approved proposed national measures granting aid to four projects related to the deployment of CCS in the UK, Germany and the Netherlands.[6]

Yet, the Commission's proclaimed intention and consequent practice do not guarantee that state aid for CCS is approved. Myhre has analysed the assessment of several notifications of state aid for CCS demonstration projects, including those mentioned above.[7] He concludes that there are significant differences in the way in which these notifications were assessed and states that 'there still is considerable scope for discretion in assessing. e.g., whether the aid measure is 'proportionate' or whether distortions of competition and effects on trade are limited, so that the 'overall balance' is positive'.[8, 9]

When seeking to support the development of CO_2 transport and storage infrastructure, Member States will be confronted with the uncertainty and loss of time related to the notification and standstill procedure under Article 108(3).[10] This uncertainty could lead to less CO_2 transport and storage infrastructure

---

[4]   Commission, 'Community Guidelines on State Aid for Environmental Protection' [2008] OJ C82/1, para 69.
[5]   Ibid.
[6]   See Commission, 'State Aid: Commission Approves UK Aid for Feasibility Studies on Carbon Capture and Storage Demonstration Projects' IP/09/555 <http://europa.eu/rapid/ pressReleasesAction.do?reference=IP/09/555> accessed 20 March 2012; Commission, 'State Aid: Commission Approves € 30 Million German Support for ArcelorMittal Eisenhuettenstadt's "Top Gas Recycling" Project' IP/10/254 <http://europa.eu/rapid/ pressReleasesAction.do?reference=IP/10/254> accessed 20 March 2012; Commission, 'State Aid: Commission Approves € 10 Million Aid for Nuon's energy-saving CO2 Capture Project in the Netherlands' IP/10/614 <http://europa.eu/rapid/pressReleasesAction.do?reference=IP/1 0/614&format=HTML&aged=0&language=EN&guiLanguage=en> accessed 20 March 2012; and Commission, 'State Aid: Commission Approves € 150 Million for Carbon Capture and Storage Project in the Netherlands' IP/10/1392 <http://europa.eu/rapid/pressReleasesAction. do?reference=IP/10/1392> accessed 20 March 2012.
[7]   Jonas W Myhre, 'Financing of CCS Demonstration Projects: State Aid, EEPR and NER funding; an EU and EEA Perspective' (2011) University of Oslo Faculty of Law Legal Studies Research Paper Series No. 2011–02 <http://papers.ssrn.com/sol3/papers.cfm?abstract_ id=1738511> accessed 20 March 2012.
[8]   As the Commission found in the four cases mentioned above.
[9]   Myhre (n 7) 74.
[10]   Under Article 108(3) TFEU, the Member States have to inform the Commission of any plans to grant or alter state aid. Until this notification procedure has resulted in a final decision, the Member State is not allowed to put the proposed measures into effect.

being developed or to the development of such infrastructure being delayed, in turn negatively affecting access possibilities for parties seeking to transport and store their captured CO_2. What is more, the classification as state aid could lead to significant (unwanted) procedural burdens.[11]

A possible way to avoid the above would be for public support for the development of CO_2 transport and storage infrastructure not to constitute state aid in the first place. Following the *Altmark* case, compensation for the discharge of a service of general economic interest[12] does not constitute state aid if a number of conditions are met.[13] As indicated earlier, Member States have a considerable margin of discretion in designating services of general economic interest.[14] I have argued before that both the transport and storage of CO_2 could, at least in the initial phase, be characterised as a service of general economic interest.[15] At first sight, it seems doubtful whether the same goes for the development of CO_2 transport and storage infrastructure. Yet, the Commission has in the past qualified public co-funding of broadband infrastructure as compensation for the provision of a service of general economic interest as it considered access to broadband services for all citizens to be a service of general economic interest.[16] In similar vein,

---

[11]   In this regard, see, for instance, Case T-233/04 *Netherlands v. Commission* [2008] ECR II-00591, paras 32–33. See also Joined Cases T-15/12 and T-16/12 *Provincie Groningen and Others and Stichting Het Groninger Landschap and Others v European Commission*, Order of the General Court of 19 February 2013, paras 33–35.

[12]   See section 6.4.3 of chapter VI.

[13]   Case C-280/00 *Altmark* [2003] ECR I-07747. The four *Altmark* conditions are further discussed in section 9.4, below. Another way for public support for the development of infrastructure not to constitute state aid is for it to qualify as public funding of 'general infrastructure'. The Commission has long considered public funding of general infrastructure such as roads and bridges not to constitute state aid. Yet, in practice, it is difficult to distinguish general infrastructure from other infrastructure. Furthermore, as private operators have, over the years, become more and more involved in major infrastructure projects funded by the state or through public-private partnerships, the Commission's stance when assessing public funding of infrastructure has become stricter. As a result, the route of having public support for the development of infrastructure qualify as public funding of general infrastructure is likewise characterised by a considerable degree of policy/legal uncertainty. See Mihalis Kekelekis, 'Recent Developments in Infrastructure Funding: When Does it Not Constitute State Aid?' (2011) 10 European State Aid Law Quarterly 433, Phedon Nicolaides and Maria Kleis, 'Where is the Advantage? The Case of Public Funding of Infrastructure and Broadband Networks?' (2007) 6 European State Aid Law Quarterly 615, and also Lambros Papadias, Alexander Riedl and Jan-Gerrit Westerhof, 'Public Funding for Broadband Networks: Recent Developments' (2006) European Commission Competition Policy Newsletter 2006–3, 13 <http://ec.europa.eu/competition/publications/cpn/2006_3_13.pdf> accessed 20 March 2012.

[14]   See section 6.4.3 of chapter VI.

[15]   Jones and Suffrin have argued that the application of the concept of services of general economic interest has been expanded significantly beyond the basic utilities. See Alison Jones and Brenda Suffrin, *EU Competition Law: Text, Cases and Materials* (Oxford University Press 2010) 602, for a list of accepted services of general economic interest. In this regard, it is interesting to note that Article 3(2) of the Electricity and Gas Directives (Directives 2009/72/EC and 2009/73/EC) allows Member States to impose public service obligations on electricity and gas undertakings relating to, inter alia, climate protection.

[16]   See Commission, 'State Aid: Commission Endorses Public Funding for Broadband Network in Limousin, France' IP/05/530 <http://europa.eu/rapid/pressReleasesAction.

public funding of CO_2 transport and storage infrastructure could constitute compensation for a service of general economic interest.

The most difficult and indeterminate of the *Altmark* criteria is arguably the fourth criterion.[17] This criterion, also known as the efficiency criterion, requires the public service provider to be selected by means of a public procurement procedure or alternatively the level of compensation for the discharge of the service of general economic interest/public service obligation[18] to be based on the costs of a typical, well-run undertaking adequately provided with the means to meet the necessary public service requirements.[19] In the following, I explore the content and meaning of the *Altmark* efficiency criterion.[20] In doing so, I answer the question of in what way Member State authorities should design public funding of CO_2 transport and storage infrastructure for it to be compatible with the *Altmark* efficiency criterion and thus to be likely to escape classification as state aid (provided that the other three *Altmark* criteria are met).

Section 9.2 briefly explores Article 107 TFEU. Section 9.3 deals with Article 107(1) TFEU and the financing of services of general economic interest. Section 9.4 examines the *Altmark* case. Section 9.5 further addresses the *Altmark* efficiency criterion and the many questions that it raises. Section 9.6 explores the European Courts' post-*Altmark* case law and the post-*Altmark* decision practice on infrastructure development of the Commission. Finally, section 9.7 draws a number of lessons for public funding of CO_2 transport and storage infrastructure.

9.2   ARTICLE 107 TFEU

Article 107(1) TFEU provides that:

> [A]ny aid granted by a Member State or through State resources in any form whatsoever which distorts or threatens to distort competition by favouring certain

---

do?reference=IP/05/530> accessed 19 March 2012. See also Natalia Fiedziuk, 'The Interplay between Objectives of the European Union's Energy policy: The Case of State Funding of Energy Infrastructure' (2009) TILEC Discussion Paper, 26 <http://papers.ssrn.com/sol3/papers.cfm?abstract_id=1386902> accessed 21 March 2012.

17   See, for instance, Jens-Daniel Braun and Jürgen Kühling, 'Article 87 EC and the Community Courts: From Revolution to Evolution' (2008) 45 Common Market Law Review 465, 476 and José Luis Buendia Sierra, 'Finding the Right Balance: State Aid and Services of General Economic Interest' in *EC State Aid Law/ Le Droit des Aides d'Etat dans la CE – Liber Amicorum Francisco Santaolalla Gadea* (Kluwer Law International 2008) 191, 199 and 214. Buendia Sierra has argued that since the four conditions must be fulfilled cumulatively, the fourth condition becomes the cornerstone of the test.

18   In the rest of this writing, I will use these two terms interchangeably.

19   Case C-280/00 *Altmark* (n 13) para 93.

20   In doing so, I only occasionally deal with EU public procurement law, a field of EU law which falls outside my expertise.

undertakings or the production of certain goods shall, in so far as it affects trade between Member States, be incompatible with the internal market.

According to settled case law, classification as state aid requires that all the conditions of Article 107(1) are fulfilled.[21] In the literature, different (numbers of) criteria are mentioned.[22] Yet, in a number of cases, the Courts indicated that Article 107(1) requires (1) an intervention by the state or through state resources, that (2) confers an advantage on the recipient, (3) is liable to affect trade between Member States and (4) distorts or threatens to distort competition.[23]

Any Member State interventions/measures meeting the above (cumulative) criteria are, in principle, to be classified as incompatible state aid within the meaning of Article 107(1). Article 107(2) provides certain exceptions for situations where the aid *will* be deemed to be compatible with the common market.[24] Article 107(3) provides certain exceptions for situations where the aid *may* be deemed to be compatible with the common market.[25] The difference between paragraphs (2) and (3) is that the latter leaves the Commission a certain amount of discretion in applying the provision, whereas the former does not.[26]

---

[21] Case C-142/87 *Belgium v. Commission* [1990] ECR I-00959, para 25; Joined Cases C-278/92 to C-280/92 *Spain v. Commission* [1994] ECR I-04103, para 20; Case C-482/99 *France v. Commission* [2002] ECR I-04397, para 68; and Case T-442/03 *SIC* [2008] ECR II-01161, para 43.

[22] See e.g. Adinda Sinnaeve, 'State Financing of Public Services: The Court's Dilemma in the Altmark Case' (2003) 3 European State Aid Law Quarterly 351, 352; Paul Craig and Gráinne de Búrca, *EU Law – Text, Cases and Materials* (3rd edn, Oxford University Press 2003) 1141–46; Braun and Kühling (n 17) 465–66; and Kekelekis (n 13) 434.

[23] Case C-280/00 *Altmark* (n 13) para 75; Case C-237/04 *Enirisorse* [2006] ECR I-02843, para 39; Case T-244/03 *SIC*, para 44; Case C 140/09 *FallimentoTraghetti del Mediterraneo* [2010] ECR I-05243, para 31; and Case T-335/08 *BNP Paribas* [2010] ECR II-03323, para 159. For an extensive discussion of the different conditions under Article 107(1) TFEU, see Leigh Hancher, Tom Ottervanger and Piet Jan Slot, *EC State Aids* (3rd edn, Sweet & Maxwell 2006) 33–88.

[24] Craig and de Búrca (n 22) 1141. Article 107(2) lists the categories of aid which are exempt from the implied prohibition of Article 107(1). Exemption is automatic if the Commission considers that the aid falls within one of the three categories listed in the sub-article. See Hancher, Ottervanger and Slot (n 23) 104.

[25] Craig and de Búrca (n 22) 1141.

[26] Hancher, Ottervanger and Slot have argued that the word 'may' ('may be considered to be compatible with the internal market') in Article 107(3) TFEU confers considerable discretion upon the Commission, the exercise of which is only subject to marginal review by the European Courts. See Hancher, Ottervanger and Slot (n 23) 109.

9.3 ARTICLE 107(1) TFEU AND THE FINANCING OF SERVICES OF GENERAL ECONOMIC INTEREST

Over the years, much ink has been spilt on the subject of the application of state aid rules to the financing of services of general economic interest. The question is whether compensation granted for the discharge of services of general economic interest confers an advantage within the meaning of the second criterion of Article 107(1). In this regard, the interplay between Articles 107(1) and 106(2) TFEU is of relevance, the latter providing for the opportunity to exempt undertakings entrusted with the operation of services of general economic interest from the rules in the Treaties, in particular those on competition.[27]

Initially, the Commission did not regard compensation for public service obligations as state aid within the meaning of Article 107(1).[28] In the *FFSA* case,[29] the Commission found that the tax exemption granted to the French post constituted a definite financial advantage, but since it did not exceed what was necessary to ensure the public service, it qualified for the derogation laid down in ex Article 90(2) of the Treaty Establishing the European Economic Community (TEEC – now Article 106(2) TFEU).[30] It concluded that the measure did not constitute state aid within the meaning of ex Article 92(1) TEEC (now Article 107(1) TFEU) since it was covered by ex Article 90(2) TEEC.[31] On appeal, the Court of First Instance (CFI) took a different approach.[32] It held that ex Article 90(2) TEEC could not prevent ex Article 92(1) TEEC from applying.[33] The compensation granted had to be qualified as state aid. Yet, ex Article 90(2) TEEC could serve to make the aid compatible with the common market.[34] In the eyes of the CFI, ex Article 90(2) TEEC was an additional basis for the compatibility of state aid with the common market, similar to ex Articles 91(2) and (3).[35]

---

27   On Article 106(2), see also section 6.4.3 of Chapter VI.
28   Sinnaeve (n 22) 352. See also Jan A Winter, 'Re(de)fining the Notion of State Aid in Article 87(1) of the EC Treaty' (2004) 41 Common Market Law Review 475, 496.
29   *FFSA* (Case NN 135/92) [1995] OJ C262/11.
30   Sinnaeve (n 22) 352. For pre-*FFSA* ECJ case law on the relationship between compensation for public service obligations and EU state aid rules, see Winter (n 28) 497 and Hancher, Ottervanger and Slot (n 23) 206.
31   Sinnaeve (n 22) 352.
32   See Case T-106/95 *FFSA* [1997] ECR II-00229.
33   Case T-106/95 *FFSA* (n 32) para 165.
34   Ibid para 172.
35   Sinnaeve has argued that the ECJ implicitly endorsed the CFI's position on the relationship between state aid rules and public service obligations in Case C-332/98 *France v. Commission* [2000] ECR I-04833. See Sinnaeve (n 22) 352.

The CFI confirmed its ruling in *FFSA* in the *SIC* case.[36] In *SIC*, it held that the fact that a financial advantage is granted to an undertaking to offset the cost of public service obligations had no bearing on the classification of that measure as aid within the meaning of ex Article 92(1) TEEC.[37] It argued that ex Article 92(1) TEEC did not distinguish between measures of state intervention by reference to their causes or aims but since the concept of aid is an objective one rather defined them in relation to their effects.[38] Referring to *FFSA*, the CFI indicated that grants intended to offset public service obligation costs could be taken into account when considering whether the aid in question is compatible with the common market under ex Article 90(2) TEEC.[39]

In *Ferring*, the European Court of Justice (ECJ) departed from the CFI's rulings in *FFSA* and *SIC* (and reaffirmed its earlier position in the *ADBHU*[40] case).[41] The ECJ argued that a tax exemption could be regarded as compensation for the public service obligations provided, and hence not as state aid within the meaning of ex Article 92 TEEC, as long as there was 'the necessary equivalence between the exemption and the additional costs incurred'.[42] According to the ECJ, the only effect of the tax exemption would be 'to put distributors and laboratories on an equal competitive footing'.[43] The Commission's policy following the *Ferring* judgment was not entirely clear.[44] While some decisions fully applied the *Ferring* approach, other decisions concluded that the measure did not qualify as state aid, but left open the possibility to nevertheless examine whether the conditions of Article 106(2) were met, thereby creating confusion.[45] Sinnaeve has argued that this was admittedly an understandable position since the criticism and divergent opinions to which the *Ferring* judgment gave rise reinforced the doubts as to whether the judgment was 'good law' or would be corrected again soon.[46]

---

[36]   Case T-46/97 *SIC* [2000] ECR II-02125. See Jean-Marc Thouvenin, 'The Altmark Case and its Consequences' in Markus Krajewski, Ula Neergaard and Johan van de Gronden (eds), *The Changing Legal Framework for Services of General Interest in Europe: Between Competition and Solidarity* (TMC Asser Press 2009) 103, 105; and Sinnaeve (n 22) 105.

[37]   Case T-46/97 *SIC* (n 36) para 84.

[38]   Case T-46/97 *SIC* (n 36) para 83.

[39]   Ibid para 84.

[40]   Case 240/83 *ADBHU* [1985] ECR 00531.

[41]   See Winter (n 28) 497.

[42]   Case C-53/00 *Ferring* [2001] ECR I-09067, para 27. This is also known as the 'compensation approach', according to which no state aid is involved if Member States merely compensate an undertaking for providing a public service; only to the extent that a measure exceeds the costs for discharging the public service obligations imposed, will it amount to state aid within the meaning of Article 107(1) TFEU. See Sinnaeve (n 22) 355.

[43]   Case C-53/00 *Ferring* (n 42) para 27.

[44]   See Sinnaeve (n 22) 353.

[45]   Ibid.

[46]   Ibid.

Part of the criticism of the ECJ's ruling in *Ferring* came from within the Court itself. In his first opinion in the *Altmark* case, AG Léger argued that the ECJ had confused the question of characterising a measure as state aid and the question of justifying state aid.[47] According to the AG, state intervention could not be assessed in terms of the objective pursued by the public authorities; those objectives may be taken into consideration only at a later stage in the analysis to determine whether the state measure is justified under the derogations provided for in the Treaty.[48]

Furthermore, AG Léger claimed that *Ferring* deprived ex Article 90(2) TEEC of a substantial part of its effect.[49] He argued that the system introduced by *Ferring* was characterised by considerable flexibility compared to the control provided for in ex Article 90(2) TEEC.[50] Finally, according to the AG, the reasoning in *Ferring* effectively removed measures for financing public services from the Commission's control.[51] Based on the above, AG Léger advised the ECJ to review the *Ferring* ruling and to return to the CFI's approach in *SIC* of considering financial compensation granted to offset the costs of public service obligations to constitute state aid within the meaning of Article 107(1) TFEU, without prejudice to the possibility of that measure being exempted under the derogations provided in the Treaty.[52]

Following AG Léger's opinion, AG Jacobs added his view to the debate and proposed in his opinion in the *GEMO* case[53] an intermediate, compromise approach.[54] AG Jacobs proposed to distinguish between two situations.[55] In cases where there is a direct and manifest link between the financing granted by the state and clearly defined public service obligations assigned to the beneficiary, such financing will not constitute state aid within the meaning of Article 107(1) TFEU. In other cases, the relevant measure can be qualified as state aid under

---

[47]   Case C-280/00 *Altmark* [2003] ECR I-07747, Opinion of AG Léger, para 76.
[48]   Ibid para 77.
[49]   Case C-280/00 *Altmark* [2003] ECR I-07747, Opinion of AG Léger, para 79.
[50]   Ibid para 89.
[51]   Ibid para 91.
[52]   Case C-280/00 *Altmark*, Opinion of AG Léger (n 47) para 98. This approach is also known as the 'state aid approach', according to which state funding granted to an undertaking for the performance of services of general economic interest constitutes state aid within the meaning of Article 107(1) TFEU, but may be justified under Article 106(2) TFEU if the conditions of that derogation are fulfilled and, in particular, if the aid does not exceed the appropriate remuneration for the costs of the service. See Sinnaeve (n 22) 354.
[53]   Case C-126/01 *GEMO* [2003] ECR I-13769, Opinion of AG Jacobs.
[54]   Sinnaeve (n 22) 353.
[55]   Thouvenin (n 36) 106.

Article 107(1). In the *Enirisorse* case, AG Stix-Hackl took a similar position.[56] Shortly thereafter, the ECJ gave its view in *Altmark*.[57]

9.4 CASE C-280/00 *ALTMARK*

9.4.1 THE RULING

The *Altmark* case concerned a preliminary reference[58] by the Bundes-verwaltungsgericht (German Federal Administrative Court). In national proceedings, a German transport undertaking (Nahverkehrsgesellschaft Altmark GmbH) had challenged the grant of renewed licences for regional bus transport services to a competitor (Altmark Trans GmbH). Nahverkehrs-gesellschaft had argued that Altmark Trans had not satisfied the requirements for the grant of a licence under national law since it needed subsidies to discharge the public service obligations deriving from the licenses granted.[59] On appeal, the Bundesverwaltungsgericht decided to stay proceedings and refer to the ECJ a question on the compatibility with EU law of the subsidies paid to Altmark Trans for the commercial operation of the licenses granted. In the first part of the question referred to the ECJ, the national court essentially asked whether subsidies intended to compensate for the deficit in operating an urban, suburban or regional public transport service would come under ex Article 92(1) TEEC in all circumstances or whether, having regard to the local or regional character of the transport services provided, such subsidies were not liable to affect trade between Member States (third criterion under Article 107(1) TFEU).[60]

---

[56]   Sinnaeve (n 22) 353–54. See Joined Cases C-34/01 to C-38/01 *Enirisorse* [2003] ECR I-14243, Opinion of AG Stix-Hackl. This approach is known as the 'quid pro quo approach', according to which one approach is applied to one category of cases and the other approach to another category. The distinction between both categories would be based on the nature of the link between the financing and the general interest obligations imposed and on how clearly these interests are defined. See Sinnaeve (n 22) 356. On the pre-*Altmark* case law on the interaction of financing of services of general economic interest and EU state aid rules, see Cesare Rizza, 'The Financial Assistance Granted by Member States to Undertakings Entrusted with the Operation of a Service of General Economic Interest: The Implications of the Forthcoming Altmark Judgment for Future State Aid Control Policy' (2003) 9 Columbia Journal of European Law 429.

[57]   Case C-280/00 *Altmark* (n 13).

[58]   Article 267 TFEU contains the preliminary reference procedure by which the ECJ can give guidance to national courts on the interpretation of the Treaties and the validity and interpretation of acts.

[59]   The relevant national provisions imposed conditions, inter alia, as to the financial solvability and reliability of the relevant transport undertaking.

[60]   Case C-280/00 *Altmark* (n 13) para 67.

The ECJ had earlier decided to reopen the oral procedure in the *Altmark* case to give the parties to the main proceedings (the Member States, Commission and Council) the opportunity to submit observations on the possible consequences of the (just delivered) *Ferring* ruling for the answer to be given to the Bundesverwaltungsgericht's question in *Altmark*.[61] In this regard, Altmark Trans, the Magdeburg regional government, Nahverkehrsgesellschaft and the German and Spanish governments had argued that the ECJ in *Altmark* had to confirm the principles of the *Ferring* judgment.[62] They considered state financing of public services to constitute aid within the meaning of ex Article 92(1) TEEC only if the advantages conferred by the public authorities exceeded the cost incurred in discharging the public service obligations.[63]

On the other hand, the Danish, French, Netherlands and UK governments had maintained that the ECJ had to adopt the approach proposed by AG Jacobs in his opinion in the *GEMO* case ('quid pro quo approach').[64] Where there was a direct and manifest link between state financing and clearly defined public service obligations, the financial support did not constitute state aid within the meaning of ex Article 92(1) TEEC. By contrast, where there was no such link or the public service obligations were not clearly defined, the public financial support would constitute state aid.

While admitting that the Bundesverwaltungsgericht's question particularly concerned the second of the four conditions under ex Article 92(1) TEEC (liable to affect trade between Member States), the ECJ held that for a state measure to be able to come under that provision, it also had to be capable of being regarded as an advantage conferred on the recipient undertaking (third condition under Article 107(1) TFEU). In this regard, the ECJ referred to, inter alia, the *ADBHU* and *Ferring* cases.[65] According to the ECJ, it followed from these judgments that, where a state measure must be regarded as compensation for public services, so that the undertakings obliged to provide those services do not enjoy a real financial advantage and the measure thus does not have the effect of putting them in a more favourable competitive position than the undertakings competing with them, such a measure would not be caught by ex Article 92(1) TEEC.[66]

---

[61] Ibid para 70.
[62] Ibid para 71.
[63] Ibid.
[64] Case C-280/00 *Altmark* (n 13) para 73.
[65] Ibid paras 85 and 86.
[66] Ibid para 87.

The ECJ held that for such compensation to escape classification as state aid, four (cumulative) conditions had to be satisfied.[67] First, the recipient undertaking is required to discharge public service obligations and those obligations have been clearly defined. Second, the parameters on the basis of which the compensation is calculated must have been established beforehand in an objective and transparent manner. Third, the compensation must not exceed what is necessary to cover all or part of the costs incurred in the discharge of public service obligations, taking into account the relevant receipts and a reasonable profit for discharging those obligations.

Finally, where the undertaking which is to discharge public service obligations is not chosen in a public procurement procedure, the level of compensation needed must be determined on the basis of an analysis of the costs which a typical undertaking, well-run and adequately provided with means of transport so as to be able to meet the necessary public service requirements, would have incurred in discharging those obligations, taking into account the relevant receipts and a reasonable profit for discharging the obligations. According to the ECJ, any measure not complying with one or more of these conditions is to be regarded as state aid within the meaning of ex Article 92(1) TEEC.[68]

9.4.2 INTERPRETATION OF THE RULING

It can be argued that the ECJ essentially confirmed the compensation approach[69] adopted in *Ferring* in the *Altmark* judgment: a mere compensation by the state for a public service obligation does not constitute state aid within the meaning of Article 107(1).[70] Yet, the ECJ's ruling in Altmark is generally seen as a compromise or even hybrid between the compensation and state aid[71] approaches.[72] While principally maintaining the approach in *Ferring*, it set two

---

[67] Ibid paras 89–93 and 95.

[68] Case C-280/00 *Altmark* (n 13) para 94.

[69] See n 42.

[70] See Erika Szyszczak, 'Financing Services of General Economic Interest' (2004) 67 Modern Law Review 982, 989, and Winter (n 28) 503. See also Lars-Hendrik Röller and Oliver Stehmann, 'Grenzen der Wettbewerbspolitik bei der Öffnung von Netzwerkindustrien' (2006) 7 Perspektiven der Wirtschaftspolitik 355, 366, and Alan Boyd and Joanne Teal, 'Interpreting the Altmark Decision: The Challenges from a Private Practitioner's Perspective' 3 <www.mcgrigors.com/pdfdocs/pl_state_aid_paper.pdf> accessed 5 March 2012.

[71] See n 52.

[72] See Sinnaeve (n 22) 362; Frederic Louis and Anne Valery, 'State Aid and the Financing of Public Services: A Comment on the Altmark Judgment of the Court of Justice' (2003) Competition Law Insight 3, 5; Michael Dougan, 'Legal Developments' (2004) 42 Journal of Common Market Studies 77, 91; Sandro Santamato and Nicola Pesaresi, 'Compensation for Services of General Economic Interest: Some Thoughts on the Altmark Ruling' (2004) European Commission Competition Policy Newsletter 2004–1, 17, 17 <http://ec.europa.eu/

(vague and broad)[73] additional criteria to accommodate concerns over too broad a discretion for Member States in financing services of general economic interest.[74] When the *Altmark* judgment was handed down, the resulting four criteria were anticipated to cause many state measures granted to offset the costs of public service obligations to qualify as compensation caught under Article 107(1).[75] The *Altmark* ruling was expected to make it more difficult for public authorities to grant state aid to undertakings entrusted with public service obligations.[76]

According to Santamato and Pesaresi, the *Altmark* criteria lead to the identification of three categories of cases: (1) the compensation is limited to the extra costs of an efficient operator; (2) the compensation does not exceed the extra costs of the (inefficient) recipient/provider and (3) the compensation exceeds the extra costs of the provider.[77] These categories of cases would respectively constitute non-aid, (possibly) compatible aid and incompatible aid.[78] Santamato and Pesaresi have argued that this framework differs from the *Ferring* case, where only the two extreme categories were foreseen: either non-aid or incompatible aid.[79]

There is a strong parallel between the *Altmark* criteria and Article 106(2) TFEU (case law).[80] The first and third *Altmark* criteria seem to be based on the latter

---

competition/publications/cpn/2004_1_17.pdf> accessed 28 February 2012; Leigh Hancher, 'The Application of EC State Aid Law to the Energy Sector' in Christopher W Jones (ed), *EU Energy Law* (Claes & Casteels 2005), 527, 618; Christopher Bovis, 'Public Service Obligations in the Transport Sector: The Demarcation between State Aids and Services of General Interest in EU Law (2005) 6 European Business Law Review 1329, 1345; Hancher, Ottervanger and Slot (n 23) 85; Quigley (n 3) 160; Thouvenin (n 36) 107; Van Bael and Bellis, *Competition Law of the European Community* (5th edn, Kluwer Law International 2010) 950; Sara Lembo and Maria Luisa Stasi, 'The Application of the Altmark Test to State Financing of Public Services in the Maritime Transport Sector' (2010) 9 European State Aid Law Quarterly 853, 853; Leigh Hancher and Pierre Larouche, 'The Coming of Age of EU Regulation of Network Industries and Services of General Economic Interest' in Paul Craig and Gráinne de Búrca (eds), *The Evolution of EU Law* (Oxford University Press 2011) 743, 760.

[73] See Sinnaeve (n 22) 362. Interestingly, Winter has argued that the four *Altmark* criteria are precise. See Winter (n 28) 503. According to Bovis, the four criteria are ambiguous as they represent the hybrid link between the compensation and the quid pro quo approach. See Bovis (n 72) 1346.

[74] These are the second and the fourth *Altmark* criteria.

[75] See, for instance, Sinnaeve (n 22) 359; Phedon Nicolaides, 'Compensation for Public Service Obligations: The Floodgates of State Aid? (2003) 24 European Competition Law Review 561, 572; Hancher (n 72) 565; and Louis and Valery (n 72) 5. In section 9.6, the post-*Altmark* decision practice and case law is examined in relation to the fourth *Altmark* criterion.

[76] See e.g. Nicolaides, 'Compensation for PSOs' (n 75) 572.

[77] Santamato and Pesaresi (n 72) 18.

[78] Ibid.

[79] Ibid.

[80] See also Sinnaeve (n 22) 357 and Hancher and Larouche (n 72) 760.

provision.[81] The first *Altmark* criterion, which requires the existence of a clear definition of the framework within which public service obligations and services of general economic interest have been entrusted to the beneficiary of compensatory payments, is consistent with Article 106(2) case law, which requires an express act by the public authority to assign services of general economic interest.[82] The third criterion, that the compensation must not exceed what is necessary to cover the cost incurred in discharging services of general economic interest or public service obligations, is compatible with the proportionality test applied in Article 106(2).[83] By contrast, the second and fourth criteria introduce new standards that appear to make the *Altmark* test more severe than that under Article 106(2).[84]

Winter has noted that if the *Altmark* criteria are not met, there seems to be no way to apply the derogation of Article 106(2).[85] Payments which are disproportionate ostensibly are not needed for the discharge of a public service obligation.[86] Similarly, if the recipient undertaking has not been properly entrusted with a task of public interest Article 106(2) could not be invoked anyway.[87] In any event, it seems clear from *Ferring* – as the Court in *Altmark* did not go back on this part of the *Ferring* ruling – that Article 106(2) may not be relied on to save state financing that would over-compensate a provider for discharging its public service obligations.[88]

9.5 THE (FOURTH) *ALTMARK* EFFICIENCY CRITERION

9.5.1 INTERPRETING THE *ALTMARK* EFFICIENCY CRITERION

The fourth *Altmark* criterion is often referred to as the 'efficiency criterion'.[89] As indicated above, it requires Member State authorities to ensure that the undertaking which is to discharge public service obligations is chosen pursuant to a public procurement procedure or alternatively to determine the level of

---

81 Hancher (n 72) 564.
82 Bovis (n 72) 1345. See also Szyszczak (n 70) 989.
83 Ibid.
84 Hancher and Larouche (n 72) 760, Bovis (n 72) 1345, and Szyszczak (n 70) 989.
85 Winter (n 28) 503.
86 Ibid.
87 Ibid.
88 Louis and Valery (n 72) 6.
89 See e.g. Case C-399/08 P *Deutsche Post* [2010] ECR I-07831, Opinion of AG Jääskinen para 55; Sinnaeve (n 22) 358; Röller and Stehmann (n 70) 366; and Hancher and Larouche (n 72) 762.

compensation needed on the basis of an analysis of the costs of a typical, well-run undertaking adequately provided with the means to be able to meet the necessary public service obligations, taking into account the relevant receipts and a reasonable profit. The wording of the fourth criterion in the *Altmark* ruling reveals the ECJ's preference for public procurement procedures/public tenders for entrusting public service obligations.[90] The underlying idea appears to be that when a public service is supplied at the least societal cost, compensation granted for discharging that service does not constitute state aid.

Yet the efficiency criterion as formulated in *Altmark* raises many questions. First, the ECJ has not given any guidance as to the design and content of the required public procurement procedure.[91] More specifically, it is not clear what is meant by the relevant public services being provided 'at the least cost to the community'. Could this, for instance, also include external (e.g. environmental) costs?[92] Can it be presumed that the services are provided at the least cost if EU public procurement legislation is respected?[93] Does the phrase refer to the cost of the most efficient producer or to the best offer available on the market?[94] The two are not necessarily the same and a tender procedure would arguably ensure the achievement of the latter only.[95]

Accordingly, a public procurement procedure would not necessarily lead to the public service obligation being provided at the least cost to society.[96] If so, would that mean that compensation granted for the discharge of the relevant public service obligation is to be characterised as state aid within the meaning of Article 107(1)? The ECJ does not indicate whether it is the instrument of public procurement or the least-cost element that is decisive. The fact that an alternative (compensation to be determined on the basis of a benchmark) is offered would seem to suggest that the latter, rather than the former, is key.

Likewise, the alternative approach of determining the level of compensation on the basis of a benchmark undertaking is far from straightforward. The obvious question is what to understand by a 'typical undertaking' 'well run and

---

[90] See also Nicolaides, 'Compensation for PSOs' (n 75) 573 and Hancher and Larouche (n 75) 763.
[91] Sinnaeve (22) 357.
[92] Sinnaeve (22) 357.
[93] Ibid.
[94] Santamato and Pesaresi (n 72) 19.
[95] Ibid. In this regard, Santamato and Pesaresi have argued that it is an acknowledged result in economics that a tender may lead to very different results depending on how it is designed and that, at any rate, a tender does not necessarily lead to the provision of a service at the least possible cost.
[96] In this regard, see also Nicolaides and Kleis (n 13) 623–24.

adequately provided' with the means to carry out its tasks.[97] As noted by Sinnaeve, in many cases, it will in practical terms hardly be feasible to make the required comparison with a hypothetical reference undertaking.[98] It is almost inherent to the concept of public services that the undertaking providing the service will often be in a situation which is very different from that of other undertakings.[99] Precisely because of the characteristics of these services, the typical undertaking, well run and adequately provided with the means to meet the necessary public service requirements, will often not exist meaning that no appropriate comparison can be made.[100] According to *Altmark*, state financing would then be deemed to constitute state aid, assuming that no public procurement procedure took place.[101]

Furthermore, what happens when the costs of the beneficiary undertaking are lower than those of potential competitors in some years and higher in others?[102] According to Nicolaides, a benchmark period will have to be established.[103] Yet it is unclear how long such period should be.[104] What happens when an undertaking has higher (than the benchmark) costs, but then commences restructuring to reduce its costs and upgrade its services during the benchmark period?[105] Should it be excluded from compensation?[106] What is more, when the costs of the public service provider are higher than those of a hypothetical market operator, the relevant measure would, according to the logic of the efficiency criterion, constitute state aid under Article 107(1). Sinnaeve has argued that it would not seem possible to authorise such overcompensation under Article 106(2) and that Article 107(3) would not seem to provide a clear basis of

---

[97]   Nicolaides, 'Compensation for PSOs' (n 75) 573. See also Hancher and Larouche (n 72) 763.

[98]   Sinnaeve (n 22) 358. Posser has argued that costs of a notional efficient undertaking will often be very difficult to calculate, particularly where there is no similar competitor whose costs can be calculated. See Tony Posser, ' 'Competition Law and Public Services: From Single Market to Citizenship Rights?' (2005) 11 European Public Law 543, 555.

[99]   Sinnaeve (n 22) 358.

[100] Sinnaeve (n 22) 358. In this regard, Santamato and Pesaresi refer to the 'more realistic [in comparison with a hypothetical example given – MH] situation in which there are only few potential providers of a certain service. In those circumstances a tender or even a survey of typical costs would not prove very useful, but skipping those procedures leads to a finding of aid. Can we really say that in all those circumstances there is advantage for the recipient?'. See Santamato and Pesaresi (n 72) 19.

[101] Sinnaeve (n 22) 358.

[102] Nicolaides, 'Compensation for PSOs' (n 75) 573.

[103] Ibid.

[104] Ibid.

[105] Ibid.

[106] Ibid. Nicolaides has argued that these questions suggest that the benchmark defined by the ECJ is essentially 'backward-looking' whereas the tender and other open selection procedures are 'forward-looking'. According to Nicolaides, the latter method of selecting undertakings is based on firmer economic foundations because what matters are the expected future costs of delivering a service.

---

compatibility either.[107] Yet in *Altmark*, the ECJ did not indicate whether and on what basis such aid can be found to be compatible with the common market.[108] In this regard, Röller and Stehmann have raised the question of whether the efficiency criterion should not play a greater role in assessing the compatibility of state aid.[109]

Further questions are raised by the notion of 'reasonable profit'. According to Santamato and Pesaresi, in the case of services of general economic interest, the ability to produce at lower cost (which generates higher profits) is often linked to the control over networks developed by the authorities or already amortised thanks to the granting of exclusive rights.[110] Can the economic rents associated with those networks be accepted as reasonable profits? Santamato and Pesaresi have argued that this question must be answered in the negative.[111] However, that would imply that even in cases where services of general economic interest are attributed through a tender, the third *Altmark* criterion (no overcompensation) might not be met and that the compensation is an incompatible aid that should have been declared.[112]

9.5.2 JOINED CASES C-83/01 P, C-93/01 P AND C-94/01 P *CHRONOPOST*

Of great relevance to the (meaning of the) *Altmark* efficiency criterion is the *Chronopost* case,[113] which was decided shortly before *Altmark*. The case concerned appeals by French postal company Chronopost, the French postal office (La Poste) and France against the judgment of the CFI in *Ufex*.[114] In the latter case, the CFI had in part annulled Article 1 of Commission Decision 98/365/EC[115] concerning alleged state aid granted by France to SFMI-Chronopost. According to the CFI, the Commission had based its decision on an incorrect interpretation of ex Article 92 TEEC by ruling out the existence of state aid without checking whether the remuneration received by La Poste for the provision of commercial and logical assistance to SFMI-Chronopost corresponded to the price that would have been asked under normal conditions.[116]

---

[107] Sinnaeve (n 22) 358. Sinnaeve has noted that when the difference in costs cannot be authorised under the state aid rules, this would de facto result in a liberalisation of all public services through the backdoor.
[108] Ibid.
[109] Röller and Stehmann (n 70) 367.
[110] Santamato and Pesaresi (n 72) 19.
[111] Santamato and Pesaresi (n 72) 19.
[112] Ibid.
[113] Joined Cases C-83/01 P, C-93/01 P and C-94/01 P *Chronopost* [2003] ECR I-06993.
[114] Case T-613/97 *Ufex* [2000] ECR II-04055.
[115] *SFMI-Chronopost* Commission Decision 98/365/EC [1998] OJ L164.
[116] Case T-613/97 *Ufex* (n 114) para 76.

The first plea submitted by Chronopost, La Poste and France turned on the concept of 'normal market conditions' used in the *SFEI* judgment[117] to determine the circumstances in which the provision of logistical and commercial assistance by a public undertaking to its subsidiaries carrying on an activity open to competition is capable of constituting state aid.[118] According to the appellants, the CFI had misinterpreted the concept of 'normal market conditions'.[119] They argued that the CFI, in referring to a private undertaking 'not operating in a reserved sector', had erred in basing its comparison on an undertaking that was structurally different from La Poste, instead of comparing the conduct of the latter with that of an undertaking in the same position (with a reserved sector at its disposal).[120]

The ECJ held that the CFI – in ruling that the Commission should have at least checked that the payment received by La Poste was comparable to that demanded by a private holding company or a private group of undertakings not operating in a reserved sector – had failed to take account of the fact that an undertaking such as La Poste was in a situation very different from that of a private undertaking acting under normal market conditions.[121] According to the ECJ, in the absence of any possibility of comparing the situation of La Poste with that of a private group of undertakings not operating in a reserved sector, 'normal market conditions' had to be assessed by reference to the objective and verifiable elements which were available.[122] It stated that the costs borne by La Poste in respect of the provision to its subsidiary of logistical and commercial assistance could constitute such objective and verifiable elements.[123]

Accordingly, there was no question of state aid to SFMI-Chronopost if, first, it was established that the price charged properly covered all the additional, variable costs incurred in providing the logistical and commercial assistance[124] and if, second, there was nothing to suggest that those elements had been underestimated or fixed in an arbitrary fashion.[125] Subsequently, the ECJ set aside the CFI judgment in *Ufex* and referred the case back to the CFI.

---

[117] Case C-39/94 *SFEI* [1996] ECR I-03547.
[118] Joined Cases C-83/01 P, C-93/01 P and C-94/01 P *Chronopost* (n 113) para 16.
[119] Ibid para 18.
[120] Joined Cases C-83/01 P, C-93/01 P and C-94/01 P *Chronopost* (n 113) para 19.
[121] Joined Cases C-83/01 P, C-93/01 P and C-94/01 P *Chronopost* (n 113) para 33.
[122] Ibid para 38.
[123] Ibid para 39.
[124] The ECJ called this 'an appropriate contribution to the fixed costs arising from use of the postal network and an adequate return on the capital investment in so far as it is used for SFMI-Chronopost's competitive activity'. See para 40 of its ruling.
[125] Joined Cases C-83/01 P, C-93/01 P and C-94/01 P *Chronopost* (n 113) para. 40.

In its second ruling (handed down five years after the *Altmark* judgment), the CFI reiterated that in the absence of any possibility of comparing the situation of La Poste with that of a private group of undertakings not operating in a reserved sector, 'normal market conditions', which are necessarily hypothetical, must be assessed by reference to the objective and verifiable elements available.[126] It held that, in those circumstances, the Commission should not, at first sight, be criticised for having based the contested decision on the only data available at the time, from which it was possible to reconstruct the costs incurred by La Poste.[127] The use of those data could be open to criticism only if it was established that they were based on manifestly incorrect considerations.[128]

9.5.3 INTERPRETING *CHRONOPOST*

As argued by AG Sharpston in her Opinion in the second *Chronopost* ruling,[129] the ECJ's test in *Chronopost* appears to be general in nature. It lays down the approach to be taken in order to assess whether the provision of commercial and logistical assistance involves state aid.[130] It does not specify the economic, accounting or financial standards to be applied.[131] In requiring 'all' additional, variable costs to be included, it does not indicate which costs should be considered variable.[132] Nor does it state what would constitute an 'appropriate' contribution to fixed costs or an 'adequate' return on the capital investment.[133]

Different views have been expressed as to the consequences of the *Chronopost* rulings for the *Altmark* judgment and its efficiency criterion. Both Szyszczak and Sinnaeve, for instance, have argued that the two rulings are contradictory.[134] According to Szyszczak, *Chronopost*, on the one hand, suggests that public service obligations are subject to special non-market rules, while *Altmark*, on the other, suggests that public service obligations should not be given any special privileges when they have not been subject to a public tender.[135] By contrast, Bartosch has noted that even though the *Chronopost* and *Altmark* rulings, at first sight, appear to be conflicting, there is no real conflict between the two judgments.[136] On the one hand, *Altmark* sets the general framework under

---

[126]   Joined Cases C-341/06 P and C-342/06 P *Chronopost* [2008] I-04777, para 148.
[127]   Ibid para 149.
[128]   Ibid para 150.
[129]   Joined Cases C-341/06 P and C-342/06 P *Chronopost*, Opinion of AG Sharpston, para 93.
[130]   Ibid.
[131]   Ibid.
[132]   Ibid.
[133]   Ibid.
[134]   Sinnaeve (n 22) 358 and Szyszczak (n 70) 990–91.
[135]   Szyszczak (n 70) 991.
[136]   Andreas Bartosch, 'Clarification or Confusion? How to Reconcile the ECJ's Rulings in Altmark and Chronopost?' (2003) CLaSF Working Paper Series Number 2, 15 <www.clasf.

which measures intended to offset the costs linked to public service fulfilment may escape the prohibition laid down by Article 107(1) TFEU.[137] On the other hand, *Chronopost* shows the limitations that the application of this general framework has in cases where for the services provided no market exists and, consequently, no comparable private operators can be found the cost structures of whom can be used as suitable benchmarks.[138] The argument appears to be that *Chronopost* stands for a 'lex specialis' which must be applied when no market exists for the service provided and consequently no comparable operator can be found as a suitable comparator.[139]

Buendia Sierra has argued that the interpretation of *Chronopost* as allowing for the benchmark comparison to be undertaken on the basis of the costs of the operator itself when no comparable operator actually exists is hard to reconcile with the logic behind *Altmark*.[140] According to Buendia Sierra, an ideal benchmark can normally be deduced from similar sectors and if this is not possible, this cannot mean that the test is considered to be fulfilled; the *Altmark* test is an exception to the principle that covering all the costs of an undertaking plus a reasonable profit is, normally speaking, an advantage.[141] As such, it cannot be interpreted in such a flexible way.[142]

Tempting as it may be to see *Chronopost* as a lex specialis to *Altmark*, I share Buendia Sierra's interpretation of the former ruling. It indeed seems very difficult to reconcile the lex specialis interpretation with the logic behind *Altmark*. The ECJ's ruling in *Altmark* was a response to the criticism that the compensation approach of *Ferring* would lead to inefficiencies being subsidised. The ECJ took a middle position between the compensation and the state aid approaches. It maintained the thrust of *Ferring* that compensation for services of general economic interest in principle does not constitute state aid, but adopted the stringent requirement that such compensation should not lead to inefficiencies being subsidised. Taking the public service provider's own costs as a point of departure for determining the level of compensation does not appear to be consistent with this approach.

Considering the likelihood of there not actually being a reference undertaking in relation to many services of general economic interest, this reading of *Chronopost*

org/assets/CLaSF%20Working%20Paper%2002.pdf> accessed 3 March 2012. See also Thomas Müller, 'Efficiency Control in State Aid and the Power of Member States to Define SGEIs' (2009) 8 European State Aid Law Quarterly 39, 41.
137 Bartosch, 'Clarification or Confusion' (n 136) 16.
138 Ibid.
139 Hancher and Larouche (n 72) 764.
140 Buendia Sierra (n 17) 199–200.
141 Ibid 200.
142 Ibid.

would probably cause compensation for those services to often automatically pass the *Altmark* efficiency test when no public tender procedure has been conducted. This, arguably, is not what the ECJ had in mind when it gave its *Altmark* ruling, which, as a matter of fact, it did some three weeks after delivering the *Chronopost* judgment.

What is more, following the lex specialis interpretation, *Chronopost* would seem to rule out the comparison with a hypothetical benchmark undertaking. If no real benchmark undertaking exists, the costs of the public service provider can be taken as a point of departure for determining the level of compensation. By contrast, the wording of the efficiency criterion does not appear to exclude the determination of the appropriate level of compensation on the basis of the costs of a hypothetical undertaking. On the contrary, by referring to the costs which a typical undertaking would have incurred, it seems to hint at the use of a hypothetical benchmark undertaking for determining the appropriate level of compensation for a service of general economic interest. The lex specialis reading of the *Chronopost* ruling would, therefore, appear to clash with the idea behind the efficiency criterion.

Considering the above, I would argue that *Chronopost* should be seen as no more than an illustration of the difficulties in applying the ideas behind the *Altmark* efficiency criterion in practice. In reality, it might often not be possible to find a benchmark efficiency undertaking. *Chronopost* illustrates that the European Courts may, in such circumstances, be tempted to deviate from the abstract approach of *Altmark* and take the costs of the public service provider as a point of departure for determining the 'efficient' level of compensation.

9.6 THE EFFICIENCY CRITERION IN POST-*ALTMARK* DECISION PRACTICE AND CASE LAW[143]

9.6.1 THE COMMISSION'S POST-*ALTMARK* DECISION PRACTICE ON INFRASTRUCTURE DEVELOPMENT[144]

To date, there have not been too many cases in which the Commission applied the *Altmark* efficiency criterion to the public funding of infrastructure

---

[143] I do not claim to give an exhaustive overview of the post-*Altmark* decision practice and case law of the Commission and Courts respectively. Rather, I try to give an overview of some of the most important issues that have arisen in the Commission's decision practice and the Courts' case law, with particular attention for the public financing of infrastructure development.

[144] So far, there have been a large number of cases in which the Commission applied the fourth *Altmark* criterion. I will, however, focus on the cases where the Commission applied the criterion in relation to the public funding of infrastructure development.

development.[145] The cases in which it did mainly dealt with the development of infrastructure for broadband internet[146] and digital terrestrial television services in rural areas. A number of these cases provide interesting insights into the Commission's application of the *Altmark* efficiency criterion in relation to the public financing of infrastructure development.

9.6.1.1 Case N 475/2003 CADA

In case N 475/2003 (*CADA*), the Commission decided on public support for the construction of new electricity generation capacity in Ireland.[147] The support was granted through long-term Capacity and Differences Agreements (CADA) that would be granted to generators undertaking the construction of new generation capacity. These agreements provided that generators received capacity payments based on their availability capacity. The Commission for Energy Regulation (CER), the Irish electricity and gas regulator, had organised a selection process in order to choose the operators of the new CADA-supported plants.

In relation to the fourth *Altmark* criterion, the Commission noted that the CADA would be awarded by means of a transparent competitive process, ensuring that the lowest price would be delivered and the fourth *Altmark*

---

145   A search in the Commission's search engine for state aid cases by (a) primary objective (services of general economic interest), (b) EU primary legal basis (Article 106(2)), and (c) all NN, C and N cases (by means of case description) results in the following cases: *Cumbria Broadband* (Case N 282/2003) Commission Decision of 10 December 2003 C (2003) 4480 fin; *Irish electricity generation capacity* (Case N 475/2003) Commission Decision of 16 December 2003 C (2003) 4488 fin; *Télécommunication Pyrénées-Atlantiques* (Case N 381/2004) Commission Decision of 16 November 2004 C (2004) 4343 fin; *Dorsal* (Case N 382/2004) Commission Decision of 3 May 2005 C (2005) 1170 fin, *DVB-T Bavaria* (Case C 33–2006) Commission Decision of 19 July 2006 C (2006) 3175 final; *DVB-T North-Rhine Westphalia* (Case C 34/2006) Commission Decision of 23 October 2007 C (2007) 5109 final; *Sociétés concessionaires d'autoroutes* (Case N 362/2009) Commission Decision of 17 August 2009 C (2009) 6506 final; *Très haut debit Hauts-de-Seine* (Case N 331/2008) Commission Decision of 30 September 2009 C (2009) 7426 final; *EstWin project* (Case N 196/2010) Commission Decision of 20 July 2010 C (2010) 4943 final; *DTT Castilla-La-Manchia* (Case CP 24/2010) Commission Decision of 29 September 2010 C (2010) 6467 final; *Spain Digital Terrestrial Television* (Case C 23/2010) C (2010) 6465 final; and *Projet T3 Est Parisien* (Case N 630/2009) Commission Decision of 26 January 2011 C (2011) 259 final.

146   There have been several state aid cases on public financing of broadband infrastructure, but only in a few of those the fourth *Altmark* criterion was applied. For an overview of cases on the public funding of infrastructure and the requirement of open, transparent and non-discriminatory tender procedures outside the framework of the *Altmark* efficiency criterion (and compensation of services of general economic interest), see Nicolaides and Kleis (n 13) 620–22. These cases will not be discussed here, but I will occasionally refer to the findings of these cases.

147   *Irish electricity generation capacity* (n 145).

criterion would be met.[148] In this regard, it mentioned several characteristics of the tender process organised to select the operators of the new power plants:

- it was launched by the independent Irish energy sector regulator;
- the public authorities had no discretional margin as to the choice of the winners;
- a notice was published in the Official Journal;
- first phase (technical) selection was based on transparent and objective criteria defined beforehand;
- second phase (price) selection was based on an objective, transparent and defined beforehand method; and
- competition had also attracted bids from generators outside Ireland.[149]

As all four *Altmark* conditions were met, the scheme involved no state aid to the generators within the meaning of Article 107(1) TFEU.[150]

9.6.1.2   Cases N 381/2004, N 382/2004 and N 331/2008 French broadband cases

In three cases on the development of broadband internet networks in rural areas in France, the Commission extensively examined the compatibility of the proposed public financing with the *Altmark* efficiency criterion.[151] These cases all dealt with the construction of broadband internet network infrastructure, to be used by telecoms operators for providing broadband internet services to residential users, businesses and public authorities. In all three cases, the construction and operation of the broadband infrastructure would be implemented by means of a public service obligation in the form of a concession under French law. Selected through a public procurement procedure, the concession holder would act as a wholesale operator providing various wholesale services to operators but not offering services to end users.[152]

The public service provider was selected using a two-phase procedure. In the first phase, the French authorities had conducted studies to examine the needs of the relevant region, determine the characteristics of the required public service

---

[148]   Ibid paras 57 and 64.
[149]   Ibid paras 58–63.
[150]   *Irish electricity generation capacity* (n 145) para 65.
[151]   *Télécommunication Pyrénées-Atlantiques, Dorsal,* and *Hauts-de-Seine* (n 145).
[152]   On the essence of these schemes see also Commission, 'Commission Approves Public Funding of Broadband Projects in Pyrénées-Atlantiques, Scotland and East Midlands' IP/04/1371, <http://europa.eu/rapid/press-release_IP-04-1371_en.htm> accessed 26 October 2012; Commission, 'State Aid: Commission Endorses' (n 15) and Commission, 'Commission Approves Public Financing Worth € 59 Million for Broadband Project in the French Hauts-de-Seine Department', IP/09/1391 <http://europa.eu/rapid/press-release_IP-09-1391_en.htm> accessed 12 December 2012.

and estimate the costs of the project and the required share of public finance. In this phase, aspects such as geographic coverage and the quality of the service provided were taken into account. In the second phase, a public procurement procedure was launched, in which the remaining candidates competed for the concession. Part of the public procurement procedure consisted of negotiations[153] with the potential providers of the service of general economic interest.[154] In this regard, the Commission noted in the *Pyrénées-Atlantiques* case that the characteristics of the public service in question did not allow for all obligations being determined beforehand and for a selection on the basis of price alone.[155]

In all three cases, the Commission concluded that the specific procedure led to the most efficient public service provider being selected and, accordingly, to the efficiency criterion being met. In the *Pyrénées-Atlantiques* and *Limousin* cases, the Commission argued that the procedure followed by the French authorities allowed for a certain amount of flexibility, the (pre-)selection of the public service provider being based on qualitative criteria and the final selection being based on the quantitative criteria of the (lowest) amount of public support required and the level of compensation to be granted.[156]

It is important to place the above findings in the context of the specific circumstances of these three cases. In its press releases on the *Pyrénées-Atlantiques* and *Limousin* cases, the Commission pointed to two important aspects in this regard. First, it underlined that the qualification of access to broadband services as a service of general economic interest was only valid for investments linked to network infrastructures and not for broadband services offered to the end user.[157] Accordingly, the public co-funding of the infrastructure could constitute compensation for the provision of a service of general economic interest. Second, the Commission indicated that it had taken into account the fact that the regions concerned consisted mainly of rural and remote areas and that the broadband services offered by existing market operators were insufficient to meet the essential needs of the people of those regions.[158]

---

[153]   Article 1(11)(d) of Directive 2004/18/EC defines 'negotiated procedures' as 'those procedures whereby the contracting authorities consult the economic operators of their choice and negotiate the terms of contract with one or more of these'.

[154]   In this regard, see also See Phedon Nicolaides, 'State Aid, Advantage and Competitive Selection: What is a Normal Market Transaction?' (2010) 9 European State Aid Law Quarterly 65, 68. Nicolaides has argued that 'the Commission Decision on the London Underground was path-breaking because it was the first time that state aid rules were applied to a public-private partnership and, more importantly, because the Commission explained that there can be no state aid even when the successful company is selected through a negotiated procedure'.

[155]   *Télécommunication Pyrénées-Atlantiques* (n 145) para. 84.

[156]   Ibid para 86 and *Dorsal* (n 145) para 77.

[157]   Commission, 'Commission Approves Public Funding' (n 152) and Commission, 'State Aid: Commission Endorses' (n 16).

[158]   Ibid.

9.6.1.3   Case C 35/2005 Breedbandnetwerk Appingedam

The importance of the latter point is illustrated by another case on the public financing of the development of broadband internet infrastructure.[159] *Breedbandnetwerk Appingedam* dealt with the public funding of a glass fibre network in the Dutch town of Appingedam. In addition to the 'regular' broadband internet services provided by two undertakings, the municipality of Appingedam wanted the town to be supplied with advanced (faster) broadband services.[160] Since the two broadband providers had shown no interest to invest in the required glass fibre network on market terms, the municipality had decided to grant public support for the development of such infrastructure.[161]

The planned network was to consist of two layers. The passive layer (rights of way, ducts, fibres, etc.) would be owned by a public foundation set up and controlled by the municipality.[162] On the basis of the passive layer, an active network would be operated which would provide wholesale services to providers, in turn offering broadband services to end users (households and businesses).[163] The active layer would be owned by an entity set up by private investors. The municipality would tender out the concession to the active layer, including the operation of the network.[164]

Even though the Dutch authorities had not explicitly invoked the existence of a service of general economic interest, the Commission did assess whether the relevant glass fibre network represented a public service.[165] In this regard, the Commission noted that the envisaged contractual relationship between the operator of the active layer and the municipality/foundation reflected a classical private-public partnership rather than the entrustment and implementation of a service of general economic interest.[166] The Commission argued that unlike, for instance, the *Pyrénées-Atlantiques* case, neither the foundation nor the operator had a clear public service mandate to enable broadband access to the general public, citizens and businesses, in rural and remote areas where no other operator provided ubiquitous and affordable broadband access.[167] In the latter case, the direct objective of the measure was to enable access to broadband services through a wholesale network to the general public in a region with limited broadband

---

[159]   *Breedbandnetwerk Appingedam* (Case C 35/2005) Commission Decision of 19 July 2006 C (2006) 3226 def.
[160]   Ibid para 7.
[161]   Ibid.
[162]   *Breedbandnetwerk Appingedam* (n 159) para 8.
[163]   Ibid para 9.
[164]   Ibid para 11.
[165]   Ibid para 38 and further.
[166]   Ibid para 40.
[167]   Ibid.

coverage.[168] According to the Commission, these conditions did not apply in Appingedam, where broadband services were already provided over two networks.[169]

Breedbandnetwerk Appingedam underlines the importance of the relevant infrastructure being used to provide a service of general economic interest.[170] If services similar to those that are to be provided by means of the planned infrastructure are already being supplied, the Commission will likely not be convinced of the public service character of the envisaged services. The *Altmark* route for having public support for infrastructure development escape state aid provisions will then be blocked.

9.6.1.4   Case N 362/2009 Sociétés concessionaires d'autoroutes

Sociétés concessionaires d'autoroutes concerned a one-year extension of motorway concessions to existing concession holders (private entities), in exchange for which the former were expected to invest in the upgrading and renovation of the motorway infrastructure under their concession.[171] Even though the case is not a classic example of direct public funding of infrastructure development, it is of relevance for the application of the *Altmark* efficiency criterion to public support for infrastructure development. This is due to the investment requirement imposed upon the concession holders, leading to the indirect public financing of infrastructure development.

In examining whether the extension of the concessions was to be seen as an advantage granted by state means within the meaning of ex Article 87(1) TEC, the Commission examined the measure's compatibility with the fourth *Altmark* criterion.[172] The beneficiaries of the envisaged French measure had not been selected by means of a public procurement procedure. According to the French authorities, the proposed funding nonetheless complied with the *Altmark* efficiency criterion as the level of compensation had been based on ratios commonly agreed by the undertakings in the sector.[173] The Commission, however, concluded that such was insufficient to meet the fourth *Altmark* criterion and declared the proposed measure to constitute an advantage within the meaning of ex Article 87(1) TEC.[174]

---

168   Ibid para 41.
169   Ibid.
170   In this regard, see also *EstWin project* (n 145), paras 42–56, and *Banda ancha* (Case N 583/2004) Commission Decision of 6 April 2005 C (2005) 981 fin, para 21.
171   *Sociétés concessionaires d'autoroutes* (n 145).
172   Ibid paras 33–36.
173   Ibid para 34. Para 34 speaks of 'ratios communément admis par les entreprises du secteur'.
174   Ibid paras 35–36.

9.6.1.5   Case N 630/2009 Projet T3 Est Pariesien

In similar vein, the Commission decided on the public funding of the construction of a city heating network in the northeast of Paris.[175] In addition to the envisaged grant to the Compagnie Parisienne de Chauffage Urbain (CPCU), the latter's existing concession for the provision of city heating services would be extended to 2024 by the municipality of Paris. In assessing the compatibility of the extension of CPCU's concession with the *Altmark* efficiency criterion, the Commission noted that the undertaking had not been selected by means of a public procurement procedure.[176] Furthermore, the French authorities had based the level of compensation granted for the public service of city heating solely on the costs borne by CPCU.[177] Accordingly, the fourth *Altmark* criterion had not been met.[178]

There are several things to take away from the above cases. First, in all of the four cases in which the Commission considered the public funding not to constitute state aid (*CADA* and three French broadband cases), public procurement procedures were followed. In this regard, it is important that the tender is EU-wide (and published in the Official Journal) and that the selection criteria are transparent, objective and defined beforehand. A tender procedure involving negotiations could be compatible with the *Altmark* efficiency criterion.[179] Furthermore, a public procurement procedure attracting bids from other Member States is seen as indicative of a competitive selection process. Also, Member States do not appear to be required to select public service providers on the basis of quantitative (financial) criteria only.

Qualitative criteria such as geographical coverage, the kind of technology used and reuse of existing infrastructure may be applied. Nevertheless, the use of such criteria must lead to a competitive tender and quantitative criteria (cost) must, in the end, be decisive. As underlined by the broadband internet cases, the Commission will have to be convinced of the public service character of the services which are to be provided through the relevant infrastructure. If not, the

---

[175]   *Projet T3 Est Parisien* (n 145).
[176]   Ibid para 37.
[177]   *Projet T3 Est Parisien* (n 145).
[178]   Ibid.
[179]   This seems to contrast with earlier cases in relation to state support for broadband infrastructure. See Hancher (n 71) 565. Hancher notes that 'a recent series of decisions in relation to state support for broadband infrastructure and related services are illustrative of the strict approach which the Commission appears to take to the application of the 'Altmark' criteria. It is particularly noteworthy that the Commission does not accept an invitation to tender and related procedure conducted on the basis of the negotiated procedure as provided for in EC Directive 92/57 in relation to public procurement of services, as sufficient to meet the standards required by the Altmark ruling'.

fourth *Altmark* criterion cannot be met. Finally, commonly agreed industry standards for compensation calculation do not guarantee an efficient supply of the specific service of general economic interest, nor do the costs borne by the public service provider itself.

9.6.2   THE COMMISSION'S 2011 ALTMARK PACKAGE

In 2011, the Commission adopted a new (legislative) package on compensation for services of general economic interest under EU state aid rules, also known as the second 'Altmark package'.[180] The package consists of two communications and a decision.[181] In December 2011, the Commission also made a proposal for a de minimis Regulation in the field of services of general economic interest, which entered into force in April 2012.[182] Commission Decision 2012/21/EU and the Communication on the framework for state aid in the form of public service compensation do not provide any guidance on the application of the fourth *Altmark* criterion and the requirements that have to be met for compensation for services of general economic interest not to constitute state; both deal with the compatibility of state aid in the form of public service compensation under EU state aid rules. However, the Communication on compensation granted for the provision of services of general economic interest and the proposed de minimis Commission Regulation could be of relevance for Member States seeking to have public service compensation escape classification as state aid under Article 107(1) TFEU.

---

180   See the Commission's website on legislation on services of general economic interest at <http://ec.europa.eu/competition/state_aid/legislation/sgei.html> accessed 28 March 2012. The first Altmark package consisted of: Commission decision of 28 November 2005 on the application of Article 86(2) of the EC Treaty to state aid in the form of public service compensation granted to certain undertakings entrusted with the operation of services of general economic interest 2005/842/EC [2005] OJ L312/67, Community framework for state aid in the form of public service compensation 2005/C 297/04 [2005] OJ C297/4, and Commission Directive 2006/111/EC of 16 November 2006 on the transparency of financial relations between Member States and public undertakings as well as on financial transparency within certain undertakings [2006] OJ L318/17.

181   These are: Communication from the Commission on the application of the European Union state aid rules to compensation granted for the provision of services of general economic interest 2012/C 8/02 [2012] OJ C8/4, Communication from the Commission European Union framework for state aid in the form of public service compensation 2012/C 8/03 [2012] C8/15, and Commission decision of 20 December 2011 on the application of Article 106(2) of the Treaty on the Functioning of the European Union to state aid in the form of public service compensation granted to certain undertakings entrusted with the operation of services of general economic interest 2012/21/EU [2012] L7/3.

182   See the Commission's website on services of general economic interest (n 182). This is the Commission Regulation of 20 December 2011 on the application of Articles 107 and 108 of the Treaty on the Functioning of the European Union to de minimis aid granted to undertakings providing services of general economic interest [2011] OJ C8/24.

As for the proposed de minimis Regulation, it is unlikely that aid granted for the development of CO_2 transport and storage infrastructure would fall within the scope of the proposed Regulation. Article 2 of the proposed Regulation determines that the proposed threshold for public service compensation not to constitute state aid within the meaning of Article 107(1) is €500,000 over any period of three fiscal years. This seems to be quite a low threshold. Even though the capture of CO_2 represents by far the largest cost component, the sums required for constructing CO_2 transport and storage infrastructure are still likely to be significant. It can therefore be questioned whether any aid granted for the development of such infrastructure will be able to meet the threshold in the proposed Regulation and escape classification as state aid within the meaning of Article 107(1) on this basis.

Of greater relevance for the application of the fourth *Altmark* criterion to public support for the development of CO_2 transport and storage infrastructure seems to be the Commission's Communication on compensation granted for the provision of services of general economic interest. It is important to note that the communication states that it is without prejudice to the relevant case-law of the European Courts. As they do not constitute one of the legal acts listed in Article 288 TFEU, communications are not legally binding. The communication does however indicate in what way the Commission intends to apply EU state aid rules to public service compensation.[183]

In section 3.6 of the communication, the Commission deals with the requirements under the fourth *Altmark* criterion. It notes that the simplest way for public authorities to meet the fourth *Altmark* criterion is to conduct an open, transparent and non-discriminatory public procurement procedure in line with EU public procurement legislation.[184] The Commission states that based on the case law of the ECJ, a public procurement procedure only excludes the existence of state aid where it allows for the selection of the tenderer capable of providing the service at the least cost to the community.[185] This is in line with the Commission's post-*Altmark* decision practice on the efficiency criterion, notably with the French cases on the development of broadband infrastructure discussed in section 9.6.1.

Concerning the characteristics of the tender procedure, the Commission notes that an open procedure[186] in line with the requirements of the public

---

[183] Commission, 'Communication on compensation granted for SGEIs' (n 181) para 3.
[184] Ibid para 63. The Commission refers European Parliament and Council Directive 2004/17/EC of 31 March 2004 coordinating the procurement procedures of entities operating in the water, energy, transport and postal services sectors [2004] OJ L134/1 and Directive 2004/18/EC.
[185] Commission, 'Communication on compensation granted for SGEIs' (n 181) para 65.
[186] Article 11(1)(a) of Directive 2004/18/EC and Article 1(9)(a) of Directive 2004/17/EC.

procurement rules is acceptable, but that a restricted procedure[187] can also satisfy the fourth *Altmark* criterion, unless interested operators are prevented from tendering without valid reasons.[188] The Commission considers a competitive dialogue[189] or a negotiated procedure with prior publication[190] sufficient to satisfy the fourth *Altmark* criterion in exceptional cases only.[191] According to it, the negotiated procedure without publication of a contract notice cannot ensure that the procedure leads to the selection of the tenderer capable of providing the relevant services at the least cost to the community.

In relation to the award criteria, the Commission notes that the lowest price[192] and the most economically advantageous tender[193] satisfy the efficiency criterion.[194] This provided that the award criteria, which can also be environmental or social, are closely related to the subject matter of the service provided and allow for the most economically advantageous offer to match the value of the market.[195] Again, this is in line with the findings of the French broadband cases discussed above, in which the Commission indicated that the award criteria may also be qualitative in nature, as long as they guarantee that the relevant service of general economic interest is provided at the lowest societal cost. In this regard, the Commission states that the awarding authority is not prevented from setting qualitative standards to be met by all economic operators or from taking qualitative aspects related to the different proposals into account in its award decision.[196]

Finally, the Commission states that there may be circumstances where a public procurement procedure cannot allow for the least cost to the community as it does not give rise to sufficiently open and genuine competition.[197] This could be the case, for example, because of the particularities of the service in question, existing intellectual property rights or necessary infrastructure owned by a particular service provider.[198] The Commission argues that in cases where only one bid is submitted the tender similarly cannot be deemed sufficient to ensure that the procedure leads to the least cost for the community.[199]

---

[187] Article 11(1)(b) of Directive 2004/18/EC and Article 1(9)(b) of Directive 2004/17/EC.
[188] Commission, 'Communication on compensation granted for SGEIs' (n 181) para 66.
[189] Article 29 of Directive 2004/18/EC.
[190] Article 30 of Directive 2004/18/EC and Article 1(9)(a) of Directive 2004/17/EC.
[191] Ibid.
[192] Article 53(1)(b) of Directive 2004/18/EC and Article 55 (1)(b) of Directive 2004/17/EC.
[193] Article 53(1)(a) of Directive 2004/18/EC, Article 55(1)(a) of Directive 2004/17/EC; Case 31/87 *Beentjes* [1988] ECR 4635 and Case C-225/98 *Commission v France* [2000] ECR I-7445; Case C-19/00 *SIAC Construction* [2001] ECR I-7725.
[194] Commission, 'Communication on compensation granted for SGEIs' (n 181) para 67.
[195] Ibid.
[196] Ibid para 67.
[197] Ibid para 68.
[198] Ibid.
[199] Ibid.

In relation to the level of compensation determined on the basis of the costs of a benchmark undertaking, the Commission first states that where a generally accepted market remuneration exists for a given service, that remuneration provides the best benchmark for the compensation in the absence of a tender.[200] In its communication, the Commission tries to shed some more light on the meaning of the different decisive components of the *Altmark* efficiency benchmark formula for those cases where no such market remuneration exists.

With regard to the concept of a 'well-run undertaking', the Commission states that the Member States should apply objective criteria that are economically recognised as being representative of satisfactory management.[201] According to it, the relevant undertaking generating a profit is not sufficient; account should be taken of the fact that the financial results of undertakings, particularly in the sectors most often concerned by services of general economic interest, may be strongly influenced by their market power or by sector-specific rules.[202] The concept of a well-run undertaking entails compliance with the national, EU or international accounting standards in force.[203] The Commission notes that the Member States may base their analysis on analytical ratios representative of productivity[204] and on analytical ratios relating to the quality of supply as compared with user expectations.[205] An undertaking entrusted with the operation of a service of general economic interest that does not meet the qualitative criteria laid down by the Member State concerned does not constitute a well-run undertaking even if costs are low.[206] Member States must take into account the size of the undertaking in question and the fact that in certain sectors undertakings with very different cost structures exist side by side.[207]

As to the reference to the costs of a 'typical' undertaking, the Commission states that this implies that there is a sufficient number of undertakings whose costs may be taken into account.[208] Those undertakings may be located in the same Member State or in other Member States.[209] The Commission expressly notes that it is of the view that reference cannot be made to the costs of an undertaking that enjoys a monopoly or receives public service compensation under conditions

---

[200]   Ibid para 69.
[201]   Commission, 'Communication on compensation granted for SGEIs' (n 181) para 71.
[202]   Ibid.
[203]   Ibid para 72.
[204]   Such as turnover to capital employed and total cost to turnover.
[205]   Commission, 'Communication on compensation granted for SGEIs' (n 181) para 72.
[206]   Ibid.
[207]   Ibid para 73.
[208]   Ibid para 74.
[209]   Ibid.

that do not comply with EU law; in both cases the cost level may be higher than normal.[210]

The costs to be taken into account are all those relating to the service of general economic interest: the direct costs necessary to discharge the service of general economic interest and an appropriate contribution to the indirect costs common to both service of general economic interest activities and other activities.[211] If the Member State can show that the cost structure of the public service provider corresponds to the average cost structure of efficient and comparable undertakings in the sector under consideration, the amount of compensation that will allow the undertaking to cover its costs, including a reasonable profit, is deemed to comply with the fourth *Altmark* criterion.[212]

Finally, the Commission states that the phrase 'undertaking adequately provided with material means' should be taken to mean an undertaking which has the resources necessary for it to discharge immediately the public service obligations incumbent on the undertaking to be entrusted with the operation of the service of general economic interest.[213] 'Reasonable profit' is understood to mean the rate of return on capital that would be required by a typical undertaking considering whether or not to provide the service of general economic interest for the whole duration of the period of entrustment, taking into account the level of risk.

The picture that emerges from the Commission's Communication on compensation granted for the provision of services of general economic interest is similar to that from its post-*Altmark* decision practice on the public funding of infrastructure development. The simplest way for public authorities to meet the fourth *Altmark* criterion is to conduct an open, transparent and non-discriminatory public procurement procedure in line with EU public procurement legislation. According to the Commission, a public procurement procedure only excludes the existence of state aid where it allows for the selection of the tenderer capable of providing the relevant service at the least cost to the community. Where the Commission's decision practice indicates that a negotiated tender procedure could be compatible with the fourth *Altmark* criterion, its communication makes clear that the Commission will deem a negotiated procedure with prior publication to suffice in exceptional cases only.

---

[210]   Ibid.
[211]   Ibid.
[212]   Commission, 'Communication on compensation granted for SGEIs' (n 181).
[213]   Ibid para 76.

What is more, a negotiated procedure without publication of a contract notice will not be enough. The Commission's communication confirms the findings of the French broadband cases in that the award criteria may also be qualitative in nature, as long as they guarantee that the relevant service of general economic interest is provided at the lowest societal cost. Under certain circumstances, a public procurement procedure cannot guarantee that the relevant service is provided at the least cost to the community as it does not give rise to a sufficiently open and genuine competition.

In relation to the level of compensation being determined on the basis of the costs of a benchmark undertaking, the Commission's communication still leaves Member States with a considerable degree of uncertainty. The Commission's statement, for instance, that the efficiency criterion is met if Member States can show that the cost structure of the public service provider corresponds to the average cost structure of efficient and comparable undertakings in the relevant sector, is of little help as it seems to do no more than reproduce the efficiency criterion's benchmark formula. Likewise, the Commission's interpretation of a 'reasonable profit' as 'the rate of return on capital that would be required by a typical undertaking considering whether or not to provide the service of general economic interest' is illustrative of the degree of legal uncertainty still left by the Commission's communication.

9.6.3 THE POST-*ALTMARK* CASE LAW

9.6.3.1 *Case T-289/03 BUPA*

The seminal post-*Altmark* judgment on the application of the efficiency criterion is *BUPA*.[214] Central to *BUPA* was the question of whether a proposed Irish scheme for equalising private medical insurance risk was compatible with EU state aid rules. Under the proposed scheme, insurers with a risk profile healthier than the average market risk profile had to pay a charge to the Irish Health Insurance Authority. Corresponding payments were then made by the Health Insurance Authority to insurers with a risk profile less healthy than the average market risk profile. BUPA Ireland, the main competitor of the incumbent insurer (the VHI) on the liberalised Irish private medical insurance market, had lodged a complaint with the Commission against the implementation of the risk equalisation scheme on the ground that it infringed ex Article 87(1) of the Treaty Establishing the European Community (TEC).

---

[214]   Case T-289/03 *BUPA* [2008] ECR II-00081.

After the Commission had decided that the proposed Irish risk equalisation scheme was compatible with EU state aid rules, BUPA brought an action for annulment of the Commission's decision before the CFI. In its first plea, BUPA argued that the Commission had misapplied ex Article 87(1) TEC. BUPA claimed, inter alia, that the fourth *Altmark* criterion had not been met as the risk equalisation scheme made no comparison with an efficient undertaking.[215] According to it, the risk profile of a private medical insurer in comparison with the average market risk profile (and with that the level of compensation granted) was determined without regard to the efficiency of the relevant insurers.[216]

The CFI first held that even though the *Altmark* judgment was delivered after the contested Commission decision had been adopted, the *Altmark* criteria were fully applicable to the factual and legal situation in *BUPA*; the ECJ had not placed any temporal limitation on the scope of its findings in *Altmark*.[217] In relation to the efficiency criterion, the CFI argued that the efficiency criterion could not be applied strictly in *BUPA*.[218] It stated that this was due to the neutrality of the compensation mechanism under the risk equalisation scheme (based on receipts and profits of the insurers) and to 'the particular nature of the additional costs linked with a negative risk profile on the part of those insurers'.[219]

In this regard, the CFI noted that the payments under the risk equalisation scheme were not determined solely by reference to the payments made by the insurer receiving compensation – a situation corresponding to that addressed by the efficiency criterion – but also by reference to the payments made by the contributing insurer, which reflected the risk profile differentials of those two insurers with the average market risk profile.[220] The level of compensation, in other words, was determined on the basis of the costs made by both receiving and contributing insurers.[221]

The CFI further argued that the Commission had been unable to identify precisely the potential beneficiaries of payments under the risk equalisation scheme and to make a specific comparison of their situation with an efficient operator due to the risk equalisation scheme not having been activated yet at the time of the adoption of the contested decision.[222] Instead of the benchmark undertaking being absent, no (potentially) benchmarked undertaking was

---

[215] Ibid para 124.
[216] Case T-289/03 *BUPA* (n 214).
[217] Case T-289/03 *BUPA* (n 214) paras 158–159.
[218] Ibid para 246.
[219] Ibid.
[220] Ibid para 247.
[221] For determining whether any compensation granted through the scheme would constitute state aid, arguably, only the efficiency of the receiving insurer was relevant.
[222] Case T-289/03 *BUPA* (n 214) para 248.

available when the Commission adopted its decision on the state aid compatibility of the risk equalisation scheme. The CFI noted that the Commission nonetheless was required to satisfy itself that the compensation granted through the risk equalisation scheme did not entail the possibility of offsetting any costs resulting from inefficiency on the part of the insurers subject to the scheme.[223]

In this regard, it stated that the Commission had expressly found that the scheme allowed the insurers to keep the benefit of their own efficiencies.[224] As the calculation of the compensation under the risk equalisation scheme depended solely on the costs not linked with the efficiency of the operators in question, that compensation was not capable of leading to the sharing of any costs resulting from their lack of efficiency.[225]

In reaction to BUPA's argument that the impact of high claims on the level of payments through the risk equalisation scheme encouraged the recipient insurer not to be efficient, the CFI noted that BUPA had not specified the nature of any lack of efficiency to which such a situation could give rise, but had essentially confined itself to invoking the absence of any comparison with an efficient operator within the meaning of the fourth *Altmark* condition.[226]

Concluding, the CFI held that the Commission had, to the requisite legal standard, taken into account the evidence which allowed it to conclude that the compensation provided for by the risk equalisation scheme was neutral by reference to any costs associated with inefficiency incurred by certain insurers.[227] Consequently, the CFI considered the Commission not to have failed to consider the *Altmark* efficiency criterion.[228]

The CFI's ruling in *BUPA* has been characterised as a retreat from the strict efficiency approach taken in *Altmark*.[229] In *BUPA*, the CFI allegedly modified the *Altmark* efficiency criterion to fit the specifics of the case.[230] Ross has argued that the clearest message that emerges from the CFI's judgment in *BUPA* is that the *Altmark* criteria are not to be regarded as a straitjacket or a one-size-fits-all

---

[223]   Ibid para 249.

[224]   Ibid.

[225]   Case T-289/03 *BUPA* (n 214) para 250.

[226]   Ibid para 252.

[227]   Ibid para 256.

[228]   Ibid para 257.

[229]   Hancher and Larouche (n 72) 764. See also Wolf Sauter and Johan van de Gronden, 'State Aid, Services of General Economic Interest and Universal Service in Healthcare' (2011) 32 European Competition Law Review 615, 618.

[230]   Andreas Bartosch, 'The Ruling in BUPA – Clarification or Modification of Altmark?' (2008) 7 European State Aid Law Quarterly 211, 211; and Hancher and Larouche (n 72) 764.

approach to services of general economic interest.[231] According to Ross, the distinctly *Altmark*-lite set of tests developed by the CFI in *BUPA* leads to doubts as to how strictly the *Altmark* tests are to be construed and applied.[232]

The question is whether the CFI's *BUPA* ruling truly represents a (principal) departure from the ECJ's *Altmark* judgment or whether it merely illustrates the way in which to overcome the practical difficulties in applying the *Altmark* criteria to complex schemes for the compensation of services of general economic interest. Buendia Sierra has noted that the *BUPA* judgment cannot be seen as a change of direction in the *ECJ Altmark* case law.[233] According to Buendia Sierra, the only reason for the CFI not to find the fourth *Altmark* criterion applicable was the very particular nature of the public service scheme concerned, which made the traditional comparison between cost and compensation not really relevant.[234] A contrario, the fourth *Altmark* criteria – as well as the other *Altmark* criteria – remains valid in the 'normal' services of general economic interest cases.[235]

Buendia Sierra is probably right in explaining the CFI's application of the efficiency criterion in *BUPA* by the specific nature of the compensation scheme at hand. When ruling that the efficiency criterion could not be applied strictly in *BUPA*, the CFI itself referred to the neutrality of the compensation mechanism under the risk equalisation scheme and the to 'the particular nature of the additional costs linked with a negative risk profile on the part of those insurers'. It basically held that the efficiency criterion could not be applied in the way foreseen in *Altmark*[236] because the level of compensation was not determined solely by reference to the payments made by the compensated undertaking, but also by reference to the payments of the compensating undertaking. The fact of level of compensation not being determined solely by reference to the costs of the compensated undertaking is arguably what made the facts in *BUPA* different from the standard *Altmark* efficiency situation. Like *Chronopost*, the *BUPA*

---

[231]   See Malcolm Ross, 'A Healthy Approach to Services of General Economic Interest? The BUPA Judgment of the Court of First Instance' (2009) 34 European Law Review 127, 138.

[232]   Ibid 138–39.

[233]   Buendia Sierra (n 17) 200.

[234]   Ibid.

[235]   Ibid.

[236]   The CFI did not hold that the fourth *Altmark* criterion was not applicable at all, as apparently suggested by Buendia Sierra. As we have seen, it held that the Commission was required to satisfy itself that the compensation granted through the risk equalisation scheme did not entail the possibility of offsetting any costs resulting from inefficiency on the part of the insurers subject to the scheme. According to the CFI, the Commission had demonstrated that no such possibility existed. In that way, the economic efficiency rationale underlying the fourth *Altmark* criterion remained intact.

ruling can be seen as an illustration of the difficulties in applying the rather abstract *Altmark* efficiency criterion.

The issue, perhaps, is not whether the fourth *Altmark* condition remains valid in the 'normal' services of general economic interest cases, but rather whether 'normal' services of general economic interest exist at all and, accordingly, how workable the efficiency criterion's benchmark comparison is in practice. As argued above, a typical characteristic of services of general economic interest appears to be that the undertaking providing such services will often be in a situation which is very different from that of other undertakings. In such circumstances, it might prove very difficult to find or hypothesise a benchmark efficient undertaking.

The conclusion to be drawn from *BUPA* should therefore arguably be that the principle behind the fourth *Altmark* criterion is still valid, but that the particular form of comparing the efficiency of the public service provider with a model efficient undertaking might be difficult to apply in practice. In that case, compensation granted for the discharge of public service obligations could still escape classification as state aid under Article 107(1), as long as it does not lead to inefficiencies being rewarded.[237]

9.6.3.2   Case T-442/03 SIC II

Another important judgment with implications for the fourth *Altmark* criterion was rendered in *SIC II*.[238] This case concerned an appeal against a decision of the Commission in which it had found, inter alia, state aid granted by Portugal to the Portuguese public television broadcaster RTP to be compatible with ex Article 86(2) TEC (now Article 106(2) TFEU).[239] According to the applicant, a commercial television company running one of the main private television channels in Portugal, the Commission had erred in law as to the conditions for the application of ex Article 86(2).[240] More specifically, SIC argued that the award of the relevant public television service to RTP without competitive tendering precluded the grant of a derogation under ex Article 86(2).[241] Referring to the *Altmark* case, it contended that the fourth criterion required the Commission, in view of the failure to follow a competitive tendering procedure, to verify that the level of compensation granted to RTP had been determined on the basis of a typical undertaking, well-run and adequately provided with

---

[237]   See also Hancher and Larouche (n 72) 764.
[238]   Case T-442/03 *SIC II* [2008] ECR II-01161.
[239]   *RTP* Commission Decision 2005/406/EC [2005] OJ L142/1.
[240]   Case T-442/03 *SIC II* (n 238) paras 129 and 133 and further.
[241]   Ibid paras 133 and 134.

means.[242] According to it, the compensation granted did not meet the *Altmark* efficiency criterion.[243]

The Commission argued that SIC confused the assessment of the compatibility of the state aid granted with the question whether the compensation granted qualified as state aid.[244] According to the Commission, it did not follow from the provisions of ex Article 86(2) or from the case law that Member States are required to follow specific procedures for choosing undertakings entrusted with the operation of services of general economic interest.[245]

The CFI held that ex Article 86(2) TEC did not include in its conditions for application a requirement on Member States to follow a competitive tendering procedure for the award of a service of general economic interest.[246] It referred to its earlier ruling in *Olson v. Commission*,[247] in which it had indicated that it is not apparent either from the wording of ex Article 86(2) or from the case law on that provision that a service of general economic interest may be entrusted to an operator only as a result of a tendering procedure.[248] The CFI argued that the applicant's claim that the Commission should, in the context of ex Article 86(2), have verified whether the television service of general economic interest had been awarded to RTP after competitive tendering had to be rejected.[249]

Furthermore, it held that even if the relevant service of general economic interest should have been the subject of a competitive tendering procedure for other reasons, the absence of this procedure could not have the result that state funding of the service of general economic interest holder's public service obligations was to be considered to be state aid incompatible with the common market.[250]

Vedder has argued that the CFI in *SIC II* held that the fourth *Altmark* criterion does not require the undertaking discharging the public service obligations to be chosen by means of a tender or public procurement procedure.[251] This arguably

---

[242] Ibid para 135.
[243] Ibid.
[244] Ibid paras 136 and 140.
[245] Ibid para 137.
[246] Case T-442/03 *SIC II* (n 238) para 145.
[247] Case T-17/02 *Olsen v. Commission* [2005] ECR II-02031.
[248] T-442/03 *SIC II* (n 238) para 145.
[249] Ibid para 146.
[250] T-442/03 *SIC II* (n 238) para 147.
[251] Hans Vedder, 'The Constitutionality of Competition. European Internal Market Law and the Fine Line between Markets, Public Interests, and Self-regulation in a Changing Constitutional Setting' in Fabian Amtenbrink and Peter AJ van den Berg (eds), *The Constitutional Integrity of the European Union* (TMC Asser Press 2010) 201, 227.

is a correct, literal reading of the CFI's judgment. Yet I would argue that it likewise follows from the ruling that the efficiency criterion does not (generally) require the level of compensation for the discharge of a service of general economic interest to be determined on the basis of the costs of a benchmark efficiency undertaking.

Even though the CFI did not directly respond to SIC's claim in this regard, this conclusion logically follows from the CFI's reasoning; the alternative option under the efficiency criterion is likewise not part of the test to be performed under Article 106(2). In *SIC II*, the CFI indicated that the *Altmark* efficiency requirement is strictly limited to the qualification of compensation granted for services of general economic interest under Article 107(1) TFEU. Only for such compensation not to be considered state aid does the level of compensation have to be determined either through a public tender procedure or by comparison to a benchmark efficiency undertaking.

Apart from the above cases, there appear to be few judgments in which the European Courts have shed more light on the application of the *Altmark* efficiency criterion.[252] Rather than dealing with the content and application of the fourth *Altmark* criterion, cases before both Courts often seem to have centred around issues of a procedural nature.

9.6.3.3 Case T-274/01 Valmont

Valmont Nederland v. Commission concerned, inter alia, financial support for the development of a non-public car park, the operator of which allowed, by way of agreement with the relevant Dutch municipality, other parties to make regular and free use of its facilities.[253] The car park operator and beneficiary of the financial support (Valmont) had brought an application for annulment of the Commission decision which had found part of the financial support granted to constitute state aid incompatible with the common market. In one of its pleas in law, Valmont argued that the Commission had wrongly concluded that part of the financing for the construction of the car park constituted a benefit. According to the Commission, half of the financing granted for the construction of the car park amounted to business costs which Valmont should normally have borne and that not doing so placed it at an advantage.

---

[252] In this regard, see also Braun and Kühling (n 17) 476. Braun and Kühling reviewed the post-*Altmark* case law between 2004 and 2007 in relation to, inter alia, the fourth *Altmark* criterion.

[253] Case T-274/01 *Valmont Nederland v. Commission* [2004] ECR II-03145.

In assessing the plea brought forward by Valmont, the CFI basically held that Valmont's burden of allowing others to use its car park constituted a service of general economic interest/public service obligation.[254] In those circumstances, the CFI argued, the Commission should examine whether or not the portion of the financing granted could be regarded as compensation for that service of general economic interest by ascertaining whether the four *Altmark* conditions were satisfied.[255] According to it, the Commission had not done so.[256] Accordingly, the CFI annulled the part of the Commission decision declaring the construction of the relevant car park to contain an element of state aid.[257] *Valmont Nederland v. Commission* shows that, if there is any possibility that the support granted compensates the costs of the discharge of a public service obligation, the Commission is under the obligation to examine whether the *Altmark* criteria are met. Without doing so, the Commission cannot rightfully come to the conclusion that the support granted led to an advantage for the beneficiary.

9.6.3.4 *Joined Cases T-309/04, T-317/04, T-329/04 and T-336/04 TV2/Danmark*

In *TV 2/Danmark v. Commission*, which concerned an application for annulment of Commission decision 2006/217/EC,[258] the CFI indicated that the Commission is under the obligation to carefully examine all conditions related to the setting of the level of compensation granted for the discharge of a public service obligation, which are brought forward by the Member State.[259] In its statement of reasons, the Commission had stated that the relevant Member State (Denmark) had not carried out *any* analysis to ensure that the level of compensation was determined on an analysis of the costs which a typical well-run and adequately provided undertaking would have incurred.[260]

According to the CFI, the Commission had ultimately done no more than reproduce verbatim the wording of the fourth *Altmark* criterion.[261] The CFI argued that such a statement of reasons could be sufficient only if it were common ground that Denmark had put nothing in place that could, in practical

---

254   Ibid para 132.
255   Ibid para 133.
256   Ibid para 134.
257   Ibid paras 137–38.
258   *TV2-Danmark* Commission Decision 2005/217/EC [2006] OJ L85.
259   Joined Cases T-309/04, T-317/04, T-329/04 and T-336/04 *TV2/Danmark v. Commission* [2008] ECR II-02935.
260   *TV2-Danmark* (n 263) para 71. Interestingly, the CFI characterised the tone of the Commission's findings in para 71 of the latter's decision as 'peremptory' (see para 224 of the judgment).
261   Joined Cases T-309/04, T-317/04, T-329/04 and T-336/04 *TV2/Danmark v. Commission* (n 259) para 232.

terms, ensure compliance with the fourth *Altmark* condition, or if the Commission had established that the analysis carried out by Denmark was manifestly inadequate or inappropriate for the purposes of ensuring compliance with that condition.[262] It stated that this was not the case as Denmark had used, inter alia, economic analyses drawn up with the help of the beneficiary's (TV2) competitors and that it was conceivable that a serious examination of all the conditions governing the setting of the amount of compensation granted, 'the analysis which the Commission should have carried out', would have led to the conclusion that Denmark complied with the fourth *Altmark* condition.[263] The CFI concluded that the Commission decision was vitiated by an inadequate statement of reasons and had to be annulled.[264]

9.6.3.5 Case T-388/03 Deutsche Post and DHL

Finally, in *Deutsche Post and DHL v. Commission*, postal companies Deutsche Post and DHL brought an application for annulment of a Commission decision[265] finding various measures adopted by Belgium in favour of the Belgium public postal undertaking La Poste not to constitute state aid.[266] The applicants argued, inter alia, that the Commission had not verified whether the relevant services of general economic interest were provided at the least cost to the community. During its preliminary examination procedure, the Commission had calculated the difference between the overcompensation as a consequence of several measures taken by Belgium and the under-compensation of the services of general economic interest which it had itself established.[267] As it considered the net additional costs incurred in discharging the relevant public service obligation to have been undercompensated, the Commission found the overcompensating measures not to constitute state aid (as they did not confer an advantage on the beneficiary).[268] By declaring the compensation not to constitute state aid since it did not exceed (and in fact remained below) the costs of the services of general economic interest, the Commission adopted an approach that is strongly reminiscent of that of the ECJ in *Ferring*.

The CFI held that it was apparent both from the contested decision and from the exchange of letters and the minutes of the meetings between the Commission and the Belgian authorities that the Commission never verified

---

[262] Ibid.
[263] Ibid.
[264] Para 234.
[265] Commission Decision of 23 July 2003 C (2003) 2508 final.
[266] Case T-388/03 *Deutsche Post and DHL v. Commission* [2009] ECR II-00199.
[267] Ibid para 22.
[268] Ibid.

that the services of general interest which La Poste provided were at a cost which would have been borne by a typical, well-run undertaking.[269] According to the CFI, the Commission merely relied on the negative balance of all the items of overcompensation and undercompensation in respect of the additional cost of the service of general economic interest.[270] Accordingly, the Commission decision had to be annulled on the basis of an incomplete/insufficient examination during the examination stage.[271] By its ruling in *Deutsche Post and DHL v. Commission*, the CFI indicated that compensation granted for the discharge of public services obligations can constitute an advantage (and thus state aid), even when the costs of those services exceed the amount of public support granted.

The most important post-*Altmark* case before the European courts arguably has been *BUPA*. Rather than representing a principal departure from the ECJ's *Altmark* ruling, the CFI's ruling in this case illustrates the way in which to overcome the practical difficulties in applying the *Altmark* criteria to complex schemes for the compensation of services of general economic interest. The conclusion to be drawn from *BUPA* should be that the principle behind the fourth *Altmark* criterion is still valid, but that the particular form of comparing the efficiency of the public service provider with a model efficient undertaking might be difficult to apply in practice. If the latter is the case, *BUPA* seems to suggest that compensation granted for the discharge of public service obligations could still escape classification as state aid under Article 107(1), as long as it does not lead to inefficiencies being rewarded.

The requirement of compensation for a service of general economic interest not leading to inefficiencies being rewarded is what sets the *BUPA* ruling apart from another case which, as we have seen, has often been characterised as conflicting with the *Altmark* judgment: *Chronopost*. The ECJ's ruling in this case has been interpreted as allowing for the efficiency criterion's benchmark comparison to be done on the basis of the costs of the undertaking itself when no comparable undertaking actually exists. Nevertheless, this lex specialis interpretation of *Chronopost* seems difficult to reconcile with the logic behind the *Altmark* ruling, in which the ECJ tried to accommodate some of the criticism of the *Ferring* judgment and sought to ensure that no inefficiencies would be subsidised. What is more, this interpretation of *Chronopost* would seem to rule out the possibility of a comparison with a *hypothetical* benchmark undertaking. This seems to conflict with the wording of and the idea behind the *Altmark* efficiency criterion.

---

[269]   Ibid para 115.
[270]   Ibid para 116.
[271]   Ibid paras 118–19.

Both *Chronopost* and *BUPA* indicate that it may be difficult to apply the efficiency criterion's second option in practice and that the Courts may, therefore, be tempted to deviate from the abstract approach of *Altmark*.[272] As *Deutsche Post and DHL v. Commission* (and *BUPA*) indicate, a creative and less strict interpretation of the *Altmark* efficiency criterion may not, however, lead to a return to the (inefficient) compensation approach of *Ferring*.

9.7 LESSONS FOR PUBLIC FUNDING OF CO$_2$ TRANSPORT AND STORAGE INFRASTRUCTURE

The post-*Altmark* decision practice, the Commission's Communication on public service compensation and the Courts' post-*Altmark* case law indicate that if Member States want public support for CO$_2$ transport and storage infrastructure to fall outside the scope of Article 107(1) TFEU, they would be wise to determine the level of compensation for the development of such infrastructure through an open, EU-wide tender procedure in line with EU public procurement rules. In this regard, it is of great importance that the transport and storage of CO$_2$ can be characterised as services of general economic interest.[273] The use of other public procurement procedures is not to be preferred. Such procedures are likely to (1) be insufficient for meeting the fourth *Altmark* criterion (negotiated procedure without publication of contract notice), (2) suffice for meeting that criterion in exceptional circumstances only (competitive dialogue or negotiated procedure with prior publication) and (3) still have to meet the requirement that all interested operators can tender (restricted procedure).

The selection criteria of any public procurement procedure will have to be transparent, objective and defined in advance. In addition to quantitative selection criteria, Member States may also use qualitative criteria, but the latter have to be closely related to the subject matter of the relevant service of general economic interest, in this case the transport and/or storage of captured CO$_2$. Furthermore, qualitative selection criteria must ensure that the CO$_2$ transport and storage infrastructure is developed at the least cost to society. It is difficult to determine the required balance between quantitative and qualitative selection

---

[272] See also Quigley (n 3) 161. Quigley, referring to, inter alia *BUPA* and *Deutsche Post and DHL v. Commission*, has noted that the CFI has stated that it is appropriate to apply the *Altmark* criteria, in accordance with the spirit and the purpose which prevailed when they were laid down, in a manner adapted to the particular facts of the case rather than to make a literal application of those criteria.

[273] See section 6.4.3 of Chapter VI on Article 102 TFEU and refusal of access to CO$_2$ transport and storage infrastructure on technical grounds.

criteria, but a selection procedure on the basis of qualitative criteria only would not seem to comply with the efficiency criterion.

Furthermore, the Commission's decision practice appears to show that Member States could have a first phase (pre-)selection by means of qualitative criteria and a second phase/final selection on the basis of quantitative criteria. First phase selection criteria such as the geographical scope of the CO_2 transport infrastructure (how many capturers will be served?) and environmental and health safety standards (crucial in relation to CO_2 storage) would seem to be allowed, as long as final selection is based on the relevant transport or storage infrastructure being developed at lowest societal cost.

There may be cases in which the Commission will not deem a public procurement procedure to lead to the relevant infrastructure being developed at the lowest cost to society. The Commission's communication is not entirely clear in this regard. Nevertheless, if only one bid has been submitted the Commission will consider the competition to have failed. By contrast, a public procurement process attracting bids from other Member States is seen as indicative of a competitive selection process.[274]

Member States are not advised to go down the road of determining the level of compensation for the development of CO_2 transport and storage infrastructure on the basis of the costs that a well-run, typical and adequately provided undertaking would have incurred. The post-*Altmark* decision practice and case law of the Commission and the Courts provide very little guidance as to the specific requirements that have to be met in this regard. Moreover, they partly seem to contradict each other. Whereas the Commission appears to follow a hard line, cases like *Chronopost* and *BUPA* indicate that the Courts may, under certain circumstances, be tempted to be more flexible. What is clear from the post-*Altmark* decision practice and case law, however, is that the simple

---

[274] Even though the storage licensing procedure under the CCS Directive could at first sight appear to be a suitable, competitive (storage operator) selection procedure (see section 2.5.1.1 of Chapter II), this procedure would likely not meet the public procurement requirements of the *Altmark* efficiency criterion. First, the Directive's storage licensing procedure does not concern the *development* of storage infrastructure, but deals with the *operation* of such infrastructure. Second, a licensing procedure is something different from a public procurement procedure. Even though any entity can apply for a storage licence, there is no bid competition as generally is the case with a public procurement procedure. In the storage licensing procedure for a particular reservoir, the first to meet the licence requirements will receive the licence. Finally, as we have seen, the Commission will deem a public procurement procedure to have failed when only one bid has been submitted, whereas a single storage licence applicant can meet the requirements of the storage licensing procedure. In relation to CO_2 transport, the Directive does not even prescribe a licensing regime.

compensation approach of *Ferring* (no advantage if compensation does not exceed (inefficient) public service costs) no longer holds.

The Commission's Communication on public service compensation provides more clarity as to the requirements to be met in relation to the alternative of benchmark compensation. Yet it still leaves Member States with a considerable degree of legal uncertainty. The Commission's statement, for instance, that the efficiency criterion is met if Member States can show that the cost structure of the public service provider corresponds to the average cost structure of efficient and comparable undertakings in the relevant sector, is of little help, as it seems to do no more than reproduce the efficiency criterion's benchmark formula. Likewise, the Commission's interpretation of a 'reasonable profit' as 'the rate of return on capital that would be required by a typical undertaking considering whether or not to provide the service of general economic interest' is illustrative of the degree of legal uncertainty still left by the Commission's communication.

Based on the above, Member States seeking to avoid the classification of public support for the development of CO_2 transport and storage infrastructure as state aid within the meaning of Article 107(1) TFEU are advised to conduct an open, transparent and non-discriminatory public procurement procedure in line with applicable EU public procurement legislation. This procedure can contain qualitative selection criteria related to, for instance, the geographical scope and safety of the infrastructure as long as the final selection is based on the relevant transport or storage infrastructure being developed at the lowest societal cost. As stated by the Commission, this is 'the simplest way for public authorities to meet the fourth *Altmark* criterion' since they will not have to go down the uncertain path of determining the level of compensation on the basis of the costs of a benchmark undertaking. In case of the development of cross-border CO_2 transport and storage infrastructure, Member States will have to cooperate in developing a suitable public procurement procedure in line with applicable EU public procurement legislation.

CONCLUSIONS

In this thesis, I researched the broader primary EU law context of a number of legal issues, related to provisions in the CCS Directive, that could have far-reaching consequences for the cross-border deployment of CCS in the EU. In the following, I first answer the question of to what extent the legal issues examined in Chapters III-IX could indeed hamper the (near-)future cross-border deployment of CCS in the EU, and explore possible solutions, if required. Second, I answer the question of in what way the legislative approach chosen by the Commission, in view of the primary EU law context, is generally likely to lead to problems for the cross-border deployment of new technologies such as CCS. Finally, I make a number of policy and legal recommendations for facilitating the cross-border deployment of CCS in the EU.

LEGAL ISSUES EXAMINED IN CHAPTERS III–IX

The legal issue raised in *Chapter III* will probably not form a significant obstacle to the future cross-border deployment of CCS in the EU. In the Introduction to this thesis, I raised the possibility of varying national implementations of Article 12 of the CCS Directive on CO_2 stream purity. Governments in Member States that have shown an interest in accommodating early CCS demonstration projects could be tempted to increase the safety standard in Article 12 by adopting CO_2 stream purity criteria that are more stringent than those in the latter provision. This could hinder the cross-border transport and storage of captured CO_2, since it would likely increase costs along the CCS value chain.

The results from Chapter III show that the risk of more stringent purity criteria hindering the cross-border deployment of CCS in the EU are minimal as the scope for Member States to adopt such criteria appears to be narrow. An analysis of the sole legal framework relevant for assessing the Member States' scope to adopt stricter CO_2 stream-purity criteria, the CCS Directive, reveals that the criteria in Article 12 probably do not constitute minimum harmonisation. Furthermore, stricter national CO_2 stream-purity criteria would seem to be in line with the broad logic of Article 12 only if such criteria would allow for a case-by-case assessment of levels of impurity. General impurity levels applying to all individual cases would seem to collide with the broad logic of Article 12, for such criteria would not respect the case-by-case approach underlying the provision. Finally, it is not certain whether the structure/spirit of the CCS Directive would allow for stricter national purity criteria, as the consequences of such criteria for the safety of CCS are unclear. In this case, the primary EU law context ensures a minimum of consistency in the Member States' implementation of Article 12 (case-by-case approach).

Likewise, the extensive case law on Article 110 TFEU would appear to ensure that the legal issue explored in *Chapter IV* of the thesis will not pose a significant hindrance to the cross-border deployment of CCS in the EU. In the Introduction, I mentioned the possibility of Member States' financial security/mechanism charges under Articles 19 and 20 of the CCS Directive hindering the free movement of captured CO_2 in the EU by making it more difficult for 'foreign' CO_2 streams to be stored than for 'domestic' CO_2 streams. Under Article 19, applicants for a CO_2 storage permit need to prove that they have the financial means to meet all future obligations, including closure and post-closure requirements. By virtue of Article 20, the storage operator has to make a financial contribution available to the competent authority before responsibility for the storage site is transferred to the latter. Article 110 TFEU prohibits Member States from imposing on imported products any internal taxation that is in excess of that imposed on similar domestic products or protects competing domestic products.

As outlined in Chapter IV, the European Court of Justice's (ECJ's) case law on Article 110 TFEU is likely to impose several requirements on the design of Member States' financial security/mechanism charges. Even though Member States, for instance, would seem to be allowed to differentiate the financial security/mechanism charge on the basis of objective criteria – for instance CO_2 stream composition – they will have to give the storage operator the opportunity to show that, by storing the specific imported CO_2 stream, he meets the criteria for being charged a particular tariff.

What is more, should it be difficult to determine the composition of the CO_2 stream (to be) imported, then the storage operator cannot be charged a flat rate duty, but should be charged the lowest tariff applicable to domestic CO_2 streams for storage. In my opinion, the possible requirements identified in Chapter IV would suffice to ensure the equal tax treatment of domestic and CO_2 streams imported for storage, particularly since Article 110 TFEU can be invoked by individuals before a Member State court.

By contrast, the legal issue examined in *Chapter V* does not seem to be remedied by the broader primary EU law context of Article 21(2)(b) of the CCS Directive. In the Introduction, I argued that Article 21(2)(b) could lead to CO_2 transport and storage capacity in some Member States being (partly) unavailable to third parties in other Member States. This could lead to problems for the latter Member States in trying to meet their national targets for the reduction of greenhouse gas emissions. Article 21(2)(b) appears to give Member States the option to require the operators of CO_2 transport and storage infrastructure to refuse to grant access to the relevant infrastructure when the capacity concerned is needed to meet part of a Member State's international and EU obligations to reduce greenhouse gas emissions. At first sight, Article 21(2)(b) appears to be at odds with Articles 4(3) TEU and 194(1)(c) TFEU. Article 4(3) TEU requires Member States to assist each other in carrying out tasks which flow from the Treaties (principle of loyalty), while Article 194(1)(c) calls for secondary EU climate and energy legislation to be drafted 'in a spirit of solidarity between Member States'.

Nevertheless, the analysis in Chapter V indicates that Article 21(2)(b) will likely not infringe these provisions. What is more, as we have seen, both provisions do not seem to narrow Member States' scope for implementing Article 21(2)(b), particularly not in view of the likelihood of individuals not being able to invoke these provisions before a national court for challenging the validity of national measures implementing Article 21(2)(b). Considering the above, Article 21(2)(b) arguably still has the potential to effectively foreclose an entire Member State's CO_2 storage market and thus to seriously hinder the cross-border deployment of CCS in the EU. As I argued in Chapter V, this could be solved by Member States requiring CO_2 transport or storage operators that refuse access on the ground mentioned in Article 21(2)(b) to expand their capacity when it is economically sensible to do so or when the potential customer is willing to pay for that expansion.

A legislative solution to this issue would be for the Commission, as part of the planned review of the CCS Directive in 2015, to propose to remove the ground for refusal of access to CO_2 transport and storage infrastructure from Article 21(2) of the CCS Directive. Alternatively, the Commission could consider developing a common EU strategy for CO_2 transport and storage, addressing the development of dedicated CO_2 transport and storage infrastructure that can be used by market players from several Member States. In this regard, it is interesting to note that a 2010 study by the Commission's Joint Research Centre (JRC) mentions advanced planning for an optimal design, which is taking into consideration the anticipated volumes of CO_2 that will have to be transported in the medium and long term and the location of CO_2 sources and sinks (reservoirs), as one of the prerequisites for the development of a trans-European CO_2 transport network.[1]

---

[1]   Joris Morbee and others, 'The Evolution of the Extent and the Investment Requirements of a Trans-European CO_2 Transport Network' (2010) JRC Scientific and Technical Reports 2

For CO_2 transport infrastructure, another way to solve this issue would be to oversize initial CO_2 pipeline infrastructure, for instance during the demonstration phase (roughly until 2020).[2] Recent research indicates that the oversizing of initial CO_2 pipelines will probably not occur without government intervention.[3] Therefore, the Commission could, as part of the review of the CCS Directive in 2015, consider proposing to add the requirement to oversize initial CO_2 pipeline infrastructure to the CCS Directive's chapter 5 on third-party access to CO_2 transport and storage infrastructure.[4] However, should such a requirement be introduced, then the Commission would also have to arrange for appropriate public means for financing the (costly) oversizing of new CO_2 pipelines. This underlines the desirability of the development of a common EU strategy for CO_2 transport and storage that provides for the necessary financial means.

Finally, the Commission could, again as part of the review of the CCS Directive in 2015, consider proposing to insert into the CCS Directive the requirement for CO_2 transport and storage project developers to run an open season and test (cross-border) market demand for capacity on the relevant infrastructure.[5] According to NERA Economic Consulting, open season procedures would ensure that prospective future developers of CCS plants have an opportunity to request additional transport capacity to exploit economies of scale.[6] Likewise,

---

<http://publications.jrc.ec.europa.eu/repository/bitstream/111111111/15100/1/ldna24565enn.pdf> accessed 4 January 2012.

[2] When a pipeline is 'oversized', it is constructed with a transport capacity greater than that required by the project for which it is constructed. In other words, the capacity of the pipeline is determined at a level which takes potential future demand for transport capacity into account.

[3] NERA Economic Consulting, 'Developing a Regulatory Framework for CCS Transportation Infrastructure (Vol. 1 of 2)' (2009) 10 <www.decc.gov.uk/assets/decc/what%20we%20do/uk%20energy%20supply/energy%20mix/carbon%20capture%20and%20storage/1_20090617-131338_e_@@_ccsreg1.pdf> accessed 4 January 2012 and CO_2Europipe, 'Developing a European CO2 Transport Infrastructure' (2011) 36 <www.co2europipe.eu/Publications/CO2Europipe%20-%20Executive%20Summary.pdf> accessed 3 January 2012.

[4] The Commission could, for instance, add an extra paragraph to Article 21(1) of the CCS Directive, which contains the general requirement for Member States to ensure that potential users are able to obtain access to transport networks and to storage sites for the purposes of geological storage of the produced and captured CO_2. Nevertheless, the consequence of requiring project developers to invest in additional capacity might be that investment is discouraged in cases where the developer would not choose to add capacity voluntarily. See NERA (n 3) 35.

[5] In an open season procedure, the name of a tender process for pipeline transport capacity, the developer makes it possible for anyone to apply to join the project. See n 20 in Chapter VII.

[6] NERA (n 3) 35. The term 'economies of scale' refers to the situation in which an increase of output can be achieved at a lower average cost due to fixed costs (costs that must be incurred even if production were to drop to zero) being spread out over more units of production. See Steven M Suranovic, International Trade Theory and Policy (2010) 80–81 and The Economist, 'Economics A-Z' <www.economist.com/research/economics/> accessed 18 April 2011.

energy consultancy Element Energy has argued that the use of open seasons provides an effective means of engaging prospective emitters with prospective pipeline developers.[7]

The findings in *Chapter VI* of the thesis show that it is difficult to give an exact answer to the question of to what extent the legal issue explored in that chapter could indeed hamper the cross-border deployment of CCS in the EU. In the Introduction to this thesis, I argued that Article 21(2)(c) of the CCS Directive has the potential to be an obstacle to the cross-border deployment of CCS as differences in national technical standards for CCS infrastructure are likely to appear. Article 21(2)(c) appears to provide the operators of CO_2 transport and storage infrastructure the option to refuse to grant access to the relevant infrastructure when the technical specifications of the specific CO_2 stream are incompatible with the required technical standard and the incompatibility 'cannot reasonably be overcome'. Yet the question is whether a refusal to grant third-party access on technical grounds is compatible with Article 102 TFEU, which prohibits the refusal of access to infrastructure which is considered indispensable to enter a certain market.

The analysis in Chapter VI indicates that the picture is complex. On the one hand, several of the prerequisites for Article 102 TFEU to be able to curb the possible effects of the application of Article 21(2)(c) seem to be fulfilled. As we have seen, there is a likelihood that undertakings active in future EU markets for CO_2 transport and storage will be in a dominant position. What is more, there seems to be a real possibility of CO_2 storage facilities being indispensable. On the other hand, there are limited means for CO_2 transport and storage operators to avoid liability under Article 102 TFEU, the most obvious, in my opinion, being a precautionary defence based on concerns about quality, safety or security.

One way of ensuring that Article 21(2)(c) cannot be abused to hinder the cross-border deployment of CCS would be for the Commission to propose, as part of the review of the CCS Directive in 2015, to insert into the CCS Directive a provision guaranteeing third parties access to CO_2 transport and storage infrastructure once captured CO_2 meets certain minimum requirements. This is the solution that the European Parliament sought when proposing to amend the text of then Article 20(1) of the proposed CCS Directive in order for it to require

---

[7]   Element Energy, 'CO_2 Pipeline Infrastructure: An Analysis of Global Challenges and Opportunities: Final Report for International Energy Agency Greenhouse Gas Programme' (2010) 85 <www.ccsassociation.org.uk/docs/2010/IEA%20Pipeline%20final%20report%20 270410.pdf> accessed 4 January 2012.

new CO_2 pipelines to be designed in such a way that they would be able to take any CO_2 stream of a given minimum quality.[8]

Such minimum technical standards could be further elaborated in national net codes, which could, like in the gas sector,[9] be declared to the Commission or in EU net codes. Preferably, these net codes would be applied by a single designated system operator, with national competition authorities guaranteeing a fair application of the relevant technical standards. Another possible solution would be for the Commission to provide (not legally binding) guidance on the question of when technical incompatibilities can no longer 'reasonably' be overcome, narrowing down the possibilities for refusal on such grounds to certain set instances. The Commission has not addressed this issue in its guidance document on CO_2 stream purity, nor, to my knowledge, in any other document.[10]

The analysis in *Chapter VII* reveals that the legal issue examined in that chapter will probably not significantly hinder the cross-border deployment of CCS in the EU. In the Introduction, I stated that the silence on the development and management of CO_2 transport infrastructure (capacity allocation and congestion management)[11] in relation to (future) cross-border capacity requirements in Article 21 of the CCS Directive (third-party access) could hinder the cross-

---

8   European Parliament, 'Report on the proposal for a directive of the European Parliament and of the Council on the geological storage of carbon dioxide and amending Council Directives 85/337/EEC, 96/61/EC, Directives 2000/60/EC, 2001/80/EC, 2004/35/EC, 2006/12/EC and Regulation (EC) No 1013/2006 (COM(2008)0018 – C6-0040/2008 – 2008/0015(COD))' A6–0414/2008 amendment 111 <www.europarl.europa.eu/sides/getDoc.do?language=EN&reference=A6–0414/2008> accessed 26 May 2011. The justification for the amendment stated that: 'this amendment makes it possible to prevent restriction of access and discrimination on the grounds of alleged technical incompatibility. It also gives every plant builder in Europe the assurance that their own CO2 – provided that they prepare it to a given quality standard – will not be excluded from the transport network on the grounds of technical incompatibility'.
9   See Article 8 (technical rules) of European Parliament and Council Directive 2009/73/EC of 13 July 2009 concerning common rules for the internal market in natural gas and repealing Directive 2003/55/EC [2009] OJ L211/94.
10   Commission, 'Implementation of Directive 2009/31/EC on the Geological Storage of Carbon Dioxide – Guidance Document 2: Characterisation of the Storage Complex, CO2 Stream Composition, Monitoring and Corrective Measures' (2011) <http://ec.europa.eu/clima/policies/lowcarbon/docs/gd2_en.pdf> accessed 26 May 2011.
11   The term capacity allocation refers to the mechanisms that are used by the network operators to allocate new capacity on EU gas and electricity transport markets. For an overview of the types of capacity allocation mechanisms in EU gas and electricity transport markets, see Commission, 'Commission Staff Working Document on Capacity Allocation and Congestion Management for Access to the Natural Gas Transmission Networks Regulated under Article 5 of Regulation (EC) No 1775/2005 on Conditions for Access to the Natural Gas Transmission Networks' SEC (2007) 822. Congestion management procedures are instruments used to allocate and re-allocate capacity allocated by the transmission system operator to market participants in a transparent, fair and non-discriminatory manner to prevent or remedy a

border deployment of CCS in the EU. Yet four EU competition cases in the gas sector appear to impose a number of obligations in this area.

The findings in Chapter VII show that CO_2 transport operators might, in spite of the lack of regulation in this regard, be under several obligations related to the cross-border capacity allocation and congestion management of such infrastructure. They could, for instance, be under the obligation to run open seasons to test (potential) market demand for CO_2 transport capacity, including (potential) demand for cross-border CO_2 transport capacity.

Furthermore, they could be required to invest in additional transport capacity if there is (cross-border) demand for extra capacity and such expansion can be profitably realised. In Chapter VI, I indicated that even though CO_2 transport operators have possibilities to avoid liability under Article 102 TFEU, these are limited. In my opinion, the mechanisms outlined in Chapter VII would go a long way to prevent the cross-border deployment of CCS in the EU from being significantly hindered. To be sure, the Commission could, as I have argued with regard to the requirement to run open seasons, propose to include these mechanisms in the CCS Directive, as part of the review of the CCS Directive in 2015. In that way, requirements can be imposed not only on those transport operators that are in a dominant position in the relevant market, but on all EU CO_2 transport operators.

By contrast, the analysis conducted in *Chapter VIII* of the thesis indicates that there might be problems in relation to the legal issue raised in that chapter. In the Introduction, I argued that the Commission, as a consequence of the little onshore storage capacity made available in the EU in recent years, might try and assume a more central role in the allocation of CO_2 storage locations. Such a role could arguably facilitate the cross-border deployment of CCS since it would create CO_2 storage opportunities for Member States with little or none of their own storage capacity. The analysis conducted in Chapter VIII nevertheless shows that if the Commission were to amend Article 4(1) of the CCS Directive in order for the provision to allow the Commission to force Member States to accept CO_2 storage in (parts of) their territory, the thus amended Article 4(1) could very well breach Article 345 TFEU. Article 4(1) currently provides that Member States retain the right to determine the areas from which storage sites are selected, including the right not to allow any CO_2 storage in parts or in the whole of their territory.

---

situation of scarcity of network capacity. See ERGEG, 'ERGEG Principles: Capacity Allocation and Congestion Management in Natural Gas Transmission Networks' (2008) 17.

On the basis of the findings in Chapter VIII, the conclusion has to be that the obstacle to the cross-border deployment of CCS caused by the right of the Member States not to allow CO_2 storage in their territory can likely not be removed by allowing the Commission to force Member States to accept CO_2 storage in (parts of) their territory.

However, there might be other ways of solving this issue. As I have argued above, the Commission could devise a common EU strategy for CO_2 storage, including the mapping and, on condition of approval by the relevant Member State, designation of suitable EU storage locations. Considering the differences in storage capacity endowments between the various Member States, the development of a common EU CO_2 storage strategy would only appear to be a natural step. For such a solution to work there would have to be strong incentives for Member States to volunteer as a host for CO_2 storage. These incentives could be financial, for instance through the granting to the storage operator of an extra premium per tonne of CO_2 stored. Alternatively, the Commission could try to persuade Member States to volunteer by offering those Member States that do provide storage options a less stringent national greenhouse gas emissions reduction target in sharing the burden of the EU's contribution to a possible future post-Kyoto climate agreement.[12]

Finally, I do not expect the legal issue raised in *Chapter IX* to form a significant obstacle to the future cross-border deployment of CCS in the EU. In the Introduction to the thesis, I mentioned the possibility of Member States providing public funding for the development of CO_2 transport and storage infrastructure. Yet, when Member States decide to do so, EU state aid provisions come into play. Article 107(1) TFEU prohibits Member States from granting aid that distorts competition and affects trade between Member States. Under Article 108(3) TFEU, Member States have to notify the Commission of (plans to grant) state aid. The uncertainty and loss of time caused by the notification and standstill procedure under Article 108(3) TFEU could delay the development of CO_2 transport and storage infrastructure, the early availability of which is of great importance for the cross-border deployment of CCS in the EU. In Chapter IX, I examined a possible route for avoiding the applicability of Articles 107 and 108 TFEU by having the public funding of CO_2 transport and storage infrastructure meet the *Altmark* criteria, in particular the fourth *Altmark* criterion (the 'efficiency criterion'). The latter provides that compensation for the

---

[12]   In such situation, host states will have an increased financial risk related to CO_2 storage. Yet, in my view, this would not necessitate financial compensation, since such increased financial burdens will eventually be paid for by foreign capturers through storage operators passing on their costs related to the financial security and mechanism to the capturers having their CO_2 stored.

discharge of a service of general economic interest is only not considered to be state aid if the service is provided at the lowest least societal cost.

The findings in Chapter IX show that there is a relatively straightforward way for Member States to meet the *Altmark* efficiency criterion. On the condition that the other three *Altmark* criteria are also met, the selection of the CO_2 transport or storage operator through a public procurement procedure in line with applicable EU public procurement law would normally lead to the public funding of CO_2 transport and storage infrastructure escaping the procedure under Article 108(3) TFEU. In this regard, it is of great importance that the transport and storage of CO_2 can be characterised as services of general economic interest.

CONSEQUENCES OF THE LEGISLATIVE APPROACH CHOSEN

In addition to the above, I posed the question of in what way the legislative approach chosen by the Commission, in view of the primary EU law context, is generally likely to lead to problems for the cross-border deployment of new technologies such as CCS. At first sight, the learning-by-doing character of the CCS Directive, reflected in the open and general framework set by the Directive, would appear to pose significant problems for an early cross-border deployment of CCS in the EU. However, the above findings suggest that, on the whole, primary EU law and the related case law and decision practice of the European Courts and the Commission seem to provide a safety net for any omissions and obstacles in the CCS Directive in relation to the cross-border deployment of CCS in the EU.

Good examples of the primary EU law context remedying the legislative approach of the CCS Directive can be found in Chapters III, IV, VI, VII and IX. Chapter III shows that when Member States seek to adopt stricter national measures,[13] the requirements that such standards have to be in line with the broad logic of the underlying provision in the CCS Directive as well as with the general spirit/structure of the Directive provide important safeguard mechanisms. Likewise, the extensive Article 110 TFEU case law explored in Chapter IV functions as a safety net for the design of Member States' financial security/mechanism charges and would normally prevent (or remedy) discriminatory or protective taxation of CO_2 streams imported for storage.

---

[13] In this respect, it is important to stress that, in my opinion, any Member State purity criterion that would either ban the presence of certain substances in the CO_2 stream or require a minimum CO_2 content, would have to be characterised as more stringent than the criteria in Article 12 of the CCS Directive. After all, the latter provision does not set those requirements.

Chapter VI reveals the significance of Article 102 TFEU in the event of a refusal to grant access to CO_2 transport and storage infrastructure on technical grounds and shows the limited possibilities for operators to avoid liability under this provision. In Chapter VII, we saw that the Commission's decision practice on Article 102 TFEU and foreclosure of EU gas markets might compensate for the lack of regulation of the development and management of CO_2 transport infrastructure.[14] Finally, Chapter IX reveals that the EU state aid regime can be bypassed by following the fairly straightforward *Altmark* path.

Nevertheless, it is important to underline that administrative or legal procedures for removing obstacles to the cross-border deployment of CCS in the EU can be costly and time-consuming. The role of the primary EU law context in remedying the legislative approach of the CCS Directive is therefore to a large extent limited to providing (costly) ex-post solutions.

As for Chapters V and VIII of the thesis, primary EU law and the related case law and decision practice do not appear to help solve the legal issues explored therein. The analysis in Chapter V shows that Articles 4(3) TEU and 194(1)(c) TFEU do not take away the possibility of Member States (completely) shutting off their CO_2 transport and/or storage markets. In a similar vein, the case law on Article 345 TFEU would generally seem to prevent the Commission from taking legislative action to tackle the problem of Member States refusing to allow CO_2 storage in their territory.

The above suggests that the legislative approach chosen by the Commission could, in relation to the development and the accessibility of crucial infrastructures, lead to problems for the cross-border deployment of new technologies such as CCS. Whereas the legal issues examined in this thesis will likely not lead to complications in the areas of (differential) product standards or the taxation of products transported cross-border, the Commission's legislative approach appears to hinder the development and accessibility of cross-border CO_2 transport and storage infrastructure. More specifically, the Commission seems to have insufficient means to play a central role in the development and accessibility of EU CO_2 transport and storage infrastructure. This illustrates the risks inherent in a legislative approach which is focused on learning-by-doing and which gives the Member States a considerable margin of regulatory discretion. This approach could lead to insufficient crucial infrastructure being developed, as a consequence of which overarching policy objectives could possibly not be met. As I have indicated above, there generally appear to be two

---

[14] Yet, as mentioned earlier, CO_2 transport operators have limited possibilities to avoid liability under Article 102 TFEU.

ways to prevent these problems from becoming obstacles to the growth of CCS in the EU.

One would be for the Commission, as part of the review of the CCS Directive in 2015, to propose to amend the Directive in such a manner that sufficient development and accessibility of CO_2 transport and storage infrastructure is ensured. This could come down to the insertion of provisions that, for instance, require operators to run open seasons when developing CO_2 transport and/or storage infrastructure, invest in capacity enlargements when there is demand for extra capacity and it can be profitable realised, and accept captured CO_2 for transport/storage when it meets certain minimum standards. Yet, as we have seen, this approach would not solve the legal issue explored in Chapter VIII of the thesis. What is more, since the availability and accessibility of CO_2 transport and storage infrastructure also importantly depends on sufficient financial means becoming available, this would perhaps not be the most suitable instrument for preventing these problems from materialising.

A second option would be for the Commission to develop a common EU strategy for CO_2 transport and storage that also addresses the crucial issue of financing the development of transport and storage infrastructure which can be used by market players from different Member States. In order to develop such a strategy, a number of steps would have to be taken.

First, as I have argued above, EU storage capacity potential would have to be mapped. For this, the work of the GeoCapacity project could be used.[15] Within the framework of this finalised EU project, the European capacity for the geological storage of CO_2 in deep saline aquifers, oil and gas structures and coal beds has been assessed.

Second, taking into account the diminished possibilities for onshore CO_2 storage in the (North-West of the) EU, possible storage locations would have to be linked to expected volumes of captured CO_2. Here, the results and the recommendations of the CO_2Europipe project, an EU project on the development of a large-scale Europe-wide infrastructure for the transport and injection of captured CO_2, could be used.[16] As I have indicated in the Introduction, the project's report on storage capacity contains two offshore (storage) only scenarios, including projections on the required lay-out of infrastructure.[17]

---

[15] EU GeoCapacity, 'Assessing European Capacity for Geological Storage of Carbon Dioxide' (2009) <www.geology.cz/geocapacity/publications/D16%20WP2%20Report%20storage%20capacity-red.pdf> accessed 13 July 2011.

[16] CO_2Europipe (n 3).

[17] Ibid 20–21.

Third, on the basis of these data, a common EU strategy that addresses the planning and the financing of the development of the required infrastructure will have to be drawn up. The final report of the CO_2Europipe project contains an example of a possible roadmap, suggesting, for instance, that Member State governments and industry jointly develop a finance model that supports the oversizing of infrastructure as well as the development of guidelines on minimum standards for CO_2 stream purity.[18]

The 2013 Regulation on guidelines for trans-European energy infrastructure (TEN-E Regulation) shows that the Commission has taken the calls for a master plan on the development of EU-wide CO_2 transport infrastructure to heart.[19] The Regulation lays down guidelines for the timely development and interoperability of priority corridors and areas of trans-European energy infrastructure, including CO_2 transport infrastructure between Member States and with neighbouring third countries.[20] The Commission aims to implement priority CO_2 transport infrastructure by streamlining permit-granting procedures and increasing public participation and acceptance, and by providing market-based and direct EU financial support.[21] The financial assistance will be provided within the framework of the 'Connecting Europe Facility', which is subject to a separate legislative proposal.[22] According to the Commission, the future development of a cross-border network for CO_2 transport requires steps to be taken now for European-level infrastructure planning and development.[23]

Yet, even though the TEN-E Regulation addresses a number of main obstacles to the development of priority CO_2 transport infrastructure, it does not seem to fully resolve the issues mentioned above. First, it predominantly deals with the development of CO_2 *transport* infrastructure and does not address the development of sufficient EU CO_2 *storage* capacity. Second, the Regulation does not appear to address the issue of the *accessibility* of priority CO_2 transport infrastructure, once this infrastructure has been realised. I would argue that for the Commission's proposal to successfully tackle the problem of possible

[18] CO_2Europipe (n 3) 51–55.
[19]   Regulation (EU) No 347/2013 of the European Parliament and Council of 17 April 2013 on guidelines for trans-European energy infrastructure and repealing Decision No. 1364/2006/EC and amending Regulations (EC) No 713/2009, (EC) No 714/2009 and (EC) No 715/2009 [2013] OJ L115/39.
[20]   Ibid, Article 1(1) and Annex I.
[21]   Commission, 'Proposal for a Regulation of the European Parliament and of the Council on Guidelines for Trans-European Energy Infrastructure and Repealing Decision No 1364/2006/EC' COM (2011) 658 final 3.
[22]   See the Commission's website at <https://ec.europa.eu/digital-agenda/en/connecting-europe-facility> accessed 12 November 2012, and Commission, 'Proposal for a Regulation of the European Parliament and of the Council Establishing the Connecting Europe Facility' COM (2011) 665 final.
[23]   Commission, 'Proposal on Trans-European Energy Infrastructure' (n 21) 3.

obstacles to the cross-border deployment of CCS in the EU, it would also have to include proposals on the development of EU CO_2 storage infrastructure as well as on the accessibility of priority CO_2 transport and storage infrastructure.[24]

In conclusion, I would advise the Commission to come up with plans for the development of CO_2 storage infrastructure and to make sure that there is sufficient access to priority EU CO_2 transport and storage infrastructure, in order to ensure that the (near-)future cross-border deployment of CCS in the EU is not hindered. Such plans would be all the more desirable given the to a large extent (costly) ex-post role of primary EU law and the related case law and decision practice of the European Courts and the Commission. Should the Commission fail to come up with plans and rules for the development and accessibility of CO_2 transport and storage infrastructure, then there is a risk of the path dependency inherent in the approach chosen for the CCS Directive hindering the cross-border deployment of CCS in the EU.[25] In fact, in the impact assessment of its Communication on Energy Infrastructure Priorities for 2020 and beyond, the Commission itself points to the risk of initial pipeline infrastructure being constructed without regard for future transport capacity needs.[26] According to the Commission, without intervention, CO_2 pipelines installed in 2014–2020 will be relatively short associated with specific projects and tailored to the (particular) needs of these projects.[27] Such risks are only aggravated if more specific rules on the access to CO_2 transport and storage infrastructure are lacking. In view of the planned review of the CCS Directive in 2015, the Commission would therefore be wise to at least explore the desirability of additional plans for ensuring a smooth (early) cross-border deployment of CCS in the EU.

---

[24] As I have argued before, the roadmap in the CO_2Europipe report provides good examples in this regard.

[25] In this regard, see, for instance, also CO_2Europipe (n 3) 20. In relation to the development of a more complex EU CO_2 transport network, the authors of the CO_2Europipe report argue that it may be challenging to change the initial organisational models 'due to the difficulty in changing established models with proven experience and momentum'.

[26] Commission, 'Accompanying Document to the Communication from the Commission to the European Parliament, the Council, the European Economic and Social Committee and the Committee of the Regions, Energy Infrastructure Priorities for 2020 and Beyond: Impact Assessment' SEC (2010) 1395 final 21.

[27] Ibid 21.

BIBLIOGRAPHY

LITERATURE/REPORTS

Ahner N, 'Final Report' (Florence School of Regulation workshop on unbundling of energy undertakings in relation to corporate governance principles, Berlin, 25 September 2009) 8 <www.florence-school.eu/portal/page/portal/LDP_HOME/ Events/Workshops_Conferences/2009/Unbundling/CorporateGovernance_AX-Programme.pdf> accessed 6 August 2012

Akkermans B and E Ramaekers, 'Article 345 TFEU (ex Article 295 EC), Its Meanings and Interpretations' (2010) 16 European Law Journal 292

Alemanno A, 'The Shaping of the Precautionary Principle by the European Courts: From Scientific Uncertainty to Legal Certainty' (2007) <http://papers.ssrn.com/sol3/ papers.cfm?abstract_id=1007404> accessed 24 May 2010

Alexiadis P, 'Informative and Interesting: The CFI Rules in Deutsche Telekom v. European Commission' (2008) 1 GCP Magazine 10

American Bar Association (ABA), Telecom Antitrust Handbook (ABA Publishing, 2005)

Amtenbrink F and HHB Vedder, Recht van de Europese Unie (4th edn, Boom Juridische Uitgevers 2010)

Anderman S, 'The Epithet That Dares Not Speak its Name: The Essential Facilities Concept in Article 82 EC and IPRs after the Microsoft Case' in A Ezrachi (ed), Article 82 EC: Reflections on its Recent Evolution (Hart Publishing 2009)

Andoura S, L Hancher and M van der Woude, 'Towards a European Energy Community: A Policy Proposal' (2010) <www.notre-europe.eu/uploads/tx_publication/Etud76-Energy-en.pdf> accessed 5 September 2011

Andrews AM, 'Picking up on What's Going Underground: Australia Should Exempt Carbon Capture and Geo-Sequestration from Part IIIA of the Trade Practices Act' (2008) 17 Pacific Rim Law and Policy Journal 407

Apotheker DF, 'The Design of A Regulatory Framework for a Carbon Dioxide Pipeline Network' (MSc thesis, Technical University of Delft 2007)

Arena A, 'The Relationship between Antitrust and Regulation in the US and the EU: An Institutional Assessment' (2011) Institute for International Law and Justice Emerging Scholar Papers 19 <www.iilj.org/publications/documents/ESP19–2011Arena.pdf> accessed 1 April 2011

Armeni C, 'An Update of the State of CCS Regulation in Europe' (2012) <www. globalccsinstitute.com/insights/authors/chiara-armeni/2012/02/13/update-state-ccs-regulation-europe> accessed 11 October 2012

— 'Key Legal Issues Arising from the Transposition of the EU CO2 Storage Directive' (presentation held at the 4th IEA International CCS Regulatory Network Meeting,

Paris, 2012) <www.iea.org/media/workshops/2012/ccs4thregulatory/Chiara_Armeni.pdf> accessed 11 October 2012

Arnull A, *The European Union and its Court of Justice* (2nd edn, Oxford University Press 2006)

Baldini C, 'Italian Competition Authority Case A358 – International Transport of Gas, ENI – TTPC (2005–2006)' (presentation at the International Competition Network Teleseminar on Abuse of Dominance in the EU Energy Sector, 8 November 2011) <www.internationalcompetitionnetwork.org/uploads/library/doc768.pdf> accessed 8 February 2012

Barnard C, *The Substantive Law of the EU: The Four Freedoms* (3rd edn, Oxford University Press 2010)

Bartosch A, 'Clarification or Confusion:? How to Reconcile the ECJ's Rulings in Altmark and Chronopost?' (2003) CLaSF Working Paper Series Number 2 <www.clasf.org/assets/CLaSF%20Working%20Paper%2002.pdf> accessed 3 March 2012

— 'The Ruling in BUPA – Clarification or Modification of Altmark?' (2008) 7 European State Aid Law Quarterly 211

Becker F, 'Market Regulation and the "Right to Property" in the European Economic Constitution' in *Yearbook of European Law* (Oxford University Press 2007)

Bengtsson C, 'Competition and Energy' (presentation at the CRA International Competition Workshop 'Competition Policy in the EU Energy Sector', Brussels, 12 February 2009) <www.crai.com/ecp/assets/Claes_BengtssonFeb09.pdf> accessed 1 February 2012

Bennaceur K and D Gielen, 'Energy Technology Modelling of Major Carbon Abatement Options' (2010) 4 International Journal of Greenhouse Gas Control 309

Benson D and A Jordan, 'A Grand Bargain or an "Incomplete Contract"? European Union Environmental Policy after the Lisbon Treaty' (2008) 17 European Energy and Environmental Law Review 280

Bergman MA, 'The Bronner Case: A Turning Point for the Essential Facilities Doctrine?' (2000) 21 European Competition Law Review 59

Berliner Informationsdienst, 'Einigung im Vermittlungsausschuss zur Solarförderung und CCS' (2012) <www.polisphere.eu/bid/einigung-im-vermittlungsausschuss-zur-solarforderung-und-ccs/> accessed 20 August 2012

Bermann GA and others, *Cases and Materials on European Union Law* (3rd edn, West Publishing Co 2011)

Beukenkamp A, 'Pipeline-to-pipeline Competition: An EU Assessment' (2009) 27 Journal of Energy and Natural Resources Law 5

Bielicki J, 'Carbon Capture and Storage and the Location of Industrial Facilities' (presentation held at the Research Experience in Carbon Sequestration 2007, Montana State University, 2 August 2007) <belfercenter.ksg.harvard.edu\> accessed 13 April 2011

— 'Returns to Scale in Carbon Capture and Storage Infrastructure and Deployment' (2008) Harvard Kennedy School Energy Technology Innovation Policy Discussion Paper 2008–04 26 <http://belfercenter.ksg.harvard.edu/files/Bielicki_CCSReturnsToScale.pdf> accessed 27 April 2011

Bielinski A, 'Numerical Simulation of CO2 Sequestration in Geological Formations' (PhD thesis, University of Stuttgart 2007) <http://d-nb.info/996784012/34> accessed 9 May 2012

Bijlsma M and others, 'Vertical Foreclosure: a Policy Framework' (2008) CPB Netherlands Bureau for Economic Policy Analysis Document No 157 27 <www.cpb.nl/en/publication/vertical-foreclosure-policy-framework> accessed 13 February 2012

Birkeland L, 'Burying CO2: The New EU Directive on Geological Storage of CO2 from a Norwegian Perspective' (2009) <www.bellona.org/filearchive/fil_Bellonas_paper_-_Burying_CO2-_The_New_EU_Directive_on_Geological_Storage_of_CO2_from_a_Norwegian_Perspective.pdf> accessed 23 March 2011

— and others, 'Improving the Regulatory Framework, Optimizing Organization of the CCS Value Chain and Financial Incentives for CO2-EOR in Europe' (2011) <http://bellona.org/ccs/ccs-news-events/publications/article/improving-the-regulatory-framework-optimizing-organization-of-the-ccs-value-chain-and-financial-inc.html> accessed 2 January 2011

Blank FT and others (CATO2), 'Permitting Cross-border Networks in Relation to Monitoring, Verification and Accounting under EU-ETS' (2011) <www.co2-cato.org/publications/publications/permitting-cross-border-networks-in-relation-to-monitoring-verification-and-accounting-under-eu-ets> accessed 11 October 2012

Blohm-Hieber U, 'The Radiaoctive Waste Directive: A Necessary Step in the Management of Spent Fuel and Radioactive Waste in the European Union' (2011) 87 Nuclear Law Bulletin 21

Borgmann-Prebil Y and M Ross, 'Promoting European Solidarity: Between Rhetoric and Reality?' in M Ross and Y Borgmann-Prebil (eds), *Promoting Solidarity in the European Union* (Oxford University Press 2010) 1

Bovis C, 'Public Service Obligations in the Transport Sector: The Demarcation between State Aids and Services of General Interest in EU Law (2005) 6 European Business Law Review 1329

Boyd A and J Teal, 'Interpreting the Altmark Decision: The Challenges from a Private Practitioner's Perspective' <www.mcgrigors.com/pdfdocs/pl_state_aid_paper.pdf> accessed 5 March 2012

Braun J and J Kühling, 'Article 87 EC and the Community Courts: From Revolution to Evolution' (2008) 45 Common Market Law Review 465

Brennan TJ, 'The Economics of Competition Policy: Recent Developments and Cautionary Notes in Antitrust and Regulation' (2000) Resources for the Future Discussion Paper 00–07 <www.rff.org/documents/RFF-DP-00-07.pdf> accessed 14 February 2012.

British Geological Survey, 'Sequestration: The Underground Storage of Carbon Dioxide' in EJ Moniz (ed), *Climate change and energy pathways for the Mediterranean: Workshop proceedings*, Cyprus (Springer 2007)

Brockett S, 'The EU Enabling Legal Framework for Carbon Capture and Storage and Geological Storage' (2009) 1 Energy Procedia 4433

Buendia Sierra JL, 'Finding the Right Balance: State Aid and Services of General Economic Interest' in *EC State Aid Law/ Le Droit des Aides d'Etat dans la CE – Liber Amicorum Francisco Santaolalla Gadea* (Kluwer Law International 2008)

Buigues P and R Klotz, 'Margin Squeeze in Regulated Industries: The CFI Judgment in the Deutsche Telekom Case' (2008) 7 CPI Antitrust Chronicle 16

California Carbon Capture and Storage Review Panel, 'Technical Advisory Committee Report: Long-Term Stewardship and Long-Term Liability in the Sequestration of CO_2' (2008) <www.climatechange.ca.gov/carbon_capture_review_panel/meetings/2010-08-18/white_papers/Long-Term_Stewardship_and_Long-Term_Liability.pdf> accessed 7 May 2012

Candeub A, 'Trinko and Re-grounding the Refusal to Deal Doctrine' (2005) 66 University of Pittsburgh Law Review 821

Capobianco A, 'The Essential Facilities Doctrine: Similarities and Differences Between the American and the European Approach' (2001) 26 European Competition Law Review 548

Carbon Capture Journal, 'CO2 Capture Project CO2 Impurities Study' (8 July 2012) <www.carboncapturejournal.com/displaynews.php?NewsID=974&PHPSESSID=mmpj7lh3tbba5mjhv5q10fkhc0> accessed 21 August 2012

—   'Report Says CO2 Leaks Could Contaminate Drinking Water' (2010) <www.carboncapturejournal.com/displaynews.php?NewsID=685> accessed 25 July 2012

Carbon Sequestration Leadership Forum (CSLF), 'Phase II Final Review from the Taskforce for Review and Identification of Standards for CO_2 Storage Capacity Estimation' (2007)

—   'Technology Roadmap' (2009)

Cardoso R and others, 'The Commission's GDF and E.ON Gas Decisions Concerning Long-term Capacity Bookings: Use of Own Infrastructure as Possible Abuse under Article 102 TFEU' (2010) European Commission Competition Policy Newsletter 2010–3, 4 <http://ec.europa.eu/competition/publications/cpn/cpn_2010_3_2.pdf> accessed 29 April 2011

CCSReg Project, 'Carbon Capture and Sequestration: Framing the Issues for Regulation – An Interim Report from the CCSReg Project' (2009) <www.ccsreg.org/pdf/CCSReg_3_9.pdf> accessed 7 May 2012

—   'Policy Brief: Compensation, Liability and Long-term Stewardship for CCS' (2009) <www.ccsreg.org/pdf/LongTermLiability_07132009.pdf> accessed 9 May 2012

Chalmers D, Davies G and Monti G, European Union Law (2nd ed, Cambridge University Press 2010)

China-UK Near Zero Emissions COAL (NZEC) Initiative, 'Summary Report' (2009) <www.nzec.info/en/assets/Reports/China-UK-NZEC-English-031109.pdf> accessed 22 March 2010

CO_2 Capture Project, 'Glossary' <www.co2captureproject.com/glossary.html> accessed 24 March 2010

Comfort N, 'RWE May Build Dutch Carbon Capture Plant Amid German Opposition' Bloomberg (22 January 2010) <www.bloomberg.com/apps/news?pid=20601100&sid=aOZGhPaq04MY> accessed 15 April 2010

Corkin J, 'Science, Legitimacy and the Law: Regulating Risk Regulation Judiciously in the European Union' (2008) 33 European Law Review 359

Craig P, EU Administrative Law (Oxford University Press 2006)

—   and G de Búrca, EU Law: Text, Cases, and Materials (4th edn, Oxford University Press 2007)

— and G de Búrca, *EU Law: Text, Cases, and Materials* (3rd edn, Oxford University Press 2003)

Dagilyte E, 'Solidarity after the Lisbon Treaty: A new General Principle of EU Law?' <www.pravo.hr/_download/repository/Dagilyte_abstract.pdf> accessed 23 October 2012

Dashwood A and others, *Wyatt and Dashwood's European Union Law* (6th edn, Hart Publishing 2011)

Davies G, 'Process and Production Method-Based Trade Restrictions in the EU' (2008) <http://papers.ssrn.com/sol3/papers.cfm?abstract_id=1118709> accessed 13 June 2012

Delbeke J, 'Enabling Legal Framework for Carbon Dioxide Capture and Geological Storage' <http://ec.europa.eu/clima/events/0006/jd_presentation_en.pdf> accessed 1 October 2012

De Búrca G, 'The Principle of Proportionality and its Application in EC Law' in A Barav, DA Wyatt QC and J Wyatt (eds), *Yearbook of European Law* 1993 (Oxford University Press 1993)

De Figueiredo M and others, 'The liability of Carbon Dioxide Storage' <http://sequestration.mit.edu/pdf/GHGT8_deFigueiredo.pdf> accessed 23 June 2010.

De Hauteclocque A, 'Long-term Supply Contracts in European Decentralized Electricity Markets: An Antitrust Perspective' (Ph.D. thesis, University of Manchester 2009)

— and V Rious, 'Reconsidering the Regulation of Merchant Transmission Investment in the Light of the Third Energy Package: The Role of Dominant Generators' (2009) Reflexive Governance in the Public Interest Working Paper Series REFGOV-IFM-67

De Sadeleer N, 'The Impact of the Registration, Evaluation and Authorization of Chemicals (REACH) Regulation on the Regulatory Powers of the Nordic Countries' in N de Sadeleer (ed), *Implementing the Precautionary Principle: Approaches from the Nordic Countries, EU and USA* (Earthscan Ltd 2006)

De Witte B, 'Institutional Principles: A Special Category of General Principles of EC Law' in U Bernitz and J Nergelius (eds), *General Principles of European Community Law* (Kluwer Law International 2000)

Del Guayo I, G Kühne and M Roggenkamp, 'Ownership Unbundling and Property Rights in the EU Energy Sector' in A McHarg and others (eds), *Property and the Law in Energy and Natural Resources* (Oxford University Press 2010)

Det Norske Veritas, 'Design and Operation of CO_2 Pipelines' (2010)

Dethmers F and N Dodoo, 'The Abuse of Hoffman-La Roche: The Meaning of Dominance under EC Competition Law' (2006) 27 European Competition Law Review 537

— and H Engelen, 'Fines under Article 102 of the Treaty on the Functioning of the European Union' (2011) 32 European Competition Law Review 86

Dixon T, 'Future Direction of CCS Directive toward Defining CO2 Quality" <www.co2captureandstorage.info/docs/oxyfuel/Discussion%20Purity/01%20-%20T.%20Dixon%20%28IEA%20GHG%29.pdf> accessed 11 March 2011

— and others, 'International Marine Regulation of CO2 Geological Storage. Developments and Implications of London and OSPAR' (2009) 1 Energy Procedia 4503

Doherty B, 'Just What are Essential Facilities?' (2001) 38 Common Market Law Review 397

Dougan M, 'Minimum Harmonization and the Internal Market' (2000) 37 Common Market Law Review 853

— 'Legal Developments' (2004) 42 Journal of Common Market Studies 77

Dutch Ministry of Economic Affairs, Agriculture and Innovation, 'Uitwerking van de afspraken voor de individuele CO2-opslagprojecten die momenteel in voorbereiding zijn' (2010) <www.rijksoverheid.nl/ministeries/eleni/documenten-en-publicaties/kamerstukken/2010/11/04/uitwerking-van-de-afspraken-voor-de-individuele-co2-opslagprojecten-die-momenteel-in-voorbereiding-zijn.html> accessed 12 December 2010)

— 'CCS-projecten in Nederland' (2011) <www.rijksoverheid.nl/documenten-en-publicaties/kamerstukken/2011/02/14/ccs-projecten-in-nederland.html> accessed 26 May 2011

Dunne N, 'Margin Squeeze: Theory, Practice, Policy: Part 1' (2012) 33 European Competition Law Review 29

Eeckhout P, *EU External Relations Law* (2nd edn, Oxford University Press 2011)

Ehlers E, 'Electricity and Gas Supply Network Unbundling in Germany, Great Britain and the Netherlands and the Law of the European Union: A Comparison' (PhD thesis, University of Tilburg 2009)

Element Energy, 'CO$_2$ Pipeline Infrastructure: An Analysis of Global Challenges and Opportunities' (2010) <www.ccsassociation.org.uk/docs/2010/IEA%20Pipeline%20final%20report%20270410.pdf> accessed 2 January 2012

— 'One North Sea: A Study into North Sea Cross-border CO$_2$ Transport and Storage' (2010) 85 <www.regjeringen.no/upload/OED/OneNortSea_Fulldoc.pdf> accessed 11 October 2012

Emch A, 'Same Same but Different? Fiscal Discrimination in WTO Law and EU Law: What Are "Like" Products?' (2005) 32 Legal Issues of Economic Integration 369

Energiebeheer Nederland (EBN) and Gasunie, 'CO2 Transport and Storage Strategy' (2010)

Engle E, 'General Principles of European Environmental Law' (2009) 17 Penn State Environmental Law Review 215

E.ON, 'Anlage A zum Schreiben vom 07.01.2010 COMP/B-1/39.317 – E.ON Gas – Zusagen an die Europäische Kommission' (2010) <http://ec.europa.eu/competition/antitrust/cases/dec_docs/39317/39317_1729_6.pdf> accessed 6 February 2012

Etteh N, 'Carbon Capture and Storage: Liability Implications' (2009) University of Dundee Centre for Energy, Petroleum and Mineral Law and Policy Annual Review 2009/10 <www.dundee.ac.uk/cepmlp/gateway/index.php?news=31264> accessed 11 October 2012

Euratom Supply Agency, 'Annual Report 2011' <http://ec.europa.eu/euratom/ar/last.pdf> accessed 8 August 2012

European CCS Demonstration Project Network, 'Thematic Report: Regulatory Development Session May 2012' (2012) <www.ccsnetwork.eu/uploads/publications/thematic_report_-_regulatory_development_session_-_may_2012.pdf> accessed 11 October 2012

European Regulators' Group for Electricity and Gas (ERGEG), 'ERGEG Principles: Capacity Allocation and Congestion Management in Natural Gas Transmission Networks' (2008)

Evrard SJ, 'Essential Facilities in the European Union: Bronner and Beyond' (2004) 10 Columbia Journal of European Law 491

Ezrachi A, 'The Commission's Guidance on Article 82 EC and the Effects Based Approach: Legal and Practical Challenges' in A Ezrachi (ed), *Article 82 EC: Reflections on its Recent Evolution* (Hart Publishing 2009)

Fairhurst J, *Law of the European Union* (8th edn, Pearson Eduction Limited 2010)

Fernando H and others (World Resources Institute), 'Capturing King Coal: Deploying Carbon Capture and Storage Systems in the U.S. at Scale' (2008) <http://pdf.wri.org/capturing_king_coal.pdf> accessed 18 March 2010

Fiedziuk N, 'The Interplay between Objectives of the European Union's Energy policy: The Case of State Funding of Energy Infrastructure' (2009) TILEC Discussion Paper <http://papers.ssrn.com/sol3/papers.cfm?abstract_id=1386902> accessed 21 March 2012

Fine F, 'NDS/IMS: A Logical Application of the Essential Facilities Doctrine' (2002) 23 European Competition Law Review 457

Fisher E, 'Is the Precautionary Principle Justiciable?' (2001)13 Journal of Environmental Law 315

— 'Opening Pandora's Box: Contextualising The Precautionary Principle in the European Union' (2007) University of Oxford Faculty of Law Legal Studies Research Paper Series Working Paper No 2/2007 4 <http://papers.ssrn.com/Abstract=956952> accessed 23 May 2011

Fleurke F and H Somsen, 'Precautionary Regulation of Chemical Risk: How REACH Confronts the Regulatory Challenges of Scale Uncertainty, Complexity and Innovation' (2011) 48 Common Market Law Review 357

Fraga FL, T Juutilainen, K Havu and J Vesala, 'Property and European Integration: Dimensions of Article 345 TFEU' (2012) University of Helsinki Faculty of Law Legal Studies Research Paper Series No 17 <http://papers.ssrn.com/sol3/papers.cfm?abstract_id=2012983> accessed 27 July 2012

Frederico G, 'The Economic Analysis of Energy Mergers in Europe and in Spain' (2011) 00 Journal of Competition Law and Economics 1

Gauer C and J Ratliff, 'EU Competition Law and Energy: Recent Cases and Issues' (presentation held at the fifty-first Lunch Talk of the Global Competition Law Centre, 18 March 2011<www.coleurope.be> accessed 30 January 2012

Gaz de France, 'Engagements Proposés Formellement par GDF Suez, GRTgaz et Elengy dans la Cadre de la Procédure COMP/B-1/39.316 – version non confidentielle' (June 2009) <http://ec.europa.eu/competition/antitrust/cases/dec_docs/39316/39316_1854_9.pdf> accessed 26 January 2012

— 'Engagements Proposés Formellement par GDF SUEZ, GRTgaz et Elengy dans la Cadre de la Procédure COMP/B-1/39.316 – version non confidentielle' (October 2009) <http://ec.europa.eu/competition/antitrust/cases/dec_docs/39316/39316_2144_9.pdf> accessed 27 January 2012

Geradin D, 'Limiting the Scope of Article 82 EC: What Can the EU Learn From the US Supreme Court's Judgment in Trinko, in the Wake of Microsoft, IMS, and Deutsche Telekom?' (2004) 41 Common Market Law Review 1519

— 'Refusal to Supply and Margin Squeeze: A Discussion of why the "Telefonica Exceptions" Are Wrong' (2010) TILEC Discussion Paper <www.ssrn.com/abstract=1762687> accessed 29 March 2011

— and R O'Donoghue, 'The Concurrent Application of Competition Law and Regulation: The Case of Margin Squeeze Abuses in the Telecommunications Sector' (2005) 1 Journal of Competition Law and Economics 355

Giannakis M, S Croom and N Slack, 'Supply Chain Paradigms' <http://fds.oup.com/www.oup.co.uk/pdf/0-19-925932-1.pdf> accessed 14 February 2012

Global CCS Institute, 'Strategic Analysis of the Global Status of Carbon Capture and Storage – Report 3: Country Studies, the European Union' (2009) <http://cdn.globalccsinstitute.com/sites/default/files/publications/8517/strategic-analysis-global-status-ccs-country-study-european-union.pdf> accessed 10 October 2012

— 'The Status of CCS Projects – Interim Report 2010' (2010) <www.globalccsinstitute.com/downloads/general/2010/The-Status-of-CCS-Projects-Interim-Report-2010.pdf> accessed 14 September 2010

— 'Accelerating the Uptake of CCS: Industrial Use of Captured Carbon Dioxide' (2011) 38 <www.globalccsinstitute.com/resources/publications/accelerating-uptake-ccs-industrial-use-captured-carbon-dioxide> accessed 22 April 2011

Goldman Sachs, 'Pipeline Financing Discussion' (presentation held at the Wyoming Natural Gas Pipeline Authority, 25 August 2003) <www.wyopipeline.com/information/presentations/2003/Goldman%20Sachs_files/frame.htm> accessed 8 February 2012

Gormley LW, Prohibiting Restrictions on Trade Within the EC (TMC Asser Instituut 1985)

— 'The Development of General Principles of Law Within Article 10 (ex Article 5) EC' in U Bernitz and J Nergelius (eds), General Principles of European Community Law (Kluwer Law International, 2000) 113

— 'Some Further Reflections on the Development of General Principles of Law Within Article 10 EC', in in U Bernitz, J Nergelius and C Cardner (eds), General Principles of EC Law in a Process of Development (Kluwer Law International,2008)

Gormsen LL, 'Why the European Commission's Enforcement Priorities on Article 82 EC Should be Withdrawn' (2010) 31 European Competition Law Review 45

Goyder DG, EC Competition Law (4th edn, Oxford University Press 2003)

Gravengaard MA and N Kjaersgaard, 'The EU Commission Guidance on Exclusionary Abuse of Dominance and its Consequences in Practice' (2010) 31 European Competition Law Review 285

Greenpeace, 'Greenpeace Q&A on Carbon Capture and Storage (CCS)' <www.greenpeace.org/international/Global/international/planet-2/report/2008/5/q-a-on-carbon-capture-storag.pdf> accessed 15 September 2011

Groussot X, General Principles of Community Law (Europa Law Publishing 2006)

Haan-Kamminga A, 'Long-term Liability for Geological Storage in the European Union' (2011) University of Groningen Centre of Energy Law Working Paper <http://papers.ssrn.com/sol3/papers.cfm?abstract_id=1858631> accessed 11 October 2012

Haigh R, 'A Framework for the Future Development of CCS' (2010) Carbon Capture Journal <www.carboncapturejournal.com/displaynews.php?NewsID=583> accessed 27 April 2011

Hancher L, 'The Application of EC State Aid Law to the Energy Sector' in CW Jones (ed), *EU Energy Law* (Claes & Casteels 2005)

— and T Ottervanger and PJ Slot, *EC State Aids* (3rd edn, Sweet & Maxwell 2006)

— and P Larouche, 'The Coming of Age of EU Regulation of Network Industries and Services of General Economic Interest' in P Craig and G de Búrca (eds), *The Evolution of EU Law* (Oxford University Press 2011)

Harbo T, 'The Function of the Proportionality Principle in EU Law' (2010) 16 European Law Journal 158

Hartley TC, *The Foundations of European Union Law* (7th edn, Oxford University Press 2010)

Hatzopoulos VG, 'The Essential Facilities Doctrine (EFD) in EU Antitrust Law' (2006) <www.concorrencia.pt/Download/Essential_Facilities-Hatzopoulos.pdf> accessed 4 May 2011

— 'Public Procurement and State Aid in National Health Care Systems' 379, 385 <www.euro.who.int/__data/assets/pdf_file/0007/138175/E94886_ch09.pdf> accessed 14 March 2012

Havercroft I, 'Long-term Liability of CCS Business' (presentation at the Global CCS Institute's regional members' meeting in Tokyo, 2012) <www.slideshare.net/globalccs/ian-havercroft-global-ccs-institute-longterm-liability-of-ccs-business> accessed 11 October 2012

Hedemann-Robinson M, 'Indirect Discrimination: Article 95(1) EC Back to Front and Inside Out?' (1995) 1 European Public Law 439

Heimler A, 'Is a Margin Squeeze an Antitrust or a Regulatory Violation?' (2010) 6 Journal of Competition Law and Economics 879

Henriksson L, 'Structural Measures in EC Competition Law: A bridge Too Far?' in U Bernitz and others (eds), *General Principles of EC Law in a Process of Development* (Kluwer Law International 2008)

Herold and others, 'Vertical Integration and Market Structure along the Extended Value Added Chain including Carbon Capture, Transport and Sequestration (CCTS)' (2010) <www.feem-project.net/secure/plastore/Deliverables/SECURE_deliverable_5%203%202.pdf> accessed 12 August 2010

Herzog H, 'CO2 Capture and Storage (CCS): Costs and Economic Potential' (presentation at the Joint SBSTA/IPCC side-event, COP11, Montreal, 30 November 2005).

Hetland K, 'Financial Security from an EU Perspective' (presentation held for the International Energy Agency (IEA) CCS Regulators Network 2010) <www.iea.org/media/workshops/2010/financialmech/Hetland.pdf> accessed 11 October 2012

Heyvaert V, 'Facing the Consequences of the Precautionary Principle in European Community Law' (2006) 31 European Law Review 185

Hinchy E, 'Abuse of Dominant Position – Telecommunications' (2011) 34 European Competition Law Review 63

Holdgaard R, *External Relations Law of the European Community* (Kluwer Law International 2008)

Holwerda M, 'Subsidizing Carbon Capture and Storage Demonstration' (2010) 4 Carbon and Climate Law Review 228

— 'Deploying Carbon Capture and Storage Safely: The Scope for Member States of the EU to Adopt More Stringent CO_2 Stream-Purity Criteria under EU Law' (2011) 2 Climate Law 37

Holyoake D and Hill C (ClientEarth), 'Final Hurdles: Financial Security Obligations under the CCS Directive' (2012) <www.clientearth.org/reports/ccsa-report.pdf> accessed 11 October 2012

Horspool M and M Humphreys, *European Union Law* (6th edn, Oxford University Press 2010)

International Competition Network, '2009 Refusal to Deal Questionnaire: Italy' (2009) <www.internationalcompetitionnetwork.org/uploads/questionnaires/uc%20 refusals/italy.pdf> accessed 6 February 2012

— 'Case Annex to ICN Unilateral Conduct Working Group: Report on the Analysis of Refusal to Deal with a Rival under Unilateral Conduct Laws' (2010) 30 <www. internationalcompetitionnetwork.org/uploads/library/doc611.pdf> accessed 6 February 2012

International Energy Agency, 'Prospects for CO2 Capture and Storage' (2004)

— 'CO$_2$ Capture and Storage: A Key Carbon Abatement Option' (2008)

— 'CO$_2$ emissions from fuel combustion – Highlights' (2009)

— 'Technology Roadmap: Carbon Capture and Storage' (2009)

— 'Energy Policies of IEA Countries: Italy 2009 Review' (2010) <www.iea.org/ textbase/nppdf/free/2009/italy2009.pdf> accessed 8 February 2012

— 'Carbon Capture and Storage – Model Regulatory Framework' (2010) <www.iea. org/ccs/legal/model_framework.pdf> accessed 18 March 2011

— 'Carbon Capture and Storage and the London Protocol – Options for Enabling Transboundary CO2 Transfer' (2011) IEA Working Paper <www.iea.org/ publications/freepublications/publication/CCS_London_Protocol.pdf> accessed 11 October 2012

— and Global CCS Institute, 'Tracking Progress in Carbon Capture and Storage – International Energy Agency/Global CCS Institute Report to the third Clean Energy Ministerial' (2012) <www.iea.org/publications/freepublications/publica-tion/IEAandGlobalCCSInstituteTrackingProgressinCarbonCaptureandStoragere-porttoCEM3FINAL-1.PDF> accessed 11 October 2012

International Energy Agency Greenhouse Gas R&D Programme, 'Capturing CO$_2$' (2007)

— 'Geological Storage of Carbon Dioxide: Staying Safely Underground' (2008) <www.co2crc.com.au/dls/external/geostoragesafe-IEA.pdf> accessed 24 March 2010

Intergovernmental Panel on Climate Change (IPCC), 'IPCC Special Report on Carbon Dioxide Capture and Storage. Prepared by Working Group III of the Intergovernmental Panel on Climate Change' (Bert Metz and others (eds) Cambridge University Press 2005)

Italian Competition Authority, 'ENI/Trans Tunisian Pipeline' (2005) <www.agcm.it/en/ newsroom/press-releases/1456-enitrans-tunisian-pipeline.html> accessed 6 February 2012

— 'Annual Report 2005' <www.agcm.it/en/annual-report/1804-annual-report-2005. html> accessed 6 February 2012

— 'ENI Fined € 290M for Abuse of Dominant Market Position in Wholesale Supply of Natural Gas' (2006) <www.agcm.it/en/newsroom/press-releases/1093-eni-trans-tunisian-pipeline.html> accessed 6 February 2012

Jans JH, 'Proportionality Revisited' (2000) 27 Legal Issues of Economic Integration 239

— and HHB Vedder, *European Environmental Law* (3rd edn, Europa Law Publishing 2008)

— and others, '"Gold plating" of European Environmental Measures?' (2009) 6 Journal for European Environmental and Planning Law 417

— and HHB Vedder, *European Environmental Law* (4th edn, Europa Law Publishing 2012)

Jones A and B Suffrin, *EU Competition Law: Text, Cases and Materials* (4th edn, Oxford University Press 2011)

Joskow PL and RG Noll, 'The Bell Doctrine: Applications in Telecommunications, Electricity, and Other Network Industries'(1999) 51 Stanford Law Review 1249

Jyllands-Posten 'CO2 Storage Protests' (5 August 2009) <http://jyllands-posten.dk/ uknews/article4187838.ece?service=printversion> accessed 5 October 2012

Kallaugher J and A Weitbrecht, 'Developments under Articles 101 and 102 TFEU in 2010' (2011) 32 European Competition Law Review 333

Kavanagh J, N Marshall and G Niels, 'Reform of Article 82 EC – Can the Law and Economics be Reconciled?' in A Ezrachi (ed), *Article 82 EC: Reflections on its Recent Evolution* (Hart Publishing 2009)

Kekelekis M, 'Recent Developments in Infrastructure Funding: When Does it Not Constitute State Aid?' (2010) 10 European State Aid Law Quarterly 433

Kellerbauer M, 'The Commission's New Enforcement Priorities in Applying Article 82 EC to Dominant Companies' Exclusionary Conduct: A Shift Towards a More Economic Approach?' (2010) 31 European Competition Law Review 175

Kirk K, 'Safety of CO2 Storage in the North Sea' <www.zeroco2.no/6-bgs-karen-kirk-storage-in-the-north-sea.pdf> accessed 9 May 2012

Kjølbye L, 'The Commission's Evolving Commitment Practice – Its Impact and the Issues that it Raises' (presentation at Florence School of Regulation EU Energy Law & Policy Workshop on Priority Access for Renewable Energy into the Grid, Florence, 28 May 2010) <www.florence-school.eu/portal/page/portal/FSR_HOME/ ENERGY/Policy_Events/Workshops/2010/EU_Energy_Law_Policy/L.Kjolbye. pdf> accessed 15 February 2012

Klass AB and EJ Wilson, 'Climate Change and Carbon Sequestration: Assessing a Liability Regime for Long-Term Storage of Carbon Dioxide' (2008) 58 Emory Law Journal 103

Klumpp T and X Su, 'Strategic Investments under Open Access: Theory and Empirical Evidence' (2011) <http://userwww.service.emory.edu/~tklumpp/docs/energy.pdf> accessed 14 February 2012

Koch O and others, 'The RWE Gas Foreclosure Case: Another Energy Network Divestiture to Address Foreclosure Concerns' (2009) European Commission Competition Policy Newsletter 2009–2, 32, 32 <http://ec.europa.eu/competition/ publications/cpn/2009_2_7.pdf> accessed 1 February 2012

Kohls M, 'Developing CCS in Germany' (presentation held at the Energy Delta Convention, Groningen, 2009) <www.rug.nl/energyconvention/EDC/Archive/EDC2009/overviewspeakers2009> accessed 10 December 2010

Koopman G, 'Exemptions from the Application of Competition Law for State Owned Companies: An EU Perspective' (presentation held at the Global Competition Law and Economics Series Conference: Competition Law and the State, Hong Kong, 18 March 2011) <www.ucl.ac.uk/laws/global-competition/hongkong-2011/secure/docs/4_koopman.pdf> accessed 1 April 2011

Kotlowksi A, 'Third-party Access Rights in the Energy Sector: A Competition Law Perspective' (2007) 16 Utilities Law Review 101

—   'Access Rights to European Energy Networks – A Construction Site Revisited' in B Delvaux and others (eds), *EU Energy Law and Policy Issues: Energy Law Research Forum Collection* (Euroconfidential 2009)

Koutrakos P, *EU International Relations Law* (Hart Publishing 2006)

Krämer L, "Environmental Protection and Article 30 EEC Treaty" (1993) 30 Common Market Law Review 111

—   *EU Environmental Law* (7th edn, Sweet & Maxwell 2011)

Kroes N, 'The Interface Between Regulation and Competition Law' 2 (speech at the Bundeskartellamt Conference on Dominant Companies, Hamburg, 28 April 2009) <http://ec.europa.eu/competition/speeches/index_theme_13.html> accessed 10 May 2011

Kuby MJ, RS Middleton and JM Bielicki, 'Analysis of Cost Savings From Networking Pipelines in CCS Infrastructure Systems' (2011) 4 Energy Procedia 2808

Lembo S and ML Stasi, 'The Application of the Altmark Test to State Financing of Public Services in the Maritime Transport Sector' (2010) 9 European State Aid Law Quarterly 853

Lenaerts K and D Arts, *Procedural Law of the European Union* (Sweet & Maxwell 1999)

—   and M Desomer, 'Towards a Hierarchy of Legal Acts in the European Union? Simplification of Legal Instruments and Procedures' (2005) 11 European Law Journal 744

—   and JA Gutiérrez-Fons, 'The Constitutional Allocation of Powers and General Principles of EU Law' (2010) 47 Common Market Law Review 162

—   and P Van Nuffel, *European Union Law* (3rd edn, Sweet & Maxwell 2011)

Lianos I, 'Categorical Thinking in Competition Law and the "Effects-based" Approach in Article 82 EC' in A Ezrachi (ed), *Article 82 EC: Reflections on its Recent Evolution* (Hart Publishing 2009)

Lonbay J, 'A Review of Recent Tax Cases – Wine, Gambling, Fast Cars and Bananas' (1989) 14 European Law Review 48

Louis F and A Valery, 'State Aid and the Financing of Public Services: A Comment on the Altmark Judgment of the Court of Justice' (2003) Competition Law Insight 3

Louka E, *International Environmental Law: Fairness, Effectiveness, and World Order* (3rd edn, Cambridge University Press 2012)

Lowe P and others, 'Effective Unbundling of Energy Transmission Networks: Lessons from the Energy Sector Inquiry' (2007) European Commission Competition Policy Newsletter 2007–1, 23 <http://ec.europa.eu/competition/publications/cpn/2007_1_23.pdf> accessed 10 January 2012

Maier-Rigaud F, F Manca and U von Koppenfels, 'Strategic Underinvestment and Gas network Foreclosure: The ENI Case' (2011) European Commission Competition Policy Newsletter 2011–1, 18 <http://ec.europa.eu/competition/publications/cpn/2011_1_4_en.pdf> accessed 9 February 2012

Majumdar A, 'Whither Dominance?' (2006) 27 European Competition Law Review 161

Mann J, J Perner and C Riechmann, 'Carbon Capture & Storage (CCS) – Supporting by Carrots or Sticks?' (presentation held at the eight Conference on Applied Infrastructure Research, Berlin, 10 October 2009) <http://wip.tu-berlin.de/typo3/fileadmin/documents/infraday/2009/presentation/09a_mann_1pdf.pdf> accessed 27 April 2011

Marias EA, 'Solidarity as an Objective of the European Union and the European Community' (1994) 2 Legal Issues of European Integration 85

Marsden P, 'Some Outstanding Issues From the European Commission's Guidance on Article 102 TFEU: Not-so-faint Echoes of Ordoliberalism' in F Etro and I Kokkoris (eds), Competition Law and the Enforcement of Article 102 (Oxford University Press 2010)

McKinsey and Company, 'Carbon Capture and Storage: Assessing the Economics' (2008) <http://assets.wwf.ch/downloads/mckinsey2008.pdf> accessed 5 October 2012

Mendelevitch R and others, 'CO_2 Highways for Europe: Modelling a Carbon Capture, Transport and Storage Infrastructure for Europe' (2010) Centre for European Policy Studies Working Document No. 340 1 <http://aei.pitt.edu/15200/1/WD_340_CO2_Highways.pdf> accessed 27 April 2011

Mikunda T and others (CATO2), 'Transboundary Legal Issues in CCS: Economics, Cross-border Regulation and Financial Liability of CO_2 Transport and Storage Infrastructure' (2011)

— and others, 'Towards a CO_2 Infrastructure in North-Western Europe: Legalities, Costs and Organizational Aspects' (2011) 4 Energy Procedia, 2409

— and H de Coninck (CO2ReMoVe), 'Possible impacts of captured CO2 stream impurities on transport infrastructure and geological storage formations – current understanding and implications for EU legislation' (2011) <www.ecn.nl/docs/library/report/2011/o11040.pdf> accessed 21 August 2012

Moen KB, 'The Gas Directive: Third Party Transportation Rights – But to what Pipeline Volumes?' (2002) 13 The Centre for Energy, Petroleum and Mineral Law and Policy Internet Journal 3

Monti G, 'Managing the Intersection of Utilities Regulation and EC Competition Law" (2008) 4 The Competition Law Review 123

Morbee J, J Serpa and E Tzimas, 'The Evolution of the Extent and the Investment Requirements of a Trans-European CO2 Transport Network' (2010) <http://publications.jrc.ec.europa.eu/repository/bitstream/111111111/15100/1/ldna24565enn.pdf> accessed 7 December 2011

Moura e Silva M, 'EC Competition Law and the Market for Exclusionary Rights' <http://homepage.mac.com/mmsilva/documents/Lumextwp5.doc> accessed 14 February 2012

Müller T, 'Efficiency Control in State Aid and the Power of Member States to Define SGEIs' (2009) 8 European State Aid Law Quarterly 39

Murphy F, 'Abuse of Regulatory Procedures: The AstraZeneca Case: Part 3' (2009) 30 European Competition Law Review 314

Myhre JW, 'Financing of CCS Demonstration Projects: State Aid, EEPR and NER funding; an EU and EEA Perspective' (2011) University of Oslo Faculty of Law Legal Studies Research Paper Series No. 2011–02 <http://papers.ssrn.com/sol3/papers.cfm?abstract_id=1738511> accessed 20 March 2012

Nederlandse Olie en Gas Exploratie en Productie Associatie (NOGEPA), 'Potential for CO_2 storage in depleted gas fields on the Dutch continental shelf' (2008) <www.nogepa.nl/LinkClick.aspx?fileticket=G11SzzTQuFc%3d&tabid=546&language=en-GB> accessed 23 June 2010.

Neele F and others (CO_2Europipe), 'Development of a Large-scale CO2 Transport Infrastructure in Europe: Matching Captured Volumes and Storage Availability' (2010) <www.co2europipe.eu/Publications/D2.2.1%20-%20CO2Europipe%20Report%20CCS%20infrastructure.pdf> accessed 5 October 2012

— 'Developing a European CO2 Transport Infrastructure' (2011) <www.co2europipe.eu/Publications/CO2Europipe%20-%20Executive%20Summary.pdf> accessed 3 January 2012

NERA Economic Consulting, 'Developing a Regulatory Framework for CCS Transportation Infrastructure, vol. 1' (2009) <www.decc.gov.uk/assets/decc/what%20we%20do/uk%20energy%20supply/energy%20mix/carbon%20capture%20and%20storage/1_20090617131338_e_@@_ccsreg1.pdf> accessed 4 January 2012

— 'Developing a Regulatory Framework for CCS Transportation Infrastructure, vol. 2: Case Studies' (2009) <www.decc.gov.uk/assets/decc/What%20we%20do/UK%20energy%20supply/Energy%20mix/Carbon%20capture%20and%20storage/1_20090617131350_e_@@_ccsreg2.pdf> accessed 3 October 2011

Nergelius J, 'General Principles of Community Law in the Future: Some Remarks on their Scope, Applicability and Legitimacy' in U Bernitz and J Nergelius (eds), *General Principles of European Community Law* (Kluwer Law International 2000)

Newcomer A and J Apt, 'Implications of Generator Siting for CO_2 Pipeline Infrastructure' (2008) 36 Energy Policy 1776

Nicolaides P, 'Compensation for Public Service Obligations: The Floodgates of State Aid?' (2003) 24 European Competition Law Review 561

— 'State Aid, Advantage and Competitive Selection: What is a Normal Market Transaction?' (2010) 9 European State Aid Law Quarterly 65

— and M Kleis, 'Where is the Advantage? The Case of Public Funding of Infrastructure and Broadband Networks?' (2007) 6 European State Aid Law Quarterly 615

— and IE Rusu, 'Private Investor Principle: What Benchmark and Whose Money?' (2011) 10 European State Aid Law Quarterly 237

Nikolinakos NT, 'Access Agreements in the Telecommunications Sector: Refusal to Supply and the Essential Facilities Doctrine under E.C. Competition Law' (1999) 20 European Competition Law Review 399

O'Donoghue R, 'Regulating the Regulated: Deutsche Telekom v. European Commission' (2008) 1 GCP Magazine 3

— and AJ Padilla, *The Law and Economics of Article 82 EC* (Hart Publishing 2006)

Oilandgaspress.com, 'China's gas pipeline length to reach 100,000 km in 2015' (22 January 2010) <www.oilandgaspress.com/wp/2010/01/22/chinas-gas-pipeline-length-to-reach-100000-km-in-2015/> accessed 23 March 2010

Oliver PJ, *Free Movement of Goods in the European Community* (4th edn, Sweet & Maxwell 2003)

— *Oliver on Free Movement of Goods in the European Union* (5th edn, Hart Publishing 2010)

Østerud E, *Identifying Exclusionary Abuses by Dominant Undertakings under EU Competition Law* (Wolters Kluwer Law and Business 2010)

Papadias L, A Riedl and J Westerhof, 'Public Funding for Broadband Networks: Recent Developments' (2006) European Commission Competition Policy Newsletter 2006–3, 13 <http://ec.europa.eu/competition/publications/cpn/2006_3_13.pdf> accessed 20 March 2012

Pielow J, and E Ehlers, 'Ownership Unbundling and Constitutional Conflict: A Typical German Debate?' (2008) 2 European Review of Energy Markets 1

— and G Brunekreeft and E Ehlers, 'Legal and Economic Aspects of Ownership Unbundling in the EU' (2009) 2 Journal of World Energy Law and Business 1

Platform Nieuw Gas, Werkgroep CO_2-opslag/Schoon fossiel, 'Beleidsrapport Schoon Fossiel'

Posser T, 'Competition Law and Public Services: From Single Market to Citizenship Rights?' (2005) 11 European Public Law 543

Quigley C, *European State Aid Law and Policy* (2nd edn, Hart Publishing 2009)

Rai V, DG Victor and MC Thurber, 'Carbon Capture and Storage at Scale: Lessons from the Growth of Analogous Energy Technologies' (2009) 38 Energy Policy 4089

Ratliff J, 'Major Events and Policy Issues in EU Competition Law, 2009–2010 (Parts 1 and 2)' (2011) International Company and Commercial Law Review 128 <www.wilmerhale.com/files/Publication/b999ed92-f1b5-4e75-b3a5-112e2c0d8d7d/Presentation/PublicationAttachment/44f5a842-d65a-4738-baf1-14e56853292e/Ratliff_offprint.pdf> accessed 30 January 2012

Rey P and J Tirole, 'A Primer on Foreclosure' (2006) <http://idei.fr/doc/by/tirole/primer.pdf> accessed 13 February 2012

Riordan MH, 'Competitive Effects of Vertical Integration' (2005) Columbia University Department of Economics Discussion Papers 0506–11, 1 <www.columbia.edu/~mhr21/Vertical-Integration-Nov-11-2005.pdf> accessed 13 February 2012

Rizza C, 'The Financial Assistance Granted by Member States to Undertakings Entrusted with the Operation of a Service of General Economic Interest: The Implications of the Forthcoming Altmark Judgment for Future State Aid Control Policy' (2003) 9 Columbia Journal of European Law 429

Roggenkamp MM, 'The Concept of Third Party Access Applied to CCS' in MM Roggenkamp and E Woerdman (eds), *Legal Design of Carbon Capture and Storage* (Intersentia 2009)

— and KJ de Graaf and JM Holwerda, 'Afvang, Transport en Opslag van CO2: De Implementatie van Richtlijn 2009/31/EG in Nederland' (2010) 37 Milieu en Recht 548

Röller L and O Stehmann, 'Grenzen der Wettbewerbspolitik bei der Öffnung von Netzwerkindustrien' (2006) 7 Perspektiven der Wirtschaftspolitik 355

Rosas A and L Armati, *EU Constitutional Law: An Introduction* (2nd edn, Hart Publishing 2010)

Ross M, 'A Healthy Approach to Services of General Economic Interest? The BUPA Judgment of the Court of First Instance' (2009) 34 European Law Review 127

RWE, 'Fall COMP/B-1/39.402 – Deutscher Gasmarkt – Zusagen an Die Europäische Kommission' (2008) <http://ec.europa.eu/competition/antitrust/cases/dec_docs/39402/39402_437_11.pdf> accessed 1 February 2012

Salerno FM, 'The Competition Law-ization of Enforcement: The Way Forward for Making the Energy Market Work?' (2008) European University Institute Working Papers RSCAS 2008/07 <http://cadmus.eui.eu/handle/1814/8108> accessed 8 February 2012

Santamato S and N Pesaresi, 'Compensation for Services of General Economic Interest: Some Thoughts on the Altmark Ruling' (2004) European Commission Competition Policy Newsletter 2004-1, 17 <http://ec.europa.eu/competition/publications/cpn/2004_1_17.pdf> accessed 28 February 2012

Santer S and others (CO$_2$Europipe), 'Making CO$_2$ Transport Feasible: The German Case Rhine/Ruhr Area (D) – Hamburg (D) – North Sea (D, DK, NL)' (2011) <www.co2europipe.eu/Publications/D4.2.2%20-%20Making%20CO2%20transport%20feasible%20-%20the%20German%20case.pdf> accessed 9 May 2012

Sauter W, 'Services of General Economic Interest and Universal Service in EU Law' <www3.udg.edu/fcee/professors/nboccard/micro2/part2/Sauter.pdf> accessed 14 March 2012

—   and J van de Gronden, 'State Aid, Services of General Economic Interest and Universal Service in Healthcare' (2011) 32 European Competition Law Review 615

Schepel H and W Sauter, *State and Market in European Law* (Cambridge University Press 2009)

Schermers HG and DF Waelbroeck, *Judicial Protection in the European Union* (Kluwer Law International 2001)

Schokkenbroek H and others, 'Elements for a National Masterplan for CCS: Lessons Learnt' (2011) 4 Energy Procedia 5810

Scholz U and S Purps, 'The Application of EC Competition Law in the Energy Sector' (2010) 1 Journal of European Competition Law and Practice 37

Schrijver N, *Sovereignty over Natural Resources – Balancing Rights and Duties* (Cambridge University Press 1997)

Schultz CR, 'Modelling Take-or-Pay Contract Decisions' (1997) 28 Decision Sciences 213

Schutze R, 'Cooperative Federalism Constitutionalised: The Emergence of Complementary Competences in the EC Legal Order' (2006) 31 European Law Review 167

Schwarze J, 'Enlargement, the European Constitution, and Administrative Law' (2004) 53 International and Comparative Law Quarterly 969

—   *European Administrative Law* (2nd edn, Sweet & Maxwell 2006)

Scott J, 'Flexibility in the Implementation of EC Environmental Law' in H Somsen (ed), *Yearbook of European Environmental Law* (Oxford University Press 2000) 37

Szyszczak E, 'Golden Shares and Market Governance' (2002) 29 Legal Issues of Economic Integration 255

— 'Financing Services of General Economic Interest' (2004) 67 Modern Law Review 982

— 'Public Services in the New Economy' in C Graham and F Smith (eds), *Competition, Regulation and the New Economy* (Hart Publishing 2004)

— 'The Survival of the Market Economic Investor Principle in Liberalised Markets' (2011) 10 European State Aid Law Quarterly 35

Sharkey WW, *The Theory of Natural Monopoly* (Cambridge University Press 1982)

Shelanksi HA and PG Klein, 'Empirical Research in Transaction Cost Economics: A review and Assessment' (1995) 11 Journal of Law, Economics and Organization 335

Sher B, 'The Last of the Steam-powered Trains: Modernising Article 82' (2004) 25 European Competition Law Review 243

Simm M, 'Institutional Report – The Interface between Energy, Environment and Competition Rules of the European Union' 19 <www.fide2012.eu/index.php?doc_id=87> accessed 25 July 2012

Sinnaeve A, 'State Financing of Public Services: The Court's Dilemma in the Altmark Case' (2003) 3 European State Aid Law Quarterly 351

Siragusa M, 'Gli Obblighi di non Discriminazione nella Regolazione Settoriale e nella Disciplina Antitrust' (contribution to the second Energy Law Conference, Rome, 6–7 April 2011)

Skougard J, 'Getting from Here to There: Devising an Optimal Regulatory Model for CO2 Transport in a New Carbon Capture and Sequestration Industry' (2010) 30 Journal of Land, Resources, and Environmental Law 357

Slot PJ, 'Harmonisation' (1996) 21 European Law Review 378

— and A Skudder, 'Common Features of Community Law Regulation in the Network-bound Sectors' (2001) 38 Common Market Law Review 87

Slotboom MM, 'Do Different Treaty Purposes Matter for Treaty Interpretation? The Elimination of Discriminatory Internal Taxes in EC and WTO Law' (2001) 4 Journal of International Economic Law 557

Somsen H, 'Discretion in European Community Environmental Law: An Analysis of ECJ Case Law' (2003) 40 Common Market Law Review 1413

Stothers C, 'Refusal to Supply as Abuse of a Dominant Position: Essential Facilities in the European Union' (2001) 22 European Competition Law Review 256

Suranovic SM, *International Trade Theory and Policy* (2010)

Suri R, 'CO_2 Compression for Capture-Enabled Power Systems' (MSc thesis, Massachusetts Institute of Technology 2009)

Szlagowski P, 'The Abuse of a Dominant Position through Strategic Underinvestment of Energy Transmission Network Interconnectors' (2010) 6 International Energy Law Review 201

Talus K, *Vertical Natural Gas Transportation Capacity, Upstream Commodity Contracts and EU Competition Law* (Kluwer Law International 2011)

— 'Long-term Natural Gas Contracts and Antitrust Law in the European Union and the United States' (2011) 4 Journal of World Energy law and Business 260

Temmink H, 'From Danish Bottles to Danish Bees: The Dynamics of Free Movement of Goods and Environmental Protection – A Case Law Analysis' in H Somsen (ed), *Yearbook of European Environmental Law* (Oxford University Press, 2000)

Temple Lang J, 'Abuse of a Dominant Position in European Community Law, Present and Future: Some Aspects' in Hawk (ed), *Fifth Fordham Corporate Law Institute* (Law and Business 1979)

— 'The Principle of Essential Facilities in European Community Competition Law: The Position since Bronner' (2000) 14 <http://lawcourses.haifa.ac.il/antitrust_s/index/main/syllabus/lang_article.pdf> accessed 5 May 2011

— 'The Duties of National Authorities and Courts under Article 10 EC: Two More Reflections' (2001) 26 European Law Review 84

— 'The Development by the Court of Justice of the Duties of Cooperation of National Authorities and Community Institutions under Article 10 EC' (2008) 31 Fordham International Law Journal 1483

Terra BJM and PJ Wattel, *European Tax Law* (2nd edn, Kluwer Law International 2012)

The Delphi Group and Alberta Research Council, 'Building Capacity for CO2 Capture and Storage in the APEC Region: A Training Model for Policy Makers and Practitioners' (2005) <http://canmetenergy-canmetenergie.nrcan-rncan.gc.ca/fichier/78892/apec_training_2005.pdf> accessed 22 March 2010

The Economist, 'Economics A-Z' <www.economist.com/research/economics/> accessed 18 April 2011

The National Mining Association, 'Carbon Capture and Storage: Status of CCS Developments' <www.nma.org/ccs/ccsprojects.asp> accessed 18 October 2012

The Parliament of the Commonwealth of Australia, 'Between a Rock and a Hard Place: The Science of Geosequestration' (2007)

Thouvenin J, 'The Altmark Case and its Consequences' in M Krajewski, U Neergaard and J van de Gronden (eds), *The Changing Legal Framework for Services of General Interest in Europe: Between Competition and Solidarity* (TMC Asser Press 2009)

Tjetland G and others, 'Incentives and Regulatory Frameworks Influence on CCS Chain Establishment' <www.sintef.no/project/ecco/Publications/Conference%2014%20June%202011/10%20-%20Regulatory%20framework%20recommendations%20-%20Goeril%20Tjetland.pdf> accessed 9 May 2012

Tobler C, *Indirect Discrimination: A Case Study into the Development of the Legal Concept of Indirect Discrimination under EC Law* (Intersentia 2005)

Trabucchi C and L Patton, 'Storing Carbon: Options for Liability Risk Management, Financial Responsibility' in *World Climate Change Report* (2008) 1

Tridimas T, 'Proportionality in Community Law: Searching for the Appropriate Standard of Scrutiny' in E Ellis (ed), *The Principle of Proportionality in the Laws of Europe* (Hart Publishing 1999)

— *The General Principles of EU Law* (2nd edn, Oxford University Press 2006)

Turocy TL and B von Stengel, 'Game Theory' (2001) CDAM Research Report LSE-CDAM-2001–09 <www.cdam.lse.ac.uk/Reports/Files/cdam-2001-09.pdf> accessed 13 February 2012

UK Department of Energy and Climate Change (DECC), DECC, 'Towards Carbon Capture and Storage: Government Response to Consultation' (2009) <http://webarchive.nationalarchives.gov.uk/+/www.berr.gov.uk/files/file51115.pdf> accessed 21 March 2011

— 'Implementing the Third Party Access Requirements of the CCS Directive – Impact Assessment' (2010)

<www.decc.gov.uk/en/content/cms/consultations/ccs_3rd_party/ccs_3rd_party.aspx>
accessed 21 March 2011

United Nations Industrial Development Organization (UNIDO), 'Carbon Capture and
Storage in Industrial Applications: Technology Synthesis Report' (2010) <www.
unido.org/fileadmin/user_media/Services/Energy_and_Climate_Change/Energy_
Efficiency/CCS/synthesis_final.pdf> accessed 18 May 2011

United States Environmental Protection Agency (EPA), 'Approaches to Geologic
Sequestration Site Stewardship after Site Closure' (2008) 8
accessed 8 May 2012

Utilityregulation.com, 'Costing Definitions and Concepts' <www.utilityregulation.com/
content/essays/t1.pdf> accessed 28 April 2011

Van Bael I and J Bellis, *Competition Law of the European Community* (Kluwer Law
International 2005)

— *Competition Law of the European Community* (5th edn, Kluwer Law International
2010)

Van Calster G, 'Greening the EC's State Aid and Tax Regimes' (2000) 21 European
Competition Law Review 294

Van Dijk M and M Mulder, 'Regulation of Telecommunication and Deployment of
Broadband' (2005) CPB Netherlands Bureau for Economic Policy Analysis
Memorandum <www.cpb.nl/sites/default/files/publicaties/download/memo131.
pdf> accessed 30 March 2011.

Van Renssen S, 'European CCS Industry Faces Moment of Truth' European Energy
Review (29 October 2012) <www.europeanenergyreview.eu/site/pagina.
php?id=3919> accessed 30 October 2012

Vandamme TA, *The Invalid Directive* (Europa Law Publishing 2005

Vangkilde-Pedersen T and others (GeoCapacity), 'Assessing European Capacity for
Geological Storage of Carbon Dioxide' (2009) <www.geology.cz/geocapacity/
publications/D16%20WP2%20Report%20storage%20capacity-red.pdf> accessed
18 October 2012

Vasques L and S Nobili, 'The Italian Competition Authority Fines ENI with the Highest
Fine Ever Imposed to a Single Company in Italy for Abuse of a Dominant Position
in the Wholesale Supply of Natural Gas on the Basis of Art. 82 EC (Trans Tunisian
Pipeline Company-ENI)' (2006) e-Competitions No501 <www.concurrences.com/
article.php3?id_article=501&lang=fr> accessed 7 February 2012

Vedder H, 'An Assessment of Carbon Capture and Storage under EC Competition Law'
(2008) 29 European Competition Law Review 586

— 'The Constitutionality of Competition. European Internal Market Law and the
Fine Line between Markets, Public Interests, and Self-regulation in a Changing
Constitutional Setting' in F Amtenbrink and PAJ van den Berg (eds), *The
Constitutional Integrity of the European Union* (TMC Asser Press 2010)

— 'Competition in the EU Energy Sector – An Overview of Developments in 2009
and 2010' (2011) 9 <http://papers.ssrn.com/sol3/papers.cfm?abstract_id=1734639>
accessed 12 May 2011

Voosen P, 'Freightened, Furious Neighbours Undermine German CO2-Trapping Power
Project' *New York Times* (7 April 2010) <www.nytimes.com/gwire/2010/04/07/07

greenwire-frightened-furiousneighbors-undermine-german-35436.html?scp=6&
sq=paul%20 voosen&st=cse> accessed 11 May 2010

Waelde TW and AJ Gunst, 'International Energy Trade and Access to Energy Networks'
(2002) 36 Journal of World Trade 191

Weatherill S, 'Recent Case Law Concerning the Free Movement of Goods: Mapping the
Frontiers of Market Deregulation' (1999) 36 Common Market Law Review 51

— 'Pre-emption, Harmonisation and the Distribution of Competence to Regulate the
Internal Market' in C Barnard and J Scott (eds), *The Law of the Single European
Market, Unpacking the Premises* (Hart Publishing 2002)

— *Cases and Materials on EU Law* (9th edn, Oxford University Press 2010)

Wennerås P, 'Towards an Ever Greener Union? Competence in the Field of the
Environment and Beyond' (2008) 45 Common Market Law Review 1645

Wiklund O and J Bengoetxea, 'General Constitutional Principles of Community Law' in
U Bernitz and J Nergelius (eds), *General Principles of European Community Law*
(Kluwer Law International 2000)

Willis P and P Hughes, 'Structural Remedies in Article 82 Energy Cases' (2008) 4 The
Competition Law Review 147 <www.clasf.org/CompLRev/Issues/Vol4Iss2Art3
WillisHughes.pdf> accessed 15 February 2012

Wilson EJ, AB Klass and S Bergan, 'Assessing a Liability Regime for Carbon Capture and
Storage' (2009) 1 Energy Procedia 4575

Winter JA, 'Re(de)fining the Notion of State Aid in Article 87(1) of the EC Treaty' (2004)
41 Common Market Law Review 475

Whish R, *Competition Law* (6th edn, Oxford University Press 2009)

Wouters J, 'Constitutional Limits of Differentiation: The Principle of Equality' (2001)
University of Leuven Institute for International Law Working Paper No 4 <https://
www.law.kuleuven.be/iir/nl/onderzoek/wp/WP04e.pdf> accessed 25 July 2011

Zhou and others, 'Uncertainty Modelling of CCS Investment Strategy in China's Power
Sector' (2010) 97 Applied Energy 2392

DOCUMENTS FROM THE EU INSTITUTIONS

Commission, 'Commission Notice on the Definition of the Relevant Market for the
Purposes of Community Competition Law' [1997] OJ C372

— 'Communication from the Commission – Services of General Interest in Europe'
2001/C17/04

— 'Commission Approves Public Funding of Broadband Projects in Pyrénées-
Atlantiques, Scotland and East Midlands' IP/04/1371, <http://europa.eu/rapid/
press-release_IP-04-1371_en.htm> (accessed 26 October 2012)

— 'State Aid: Commission Endorses Public Funding for Broadband Network in
Limousin, France' IP/05/530 <http://europa.eu/rapid/pressReleasesAction.
do?reference=IP/05/530> accessed 19 March 2012

— 'DG Competition Report on Energy Sector Inquiry' SEC (2006) 1724 <http://
ec.europa.eu/competition/sectors/energy/inquiry/full_report_part1.pdf> accessed
7 November 2011

— 'Sustainable Power Generation from Fossil Fuels: Aiming for Near-zero Emissions from Coal after 2020' COM (2006) 843 final

— 'Commission Staff Working Document on Capacity Allocation and Congestion Management for Access to the Natural Gas Transmission Networks Regulated under Article 5 of Regulation (EC) No 1775/2005 on Conditions for Access to the Natural Gas Transmission Networks' SEC (2007) 822

— 'Antitrust: Commission Initiates Proceedings against the ENI Group Concerning Suspected Foreclosure of Italian Gas Supply Markets' MEMO/07/187 <http:// europa.eu/rapid/pressReleasesAction.do?reference=MEMO/07/187&format=HTM L&aged=1&language=EN&guiLanguage=en> last accessed 6 February 2012

— 'Opening of Proceedings' (2007) <http://ec.europa.eu/competition/antitrust/cases/ dec_docs/39402/39402_43_10.pdf> accessed 1 February 2012

— 'Antitrust: Commission Initiates Proceedings against RWE Group Concerning Suspected Foreclosure of German Gas Supply Markets' MEMO/07/186 <http:// europa.eu/rapid/pressReleasesAction.do?reference=MEMO/07/186&format=HTM L&aged=0&language=EN&guiLanguage=en> accessed 1 February 2012

— 'Antitrust: Commission Opens Formal Proceedings against Gaz de France Concerning Suspected Gas Supply Restrictions' MEMO/08/328 <http://europa.eu/ rapid/pressReleasesAction.do?reference=MEMO/08/328&format=HTML&aged=0 &language=EN&guiLanguage=en> accessed 26 January 2012

— 'Questions and Answers on the Directive on the Geological Storage of Carbon Dioxide' (2008) <http://europa.eu/rapid/pressReleasesAction.do?reference=MEM O/08/798&language=EN> accessed 7 December 2010.

— 'Accompanying Document to the Proposal for a Directive of the European Parliament and of the Council on the Geological Storage of Carbon Dioxide: Impact Assessment' COM (2008) 18 final

— 'Antitrust: Commission Accepts Commitments by GDF Suez to Boost Competition in French Gas Market' IP/09/1872 <http://europa.eu/rapid/pressReleasesAction. do?reference=IP/09/1872&format=HTML&aged=0&language=EN> accessed 30 January 2012.

— 'Antitrust: Commission Confirms Sending Statement of Objections to ENI Concerning the Italian Gas Market' MEMO/09/120 <http://europa.eu/rapid/ pressReleasesAction.do?reference=MEMO/09/120&format=HTML&aged=1&lang uage=EN&guiLanguage=en> accessed 6 February 2012

— 'State Aid: Commission Approves UK Aid for Feasibility Studies on Carbon Capture and Storage Demonstration Projects' IP/09/555 <http://europa.eu/rapid/ pressReleasesAction.do?reference=IP/09/555> accessed 20 March 2012;

— 'Summary of Commission Decision of 18 March 2009 Relating to a Proceeding under Article 82 of the EC Treaty and Article 53 of the EEA Agreement Case COMP/B-1/39.402 RWE Gas Foreclosure' [2009] OJ C133/08

— 'Antitrust: Commission Welcomes E.ON Proposals to Increase Competition in German Gas Market' MEMO/09/567 <http://europa.eu/rapid/pressReleasesAction. do?reference=MEMO/09/567&format=HTML&aged=0&language=EN&guiLangu age=en> accessed 6 February 2012

— 'Antitrust: Commission Opens German Gas Market to Competition by Accepting Commitments from RWE to Divest Transmission Network' IP/09/410 <http://

europa.eu/rapid/pressReleasesAction.do?reference=IP/09/410&format=HTML&ag
ed=0&language=EN&guiLanguage=en> accessed 1 February 2012
— 'Communication from the Commission – Guidance on the Commission's
Enforcement Priorities in Applying Article 82 of the EC Treaty to Abusive
Exclusionary Conduct by Dominant Undertakings' 2009/C 45/02
— 'Antitrust: E.ON's Commitments Open up German Gas Market to Competitors'
IP/10/494 <http://europa.eu/rapid/pressReleasesAction.do?reference=IP/10/494&fo
rmat=HTML&aged=0&language=EN&guiLanguage=en> accessed 6 February
2012
— 'Antitrust: Commission Welcomes ENI's Structural Remedies Proposal to Increase
Competition in the Italian Gas Market' MEMO/10/29 <http://europa.eu/rapid/
pressReleasesAction.do?reference=MEMO/10/29&format=HTML&aged=0&langu
age=EN&guiLanguage=en> accessed 8 February 2012
— 'Antitrust/ENI Case: Commission Opens up Access to Italy's Natural Gas Market'
IP/10/1197 <http://europa.eu/rapid/pressReleasesAction.do?reference=IP/10/1197&
format=HTML&aged=0&language=EN&guiLanguage=en> accessed 8 February
2012
— 'Building a post-2012 global climate regime: the EU's contribution' (2010) <http://
ec.europa.eu/environment/climat/future_action.htm> accessed 23 March 2010
— 'Case COMP/39.317 E.ON Gas Initiation of Proceedings' (2010) <http://ec.europa.eu/
competition/antitrust/cases/dec_docs/39317/39317_1713_8.pdf> accessed 3 February
2012
— 'State Aid: Commission Approves € 30 Million German Support for ArcelorMittal
Eisenhuettenstadt's "Top Gas Recycling" Project' IP/10/254 <http://europa.eu/
rapid/pressReleasesAction.do?reference=IP/10/254> accessed 20 March 2012
— 'State Aid: Commission Approves € 10 Million Aid for Nuon's energy-saving CO2
Capture Project in the Netherlands' IP/10/614 <http://europa.eu/rapid/
pressReleasesAction.do?reference=IP/10/614&format=HTML&aged=0&language=
EN&guiLanguage=en> accessed 20 March 2012)
— 'State Aid: Commission Approves € 150 Million for Carbon Capture and Storage
Project in the Netherlands' IP/10/1392 <http://europa.eu/rapid/pressReleases
Action.do?reference=IP/10/1392> accessed 20 March 2012
— 'Commission Staff Working Document: Accompanying Document to the Revised
Proposal for a Council Directive (Euratom) on the management of spent fuel and
radioactive waste' SEC (2010) 1289 final 10
— 'Questions & Answers: Safety Standards for Nuclear Waste Disposal'
MEMO/10/540 <http://europa.eu/rapid/pressReleasesAction.do?reference=MEMO
/10/540&format=HTML&aged=0&language=EN> accessed 7 August 2012
— 'Commission Staff Working Document: Summary of the Impact Assessment
Accompanying Document to the Revised Proposal for a Council Directive
(Euratom) on the Management of Spent Fuel and Radioactive Waste' SEC (2010)
1290 final 3 <http://eur-lex.europa.eu/LexUriServ/LexUriServ.do?uri=SEC:2010:12
90:FIN:EN:PDF> accessed 7 August 2012
— 'Antitrust: Commission's Commitment Decision Opens German Gas Pipelines to
Competitors – Frequently Asked Questions' MEMO/10/164 <http://europa.eu/

rapid/pressReleasesAction.do?reference=MEMO/10/164&format=HTML&aged=0 &language=EN&guiLanguage=en> accessed 6 February 2012

— 'Implementation of Directive 2009/31/EC on the Geological Storage of Carbon Dioxide –Guidance Document 2: Characterisation of the Storage Complex, CO2 Stream Composition, Monitoring and Corrective Measures' (2011) <http://ec. europa.eu/clima/policies/lowcarbon/docs/gd2_en.pdf> accessed 26 May 2011

— 'Implementation of Directive 2009/31/EC on the Geological Storage of Carbon Dioxide – Guidance Document 4: Article 19 Financial Security and Article 20 Financial Mechanism' (2011) <http://ec.europa.eu/clima/policies/lowcarbon/ccs/ implementation/docs/gd4_en.pdf> accessed 7 May 2012

Council of the European Union, 'Proposal for a Directive of the European Parliament and of the Council on the geological storage of carbon dioxide and amending Council Directives 85/337/EEC, 96/61/EC, Directives 2000/60/EC, 2001/80/EC, 2004/35/EC, 2006/12/EC and Regulation (EC) No 1013/2006' 14532/08 <http:// register.consilium.europa.eu/pdf/en/08/st14/st14532.en08.pdf> accessed 3 November 2011

— 'Pursuing the Treaty Reform Process' 10659/07 POLGEN 67 <http://register. consilium.europa.eu/pdf/en/07/st10/st10659.en07.pdf> accessed 1 September 2011

European Council, 'Brussels European Council 21/22 June 2007 – Presidency Conclusions' 11177/1/07 REV 1 <http://register.consilium.europa.eu/pdf/en/07/ st11/st11177-re01.en07.pdf> accessed 1 September 2011

— 'Brussels European Council 8/9 March 2007: Presidency Conclusions' 7224/1/07

European Parliament, 'Draft report on the proposal for a directive of the European Parliament and of the Council on the geological storage of carbon dioxide and amending Council Directives 85/337/EEC, 96/61/EC, Directives 2000/60/EC, 2001/80/EC, 2004/35/EC, 2006/12/EC and Regulation (EC) No 1013/2006 (COM(2008)0018 – C6–0040/2008 – 2008/0015(COD)) A6–0414/2008 <www. europarl.europa.eu/oeil/FindByProcnum.do?lang=en&procnum= COD/2008/0015> accessed 26 May 2011

— 'Report on the proposal for a directive of the European Parliament and of the Council on the geological storage of carbon dioxide and amending Council Directives 85/337/EEC, 96/61/EC, Directives 2000/60/EC, 2001/80/EC, 2004/35/ EC, 2006/12/EC and Regulation (EC) No 1013/2006 (COM(2008)0018 – C6–0040/2008 – 2008/0015(COD))' A6–0414/2008 <www.europarl.europa.eu/ sides/getDoc.do?language=EN&reference=A6–0414/2008> accessed 26 May 2011

WEBSITES

CO_2 Capture project website

CO_2 ReMoVe project website accessed 22 October 2012

Commission website on carbon capture and geological storage <http://ec.europa.eu/ clima/policies/lowcarbon/ccs/index_en.htm> accessed 20 July 2012

Commission website on the EU climate and energy package <ec.europa.eu/environment/ climat/climate_action.htm> accessed 8 March 2010

Commission website on services of general economic interest <http://ec.europa.eu/competition/state_aid/legislation/sgei.html> accessed 28 March 2012>

Europa, 'Summaries of EU legislation: Guidelines on Vertical Restraints' <http://europa.eu/legislation_summaries/other/l26061_en.htm> accessed 13 February 2012

EU website on the Schuman declaration <http://europa.eu/about-eu/basic-information/symbols/europe-day/schuman-declaration/index_en.htm> accessed 3 August 2012

IPCC website on technology research, development, deployment, diffusion and transfer <www.ipcc.ch/publications_and_data/ar4/wg3/en/tssts-ts-2-6-technology-research.html> accessed 9 June 2010>